Content With My Wages

A Sergeant's Story

Book II-The War On Drugs

Also by the author:

Content With My Wages-A Sergeant's Story: Book I-Vietnam
Content With My Wages-A Sergeant's Story: Book III-Afghanistan

Content With My Wages

A Sergeant's Story

Book II-The War On Drugs

Gregory H. Murry
Master Sergeant, U.S. Army (Retired)

No End To Publishing Company
Austin, Texas, USA

Library of Congress Cataloging-in-Publication Data

Murry, Gregory H.
Content with My Wages-A Sergeant's Story: Book II-The War On Drugs/Gregory H. Murry
Includes index.

ISBN - 978-0-9903976-5-6

Uncorrected Proof #5 revised

No End To Publishing Company LLC
Austin, Texas, USA
Manufactured in the United States of America

And some serving in the military also questioned him,
saying, and we, what shall we do? And he said to them,
"Extort nothing from anyone by force, nor take any-
thing by false accusation,
and be content with your wages."
Luke 3:14

"You'll get over it Joe. Oncet I was gonna write a book exposin the army after the war myself."

SSG (T) Bill Mauldin, 45th Infantry Division

To

Thomas Payne, George Huseman, 'Steve', Chris Jay,
Tom Stephens, Larry Rayburn, James F, Clayton Hardee, Bill S: the original members of the Special Operations Detachment (SOD)
Carpe Per Diem!

And

The members of SOD and the Texas National Guard's
Joint Counterdrug Task Force

And

The members of the following units:

Troop D, 1-124 Cavalry, 49 Armored Division
Company G, (Airborne-Ranger), 143d Infantry (LRRP)
LRSD, 1-124 Cavalry, 49 Armored Division
143d Infantry Detachment (LRS), 49Armored Division
G2, 49 Armored Division
RCPOC, 49 Armored Division
136 Regional Training Academy

And Especially

The wives and children that held things together on the home front.

In Memoriam

William (Bil) Enney
Gregory Dembowski
Larry Rayburn
James (Mike) Hargett

Master Sergeant William (Bil) Enney
United States Army Special Forces

Table of Contents

Spring Break

It was Spring Break on South Padre Island and we were there for the party. Tonight, we were going to get down on Boca Chica Beach. We bailed out of the Blazer just after the sun went down. There was no one on the beach near us. We grabbed the LORIS and the batteries and moved them into the dunes. The Blazer drove off and the three of us were alone.

We struggled over the first line of dunes and found a small valley where we could set up the monitor. We placed the LORIS head on the top of a dune, camouflaged it and ran the cables down to the monitor, hooked them up and powered up the LORIS. The monitor brightened and Pat began adjusting the controls. Dan was up on a dune next to the one with the LORIS head on it. He was watching the beach with a pair of PVS-7 Night Vision Goggles.

I was walking around videotaping the activities of Pat and Dan when Pat said, "Hey, come here and look at this." I walked over to the monitor and saw a Shark Boat, low in the water, going north, just outside the surf line. I called the CP and said "We have a shark boat approximately 100 meters in front of our position, heading north at this time." Pat was panning slowly and switching the near and far views and going from white hot to black hot. He was annoyed that I wouldn't let him talk on the radio while he operated the LORIS. He's like that. He was also the best operator I ever worked with and capable of doing everything on this mission at the same time.

The CP called back and asked a series of questions that we tried to answer as we watched the monitor. There were 5 men in the back and there appeared to be a large box covered with a tarp in the front. It was around 7:30 PM, pitch dark, one/half mile from the border, and it would be hard to believe that these guys had any legitimate business here. We were fascinated as we watched them turn toward shore a ways north of our position and navigate their way through the three lines of surf to beach the boat. We watched as the smugglers lifted bundles out of the boat and run up the beach with them where they disappeared from our sight in the dunes. We had Dan look up the beach to check for vehicle lights and he didn't see any so we figured that they were burying them in the dunes like the load we had found a few days before. These guys were good.

Author's Note

I wrote this for operations and intelligence sergeants. If you are neither, read on at your peril. The officers I served with were able to operate in this career-risk environment. For that alone I salute them. This is both a memoir and a history at the squad level. It covers a lot of ground but focuses on the activities of the Special Operations Detachment (SOD) of the Texas National Guard's Counterdrug Task Force.

I'm proud to have served there with some of the best soldiers I ever met. In writing this I didn't spare myself from my failures because they are teaching points for military/life lessons. To my comrades who made mistakes while serving, I hope you learned from yours; I know I did from mine.

Sergeant First Class Thomas Payne and I founded SOD and selected its first members. Some were experienced special forces and ranger soldiers, and when they incorporated what we had learned from running missions on the border, developed very professional teams.

During my ten years with SOD we conducted hundreds of missions. I was on all of them in the early years but as the unit grew and we decentralized our command and control, many missions took place that I was not a part of so except for some anecdotes they are not mentioned.

I included my experiences as an instructor and administrator in the Army's Noncommissioned Officer Education System. It is our duty to use what we learned from our failures to train the ones coming up. Wisdom is the compound summation of all our previous learning from our mistakes and the ability to expound on those subjects for the benefit of younger people who have little or no experience yet think they know everything.

Regretfully, the continual use of the letter 'I' makes it appear that this story is all about me. General Creighton Abrams Jr. said he would never write a memoir because of it. The Armed Forces is a team effort; what I have written is the view of an old soldier who was proud serve with these men and be part of the effort to improve the team.

For the sake of privacy, I have not used the names of many of the men and women that were part of this story. Most of the pictures are frame captures from video I took while serving with LRSD, SOD, and the CD Task Force. All other photos are credited.

Proclaim this among the nations: Consecrate for war; stir up the mighty men. Let all the men of war draw near; let them come up. Beat your plowshares into swords, and your pruning hooks into spears; let the weak say, "I am a warrior."
Joel 3:8-9

Preface

Each state and each territory of the United States has a military force called the National Guard. The National Guard traces its origins to the tradition of the British militia where every able-bodied man was to be ready to defend his town or country when needed. Today, the Guard is simply a reserve component of the U.S. Armed Forces. The Army Guard is the state's ground force and the Air Guard is its air force. During peacetime the Guard is commanded by the governor of the state and can be used to assist the civil authorities with natural disasters and law enforcement situations. During national emergencies it may be called to Federal service by the President.

Militias were difficult to use as an offensive force; it took regulars to defeat the British at Yorktown during our revolutionary war; but they did provide a rapidly assembled military force when needed. The two world wars showed that with training, National Guard troops could fight alongside the regular army. On D-Day, a National Guard Division landed on Omaha Beach next to the 1st Infantry Division and took their objectives at a heavy cost.

The Texas National Guard

The Texian Militia was organized by Stephen F. Austin in 1823 while Texas was still part of Mexico. The militia defended the early settlers from attacks by the Comanches and other tribes in the area. The most famous of the early militia forces were the Texas Rangers. They were volunteers that quickly formed small units out of necessity while Texas was still part of Mexico. They fought in the war of 1846 with Mexico and with the Comanches and other tribes along the frontier for the rest of the 1800s.

The Guard is still the state's militia and it is a hometown tradition in Texas with fathers serving with their sons and preparing to go where the country sends them in time of war. After a long hiatus of fifty-six years, the Texas Guard deployed to Afghanistan and Iraq after 911.

The Texas National Guard's Joint Counterdrug Task Force

The Texas National Guard's Joint Counterdrug Task Force is an ad hoc organization hastily formed in 1988 to support law enforcement agencies investigating drug trafficking activities in Texas. After the task force was formed, it began sending troops to the border to help U.S. Customs with the inspection of trucks coming into Texas from Mexico. Border surveillance missions followed and by 1991 the task force was running hard.

Within the Counterdrug Task Force were several special ad hoc units: The Reconnaissance and Interdiction Detachment (RAID), an Army Guard aviation unit flying specially equipped helicopters. The Air Guard flew aerial photograph missions using a RC-26, a twin-engine reconnaissance and surveillance airplane. The Special Operations Detachment (SOD) was a ground surveillance and reconnaissance unit composed of both Army and Air Guard personnel.

The RAID unit came out of Operation Green Thunder, a domestic marijuana eradication program, that utilized Army National Guard helicopters, pilots, crew chiefs, and refuelers with law enforcement officers as spotters. These operations were in progress when the CDTF entered the drug war in 1988. The pilots and crew chiefs were assigned to Army National Guard aviation units. Some of these personnel applied for full-time positions in the newly formed RAID unit.

Ten states, including Texas were given the RC-26, a twin engine turbo-prop Fairchild Metroliner. Texas pilots were selected from a F-16 fighter wing and trained by U.S. Customs for aerial interdiction. Cameras for aerial photographs were added for the primary mission.

The Special Operations Detachment was formed with volunteers from the two Long Range Surveillance units in Texas. Manned by special forces veterans and rangers, SOD, like their Texas Ranger forebearers soon became notable along with the two aviation units for using tactics, techniques, and procedures not found in the manuals.

Acknowledgments

I have to thank the following people who reviewed the manuscript:

Colonel William 'Buckwheat' Pettit (USAF Retired)
Colonel Russ Malesky (USAF Retired)
CW3 Jim Walker (USA Retired)
SGM 'Steve'(USSF Retired)
SFC Thomas Payne (USSF Retired)

Amy S who said, "I know why you call it Special Operations, it's because you can do anything you want to do."

There was another person who helped me through the entire ordeal of writing this story and desires to remain anonymous. For her I thank our God.

Abbreviations
(And other useful information)

AAR - After Action Report or Review
AGR - Active Guard and Reserve -
AK-47 - Russian Assault Rifle-7.62mm - An actual assault rifle.
AN/PVS-5/7 NVG - Night Vision Goggles.
AT - Annual Training/Summer Camp
A-Team - Twelve man Special Forces unit.
BAMC - Brooke Army Medical Center - San Antonio, TX
BMNT - Begin Morning Nautical Twilight
BP - Border Patrol
CAB - Combat Action Badge
CDC - Counterdrug Coordinator
CDTF - Counterdrug Task Force
CG - Commanding General
CI - Confidential Informant
CIB - Combat Infantryman Badge -
CMB - Combat Medic Badge
CMIS - Counterdrug Management Information System
CO - Commanding Officer
CP - Command Post
DD-214 - Official Record of Military Service
DFAC - Dining Facility or Mess Hall
DOD - U.S. Department of Defense - The Pentagon.
DZ - Drop Zone for parachute landings
ERO - Engine Running On/Off Load
FSE - Forward Support Element
HF - High Frequency Radio
HQ - Headquarters - Military Commander's Location
HUMVEE - (Hummer) Wheeled vehicle that replaced the Jeep.
IG - Inspector General
Intel - Intelligence
IPB - Intelligence Preparation of the Battlefield
LRRP - Long Range Reconnaissance Patrol
LRS - Long Range Surveillance (Detachment or Company)
LZ - Landing Zone-Helicopter
M16A1 - U.S. Battle Rifle - 5.56mm

MFR - Memorandum For Record
MOS - Military Occupational Specialty - Army job designator
MRE - Meals, Ready to Eat - U.S. military field rations
MSS - Mission Support Site
MTOE - Unit Organization and Equipment
NCO - Noncommissioned Officer-A sergeant
NCOER - NCO Evaluation Report
NCOES - Noncommissioned Officer Education System
NCOIC - Noncommissioned Officer in Charge
NGB - National Guard Bureau
ODA - Operational Detachment Alpha - Special Forces A-Team
OER - Officer Evaluation Report
OPSEC - Operations Security
OPTEMPO - Operations Tempo
PAIC - Border Patrol Agent In Charge
PX - Post Exchange-Retail shopping store on an Army Base.
REMF - Rear Echelon Military Forces (or words to that effect
RIP - Relief in Place - A new unit takes over from the old unit
R+R - Rest and Recuperation
RTI - Regional Training Institute
S1 - U.S. Army staff officer responsible for personnel
S2 - U.S. Army staff officer responsible for Intelligence
S3 - U.S. Army staff officer responsible for operations
S4 - U.S. Army staff officer responsible for logistics
SATCOM - A military radio utilizing a satellite repeater
SF - Special Forces - The U.S. Army Green Berets
SITREP - Situation Report
SMU - Special Mission Unit
SOF - Special Operations Forces
SOP - Standing Operating Procedures -
SUV - Sports Utility Vehicle
TF - Task Force
TOC - Tactical Operations Center - Military Command Post
UCMJ - Uniform Code of Military Justice - Military Penal Code

Time

Coordinated Universal Time (UTC)-12:00AM-12:00PM

The Road to War

People have sought to escape anxiety and the depression that follows by ingesting substances that make them feel good and takes their mind off the troubles of this life. Alcohol and various organic substances have been used since the beginning of man's history to facilitate this escape. Unfortunately, not everyone is able to function in a responsible manner while indulging in this escapism and their irresponsibility has resulted in injury and death to themselves and those around them. Another effect of this activity has been an eroding of the will power necessary in every person to fulfill their obligations with regard to paying their own way by working.

Since the beginning of human government there have been measures taken to restrict the use of alcohol and these substances when it became apparent that their unrestricted use was causing a noticeable increase in accidents and untimely deaths by misadventure. The measures taken by the U.S. government over the years began to unravel in the 1920's with the rise of the Jazz Age.

With the music came the alcohol and substance abuse. The first attempt at dealing with the effects of alcohol abuse was Prohibition, a Constitutional amendment that banned the manufacture and sale of alcohol in the U.S. By the time it was repealed in 1933, there was more alcohol consumption in the country than there was before the ban.

After WWII, the country was ready for some good times after having come out of the Great Depression and the war. The music was changing and the unrestricted flow of alcohol livened up the party but people wanted more. The Beatniks introduced them to marijuana and heroin. By the end of the 50s there was a small network of people that had found ways to avoid the inevitable 9 to 5 and they were congregating on the east and west coasts where they created small colonies of full time escapees.

When a Swiss scientist accidentally discovered the psychological effects of a substance he was working on, he named it LSD. In the 1950s it was tested by the CIA as a truth serum and by the Army as a chemical warfare weapon. Somehow it got out of the labs and that had unintended consequences that continue to this day.

On the east coast, Timothy Leary, a Harvard University professor took some and proclaimed the millennium. He and some colleagues built a religion around psychedelic drugs and went forth advising people to "Turn on, tune in, and drop out." On the west coast, a successful novelist, Ken Kesey, who was a guinea pig for Veterans Administration tests of psychedelic drugs, financed a series of 'Acid Tests.' Calling themselves the 'Merry Pranksters,' they poured large quantities of LSD into barrels of Kool Aid and gave it away to crowds of young people while a loud rock band named the Grateful Dead blasted away at the audience's efforts to hold onto their sanity. The sixties had arrived.

The war in Vietnam brought home a generation of young men who were ready for anything after having escaped death and returned to a new world. The music changed, an anti-war movement had sprung up fueled by fear of conscription and huge crowds of young people were thumbing their noses at their parents and the restrictions imposed upon them by their society; revolution was in the air.

President Nixon was elected in 1968 and promised to bring 'law and order' to the country and a War on Drugs. It is easy to see why he did, in retrospect, because of the collateral damage caused by rampant drug abuse. It was on a scale unheard of in our history and was a major contribution to the chaos of the great rebellion of the 1960's by the young people of the U.S. and other western nations.

President Nixon first used the term in 1971 when he described illegal drugs as public enemy number one. He initiated Operation Intercept which forced the Mexican government to allow the spraying of herbicides on the marijuana plantations south of the border. Alcohol was now passé to the youth. Marijuana was refined into hashish, Peyote was synthesized and called Mescaline, Magic mushrooms became Psilocybin, LSD, Speed, Cocaine, Crack; the list is now endless as new designer drugs, permutations of older discoveries, popped up so fast that the government couldn't keep up with them.

Nixon created the Drug Enforcement Administration in 1973 in an attempt to focus law enforcement efforts on the large drug trafficking organizations that were supplying the demand for more and better drugs. The Vietnam War ended, the music changed and now people needed something new.

The Mexican marijuana market had dried up due to health problems the smokers were experiencing. The DEA was in Mexico spraying paraquat, an 'agent orange' type herbicide that was being used to try and stop marijuana production in Mexico. Congress felt the pressure from their constituents and cut funding for the spraying operations in Mexico. With the Mexican weed out of favor, the Colombians stepped in with high grade 'Santa Marta Gold' to replace the 'Acapulco Gold' that had been the 'gold' standard. When the music changed again, the Colombians used their marijuana distribution network to flood the U.S. with cocaine to go along with the 'disco' craze.

President Reagan restarted the war after President Carter had ignored the problem. He called for renewed efforts and additional funding for law enforcement. The Mexican drug traffickers fought back. They kidnapped, tortured, and murdered a DEA agent named Enrique (Kiki) Camarena. Mexico now became a trans-shipment point for Colombian cocaine but the main activity was in the Caribbean with South Florida now becoming the place to go for coke.

Meanwhile, the Cold War was still on and one of its battlefields was in Central America. The government of Nicaragua had been overthrown by the Sandinista revolutionary movement and the new government was being supported by the Soviet Union and its surrogate, Cuba. The U.S. was fighting communist subversion in El Salvador and we were supporting counter-revolutionaries in Nicaragua known as the 'Contras.' The U.S. Congress cut off funding for the Contras and so they, along with the Sandinistas, got involved with cocaine trafficking in order to pay for their war.

The U.S. was conducting what Congress later called an illegal covert operation, now known as Iran/Contra. In the freedom of the gray area that existed, some of the more pragmatic intelligence officers that considered the communist threat more dangerous than the drug threat, turned a blind eye to their more entrepreneurial helpers. Some of the contract pilots, who didn't like flying an empty airplane back to the states after delivering a shipment of weapons and supplies, were implicated in the flood of cocaine that was cooked into crack, which inundated South Central Los Angeles.

In 1986, President Reagan added drug trafficking to the list of threats to our national security. This directed the military to get involved in drug interdiction. In 1988, Colors, a cop movie about gangs and crack came out and it portrayed a situation that was unimaginable for mainstream Americans.

By 1989 President Bush and the congress enacted a National Drug Control Policy. Among the many initiatives spawned by this legislation was allocated funds for counterdrug operations by the National Guard and all 50 states established programs.

The best history that I ever read about drug abuse and trafficking is "Desperados" by Elaine Shannon. It details the convoluted thinking by our society about the problems of drug abuse and drug trafficking. The corrupting influence of drug money became so powerful that countries were being run by men who were beholden to the drug lords.

My personal experiences in the 60's had made me aware that drug smuggling and trafficking existed but I didn't really know the extent of either. The media had given us some great movies like the French Connection, Serpico and a long running TV show called Miami Vice. These movies fired the imagination of patrol cops like me and when the military got involved, the idea of going toe to toe with Colombian drug lords with Mac 10 submachine guns on the Rio Grande was an enticing one. 'It ain't much of a war but it's the only war we got,' was a popular joke during the early years of the Vietnam War, and some of us who were in the military at this time, when we were invited to the drug war, quickly saw the logic and signed up. A few years after I raised my hand, the Soviet Union collapsed and the drug war was the only game in town.

No one serving as a soldier entangles himself with the affairs of this life, that he may please the one who enlisted him.
2 Timothy 2:4

Chapter 1-1977

Back in the National Guard

I joined the Texas Guard, probably for a kind of catharsis. While I was in high school, I had joined the Arizona National Guard's 158th Infantry Regiment in 1963 when I was 17. It was a great experience. In 1965 I joined the regular army and served in Germany which was also a great experience before volunteering for Vietnam. After serving in the 3rd Infantry (Old Guard) in Washington DC, I was compassionately reassigned to Fort MacArthur, California because my father was dying. I took part in Project Transition and partied in San Diego until my enlistment was up. Then I returned to Fort Mac and was honorably discharged there in 1969.

After living in San Diego, I returned to Corona del Mar where I had discovered surfing before I joined the Army and tried college on the GI Bill. After bombing in college, I got married and we moved to Marin County in northern California where I worked as a horse wrangler on a ranch. After my wife left, I drifted down to Mexico, Central and South America for several years. Upon my return to the states in 1976, I hitchhiked to Austin, Texas to see Rodney, my closest friend in Vietnam. Rodney had gotten into the drug trade and offered to bring me in but I found work on a drilling rig.

By 1977, I was in the throes of PTSD as a result of my experiences as an infantry sergeant in Vietnam. In 1967 when I returned to the states, I occasionally thought about the orders I had given and the things that had happened as a result. At random times these film clips would pop into my mind and I would relive those moments in combat. I would clench my teeth while trying to will an alternate ending to the grim consequences of those orders. I could be with a group of people or by myself when these images would push aside whatever my mind was involved in and I would find myself back in Vietnam.

I was working as a fuel truck driver for a construction company when the mechanic told me I could make a lot more money if I could work on diesel engines. I went to a National Guard recruiter and he assured me that they had a lot of diesel engines and that they would train me to be a mechanic.

The Guard was offering one-year enlistments so I raised my hand. I came in as a private E-2, my rank at discharge in 1969. I had been a sergeant then but due to some of my unsanctioned behavior, the UCMJ provided the remedy but I left active duty with an Honorable Discharge. My first unit was a heavy maintenance battalion. When I went with them to Fort Hood for two weeks of annual training (AT) or summer camp, I realized that I didn't belong there. The real mechanics were in the ordnance depots; all I did was undo the motor-mounts so that we could lift the motor out and ship it to the depot.

During the second week, we went to the field and did nothing. I was sitting on a fallen tree feeling sorry for myself. A man in a jeep stopped and asked what I was doing. I saw that he was an officer with a cross on his collar, a chaplain. He insisted that I talk to him so I poured out my tale of woe. He reached in his pocket, pulled out a business card, punched a hole on the edge and handed it to me. It was a TS (Tough Shit) Card. I had read about them in books on WWII. He gave me a smile and drove off leaving me smiling too.

The Blues (Recon)

After a dismal year with the heavy maintenance battalion where I was a fish out of water, I transferred to the Blues; the Recon Platoon of Troop D (Air), 1st Squadron, 124th Cavalry. The air cavalry troop was a Vietnam War legacy unit. The Blues of the 1st Squadron, 9th Cavalry of the 1st Cavalry Division were legendary in Vietnam and were portrayed in the movie "Apocalypse Now."

There were many Vietnam vets in the troop. We would get drunk and tell stories about the strange and funny things that happened to us in the war. We never talked about the other thing, except in an offhanded manner. We all agreed though that it was easy to get killed in strange ways.

The Recon Platoon prided themselves on helicopter rappelling. Standing on the skid of a Huey helicopter hovering a hundred feet above the ground, hooked in to ropes anchored to the helicopter before stepping off is a piece of cake if everyone knows what they are doing.

Once, when we were performing a rappelling demonstration for the air show at Bergstrom Air Force Base, the rappel master inserted the ropes into my snap-link backwards in what is known as a 'fatal hookup.' When I pushed off the skid and started to drop, the ropes bunched up in my snap-link, abruptly stopping my descent. I swung under the skid and up toward the bottom of the helicopter. I stopped my upward swing by putting my boots on the bottom of the chopper and pushing. As I fell backwards I looked down at my snap-link and saw the ropes were coming out so I grabbed them with my gloved hands and slid 100 feet to the ground. I showed my snap-link to the belay man and then pulled the smoking gloves off my hands before moving off the field.

My first summer camp at Fort Hood was fun and the beer was more than plentiful. Training was easy; we did some patrolling and worked a lot with the four Huey helicopters that were our primary means of transportation. Back in Austin, after every weekend drill, we would assemble at the Saddle Club and get as drunk as we could before closing time. My second summer camp was less satisfactory. We didn't do much beside drinking a lot of beer. One of the guys got a jeep stuck in a creek and the platoon spent a week getting it out.

After two years with recon, I was promoted to sergeant. As a sergeant it was my duty to train the men in my squad. There were published training standards for a recon platoon but our leaders seemed to be ignoring them. My experiences in combat had taught me that realistic training was the only way to prepare for it. Remembering my own failures to correct my men when they failed to keep the machine gun ammunition with the gun whenever we set up, costing them their lives, the only positive thing that I could see to do was to try and redeem myself by being a better sergeant.

I pushed my way into planning the training of the Recon Platoon, hoping to use lessons I had learned to train the next generation of scouts. My recommendations were deemed extreme and I soon found myself standing in front of the troop commander. After trying to placate my demands for more realistic training by telling me that he was going to promote me to SSG and then, when I persisted, offering to send me to OCS, the Major's face got red when I replied, "that's fine sir, but what about the training?" He then said, "I want you out of my unit, Sergeant Murry, you're too good a soldier and you don't belong here!"

Now I had a new mantra, “I want you out of my unit; you’re too good a soldier.” This was the ultimate conundrum for a soldier. Good soldiers were held at a premium and a commander would never get rid of someone for being too good. With those words ringing in my ears, I went to Camp Mabry and filed a complaint with the Inspector General. He listened sympathetically and offered to get me into another unit, while not touching the basis for my complaint.

I decided to fight the Major and let it be known that if he wanted me out of the unit he would have to throw me out. I was transferred from Recon to the refuelers in the headquarters platoon. The next drill was pretty lonely until the refueler NCOIC, a black sergeant, took me aside and welcomed me to his unit. All the refuelers were black. He knew about the background of my case and decided that I was being squashed by the system.

Salvation

On the Monday following drill I was sitting in my tumbled down house out in the country looking at a couple of paperback books. One was called “Franny and Zooey,” by J.D. Salinger. I can’t tell you why I bought that book because I only read military history. In the story about Franny, a college girl, she talks about reading a book called “The Way of a Pilgrim.” She tells her boyfriend that the book is about a Russian peasant who wandered across Russia until he found the ‘Jesus Prayer,’ “Lord Jesus Christ, have mercy on me a sinner.” Her boyfriend is interested in other things and her behavior is disconcerting. The story ends with her praying and her boyfriend leaving. I can’t say how I managed to read the whole story but I did

The other book I had found in the Service Club at North Fort Hood during our last Annual Training. It was “The Way of a Pilgrim,” (Translated by R.M. French). When I saw it I was amazed. I thought the book mentioned in Salinger’s story was some kind of literary device. The Service Club was giving these books away so I grabbed it. I read the book and found the Jesus Prayer. For the past year I had been considering a run at Ranger School. I had a 5 mile running course that took me through the countryside of north Austin. I ran that course three times a week and after I found the book, as I ran, I said the Jesus Prayer to the cadence of my feet hitting the ground.

That morning I was feeling pretty discouraged. There were no slots for Ranger school in the refuelers and today was a running day. I heard the sound of a car driving into my yard and when I looked through the screen door I saw my First Sergeant get out of the car and walk toward the house. "What are you doing here, Top?" I said as I held the door open for him to come in. He didn't say anything as we walked over to some chairs. I sat down and looked up at him.

The atmosphere had changed and I was like one in a trance. He stood there, gently swaying. "Murry, you ever been saved?" he asked, slurring his words slightly. "No," I said, not knowing what he was talking about but knowing. "Do you want to get saved?" Immediately I said "Yes," and like a mighty rushing wind, the Lord Jesus came in!

All at once I began to weep and eventually I began saying, "Thank you, Lord, thank you, Lord," without any conscious realization about who I was thanking or why. The First Sergeant didn't seem to consider my behavior unusual and after a few moments said, "Well, come along with me and I'll get you saved."

He led me out to his car and we drove towards Austin. I was weeping and thanking the Lord while he was driving us to some place. Eventually, I was able to discern the fact that he was actually pretty drunk. Normally I might have offered to drive or at least remonstrated with him as he careened through the city streets, but today was different and it was if I was above those petty concerns.

We reached our destination and pulled into the parking lot of a Baptist Church. There was a lot of activity for a Monday afternoon. Groups of people were walking toward the meeting place. The First Sergeant told me to follow him and he led me to a man in a suit standing by the entrance and introduced me to the pastor, saying, "I brought this man here to get saved." I was still weeping and the pastor probably knew what was going on with me. The fact that the First Sergeant was drunk didn't seem to upset him and he told him to take me up to the front which he did and he told me to sit on the front pew. I had no idea what was going on but I wasn't alarmed.

Something had happened to me that I couldn't explain but whatever it was, it was beyond the realm of good and evil. I sat there gently weeping while some men gave me forms to fill out. They were applications to join their congregation. One form offered me some choices regarding my 'ministry,' whatever that was, and another was for my address so they could send me some collection envelopes. I filled them out, weeping and thanking the Lord, still clueless as to what happened.

The pastor started the service and I found out that I was in the first meeting of a week-long revival. He introduced a speaker who proceeded to entertain the congregation with story after story from the Bible to which he gave a humorous slant before turning it around and slapping us upside the head with the facts of our sinful condition. At the end there was a call to come forward and one of the men escorted me up to join a small group of men and women. I was still weeping and I repeated the prayer of the speaker along with the others.

Then the whole congregation was invited to come forward to welcome us into the fellowship. Several pilots from the Air Cavalry Troop were among those who came forward and shook my hand. One of them, Bobby Boyd, was a pilot in my platoon and he appeared to be delighted to see me, weeping my eyes out in front of this large crowd of people.

The First Sergeant took me home after the service. He bought me a chicken dinner and dropped me off at my house. I continued to weep until I fell asleep that night. The next morning when I woke up I decided to return to the Baptist Church and for the rest of the week I listened to speaker after speaker as they attempted to revive the congregation. I don't know about the rest, but on Friday the honeymoon was over for me and I was as dead as a door nail.

G Company

I came to the realization that I needed to drop my feud with the Air Cavalry. I went back to Camp Mabry and transferred to G Company, (Airborne/Ranger), 143 Infantry (LRRP). The catharsis I was looking for when I joined the Guard wasn't there. When I joined G Company in 1980, PTSD had pretty much taken over my mind and there were few moments that I wasn't back in the Nam trying to undue my mistakes, my decisions, and especially the orders I gave that resulted in the deaths of some men.

Salvation Continues

After a couple of drills, I met Alan Waters, a Vietnam veteran of the 173rd Airborne Brigade. He was working as a parachute rigger and had recently been accepted as a recruit for the Austin Police Department. We were talking and he asked me if I was a Christian. I remembered Jimmy Carter saying that he was a born-again Christian and I now realized that after my conversion experience of a few months before, I was too. “Yes,” I said, and he replied, “Would you like to go to a Christian meeting sometime?” "Yes," I said again. Again, he asked, “Would you like to go to one tonight?” I answered, “yes,” again and was soon sitting on the back row of a meeting in south Austin.

The man in the front was speaking in a gentle way. He said that Christians needed to be people who called on the name of the Lord. I was sitting in the back with every flashback I had ever entertained flowing through my mind. It was like an endless barrage of direct hits, each one worse than the one before, with the main theme being that whatever happened it was my fault. I sat there and quietly said, “Oh Lord Jesus.” Suddenly the flashbacks were gone and I was experiencing peace.

I found the Lord again, like that day when I wept and couldn’t stop. I wasn’t weeping this time but I was enjoying Him. From then on, every time I had a flashback, I called out, “Oh Lord Jesus!” and in about three years my wounded heart was healed from what is now known as PTSD.

The people meeting as the church in Austin welcomed anyone who confessed that Jesus Christ was their Lord. They had received much help in understanding the Bible from two Chinese Christians named Watchman Nee and Witness Lee. I marveled at the irony of the Lord’s way to send these men’s ministry of the Word from the ‘backward’ land of China to the super power of this age so that we might experience God’s full salvation through the exercise of our human spirit that was one with the Holy Spirit.

G Company

I was placed on a recon team with Gary Parrish, the team leader. He was a former artillery captain and 101st Veteran who had been ‘rifed’ after serving as a forward observer in Vietnam. Ralph Carr, the assist team leader was a Dallas police officer. John Kulas was the RTO. He had just returned from his basic and MOS training. Gary was a very competent leader and his map reading skills kept us from traveling farther than was needed. I remained in G Company for almost a year. The training was much better. We practiced patrolling and immediate action drills. One very dark night while walking point for my team at Camp Swift, I stepped into a gully and dropped about five feet, landing on my rucksack. I was stunned but managed to whisper "stop" to the man following me so he wouldn't land on top of me while I lay there for a moment, wondering what I was doing back in the military.

High-Frequency radio communications was new to me. We used long wire antennas and code-burst devices. We were expected to learn Morse Code and I endeavored to get proficient. Texas had had its own jump school but it shut down a month before I joined G Company, so getting airborne qualified would have to wait until Fort Benning.

During one of the drills at Camp Swift we were taking a break in a platoon formation when one of the men on my team said to me, "What's with you, Murry? You look different." I turned to him and noticed that most of the platoon was looking at me, "I found the Lord Jesus," I replied. There was a long silence while the others stared at me until the platoon sergeant shouted "fall in!" As I continued to call on the Lord and pursue Him I began to sense that my time in the military was coming to an end. My enlistment was up in 1981 and while I considered reenlisting, I felt that the Lord had another path for me so I turned in my gear and took my discharge.

Alan got me a job at Camp Mabry doing landscape maintenance. Then some men in the church hired me to work in their landscaping business. I mowed lawns for the next few years until Alan suggested I become a police officer. I thought that would be interesting. It reminded me of the song "Clouds" by Joni Mitchell. After completing the peace officer course at a regional academy, I was hired by the Austin Police Department to work at the city hospital.

June 1985: Police Work: It's Not For Everyone!

On my first day as a police officer, I was confronted by my shift sergeant and two others who were all former Special Forces soldiers of the post-Vietnam Army. Each one wore the Special Forces crest for a tie clasp on his uniform. They wanted me to be impressed with the fact that they were the thin blue line that had cleaned up Dodge and they wanted to know, if I was sure that I had what it takes to stand shoulder to shoulder with them?

I laughed and told them that I had fought toe to toe with the North Vietnamese Army so I wasn't too worried about what the criminals of Austin, TX had to offer. That was a mistake. They had spent the last ten years listening to Vietnam vets in the Army telling them that since they hadn't been there they needed to shut up.

The second day, as I left the Police Office and was crossing the street on my way to the ER, a call came out on the radio that a woman in the ER checkout office had been assaulted by a patient. The suspect description followed: a black male carrying a child. Coming towards me about twenty feet in front of me was a man matching that description and he looked angry.

I stopped and asked him if I could talk to him. He was angry and began shouting that the woman had insulted him in front of his kid. I assured him that we could work something out and he soon calmed down. He had just started again to explain what had happened when I heard the sound of running feet coming from behind me. Before I could say something, two of my fellow officers jumped between me and him. One grabbed the child and the fight was on.

I had to help my fellow officers subdue the man and he was taken to jail. I took the report from the victim who told me that the man had thrown a pen on the counter and it had bounced through a small slot at the bottom of the window and landed in her lap. I talked to another employee who witnessed the event. She told me that the checkout clerk was ragging the man about a previous unpaid bill.

The story didn't end there. A few days later the man held a press conference in front of APD Headquarters where he had handcuffed himself to a banister and railed about police brutality. There is a lesson here that all police officers should know: the first officer on the scene handles the situation and the others stay out of it and back him up.

A few weeks after I started working, a money bag from the cafeteria was stolen from the police office during the night shift. All the night shift officers including myself were sent to a polygraph examiner. He must have been new because after I was wired up, the first thing he said was "Have you ever stolen anything?" When I answered "Yes" he got flustered and asked, "Did you steal the money bag from the police office?" I answered "No." After a minute of his looking at the printout he said, "OK, you can go."

My probation was for six months. During that time I had been reprimanded several times for preaching the gospel while on duty to people I thought would be interested. This caused my sergeant to recommend that I be fired at the end of my probation. The night before my probation was up I was walking around the outside of the hospital making security checks. I stepped through a gate that led to a door for the Alternative Birthing Center and saw a plastic garbage bag on the ground at the bottom of some stairs. Trash was not that unusual around the hospital but I decided to look inside and found a zippered money bag like ones we stored in the police office. I radioed my sergeant who was in the office typing up a form with his recommendation that I be terminated. I recommended that he come through the hospital to meet me. He did and quickly determined that the money bag had been stolen from the cafeteria.

The afternoon shift officers were still in the police office and two of them volunteered to watch the area from across the street. Another officer was assigned to lie on the roof above the area and I was assigned to watch the bag from inside the hospital. About two hours later the man on the roof got tired of laying on the gravel roof and stood up just two men were parking their car near the entrance to the Alternative Birthing Center.

They saw the officer stand up and drove away. The officers across the street gave chase and pulled them over. They were cooks in the cafeteria but since they hadn't touched the bag they had to let them go.

My sergeant was pretty happy with my part in the event and at the end of my shift he showed me the form recommending that I be fired before he tore it up and had me sign the new one that recommended that I be retained at the end of my probation which was that morning. It took a year before I was finally accepted by the other officers and when I was, they even recommended me for the annual Pride of the City award which I was given.

Working the city hospital was not the same as working the street. Every night I went to work and spent 8 hours dealing with the aftermath of human interaction with anger, drugs, and alcohol. These contributing factors blurred what good judgment might have been available to these battered and broken souls who were rolled into the ER by EMS, or who were dropped off on the ambulance dock by people who were in too much of a hurry to tell us what had happened.

Murder, assault, rape, child abuse, and violent mental meltdowns were among the crimes I was expected to investigate, photograph, and report while restraining suspects and maintaining order among the drunken and drugged friends and family who sat in the waiting room.

Occasionally I had to tell some hopeful family member that there was no hope for the survival of their loved one and endure their screams until I could walk them outside where they wouldn't disturb the living. We made lots of arrests and when it was slow we wrote parking tickets or made traffic stops for speeding on the streets near the hospital.

While I was on the night shift, the department sent me to several law enforcement continuing education courses at the Texas Department of Public Safety headquarters in Austin. The Criminal Investigator's course was very interesting and useful for our work but I was fascinated by the Police Photography course. Beside the routine photographs we took in the ER, I began to experiment with telephoto lens using a 35mm camera that I bought from another officer at the department. Tom Stephens had a lot of knowledge about photography that he had gained from Special Forces and film school at UT. We tried different things on slow nights and that helped me understand how difficult it was to get facial recognition photos in low light using telephoto lenses.

Marriage

A year after I became a police officer I married a women in the church. Before we got married I said that the only thing I expected from her was to keep the house quiet while I slept during the day because I worked the night shift at the city hospital. She told me she couldn't do it but I married her anyway.

We were 40 year-old battered refugees from the 60's; we had lots of baggage as we began our married life. She took good care of me and always had a good meal on the table, clean clothes in my drawer, and a clean house to live in. A year after we were married she brought forth our son Richard.

My wife was from Rochester, New York where they thought and did things completely opposite of what I, from Corona del Mar, California, believed was the best approach to the things of this life. Richard began to function at an early age as the peacemaker and when that wasn't possible, the referee, as we worked through our differences.

The only thing that kept us from following the way of the world and divorcing was our vision of God's eternal purpose and His economy for carrying it out. We learned these things from the ministry of the age that was brought to us by Watchman Nee and Witness Lee, two brothers from mainland China. They showed us the truth in the Bible regarding Christ and the church and how we might experience all the promises in the New Testament by the exercise of our human spirit, made alive when we were born again.

1987: Austin Police Department/Hospital Police (Day Shift).

For two years I worked the 11 PM to 7 AM shift. When I transferred to the day shift at the request of my wife, the night shift cops told me to take the bulletproof panel in the front of my vest and put it in the back. When I asked why, one of them said, "To protect you from knives; you're on the day shift!" How right they were.

Another night shift patrolman was assigned to the new position of investigator. One of the first things he did was to make a new photography policy and as a result we would never see the pictures that we took in the ER. I had gone to a Police Photography school at the State Police HQ so I told him that we should be able to see the photos we took and learn from our mistakes. He decided that this was a challenge to his new status as an investigator and some heated arguments ensued. After one of them, in front of several other officers, when my logic had trumped his vanity, he stepped up behind me as I was leaving and whispered, "I'll get even!" and he did.

One morning, a few weeks later, we had two homicide/suicide victims in the ER and I took the photos. I got busy with the increased activity and forgot to fill in the information for them. The investigator discovered my error and accused me of purposely interfering with a homicide investigation. The chief wrote me a letter of reprimand which I rebutted. Most of the other officers had made an occasional mistake with the photo information and they didn't get a reprimand, so I felt like the investigator was making good his threat. I explained all of it to the chief but he chose to let the reprimand stand. It didn't affect my employment but it did reflect on my integrity; I knew I would never get a promotion in the department.

Back In The Guard

After three years with Hospital Police, my former night shift supervisor, Sergeant Mike Hargett, who had been a Captain in an Army Reserve Special Forces unit and had recently transferred to the Texas National Guard, told me that he was getting command of the Long Range Surveillance Detachment (LRSD). It was a recently activated unit that was rising from the ashes of my old outfit, the Blues (Recon) Platoon of the Air Cavalry Troop that I had served with 8 years before.

Hargett asked me if I would like to reenlist in the Guard and be his driver in the LRSD. I thought about it for a short time and that evening I told my wife that I was going back in the Guard, part time. It was something I wanted to do and it would give us a little more money. She didn't have any objections so the next time I saw Hargett, I told him, "I'm in." There were no overt wars that the U.S. was fighting but one never knew. Since I didn't have any premonitions or overwhelming feelings of patriotism, I guess I was just bored. In 1981 when I got out I was sure I would never wear the uniform again but, like malaria, when it gets in your blood, it comes back every now and then.

Recon, Troop D (Air), 124 Cavalry

Company G (Airborne-Ranger) 143 Infantry (LRRP)

Author with family and at work with Hospital Police

ER Ambulance Dock

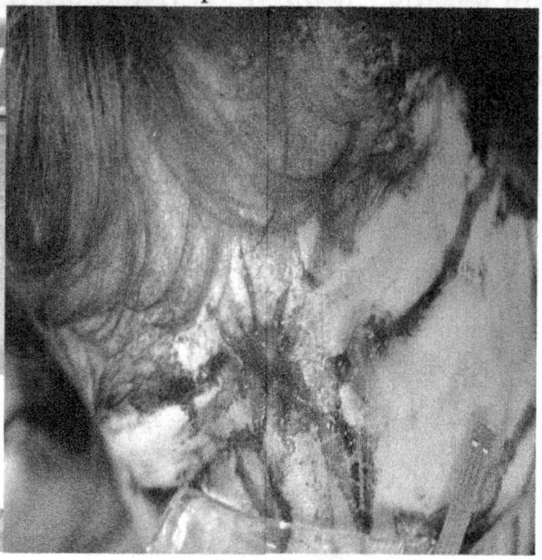

GSW to the head (she survived)

LRSD-SFC Payne-Airborne Refresher-Rigging Rucks For Airborne Operation

And you will be hearing of wars and rumors of wars. See that you are not alarmed, for it must happen; but the end is not yet.
Matthew 24:6

Chapter 2 1988

Long Range Surveillance Detachment (LRSD)

After I told Sergeant Hargett I was in; I went to the recruiter and signed up, took a physical in San Antonio and was cleared for enlistment in the Texas Army National Guard. I had to take the ASVAB test which is used for job qualification and when I asked the recruiter how I did, he said, "You can do anything you want." I asked to see my scores and was surprised to see that they had all dramatically improved. I took the same test 12 years before and I knew I wasn't any smarter so I had to thank the Lord for his word in Romans 12:1, "be renewed in the spirit of your mind."

The LRSD was being formed as a unit of the 124th Cavalry by disbanding the Recon (Blues) Platoon of the Air Cavalry Troop and forming an Airborne Ranger type reconnaissance unit. The army called it a surveillance unit because they were trying to keep it from being associated with the Long Range Reconnaissance Patrol (LRRP) units of the Vietnam War. This was the final stretch of the Cold War and some army leaders felt that an 'eyes behind the enemy lines' capability would counter any anti-satellite threat for the big war in Europe.

Captain Hargett decided to have the NCOs of the Blues Platoon participate in the selection of the Detachment Sergeant. Since most of them were neither airborne nor ranger qualified, their vote would be weighted by their own criteria which would contain a good deal of 'what's in it for me?' considerations. I was the only combat infantry veteran in the unit at that time and since no one asked me, I kept my thoughts to myself because I wanted to get along. By telling them that since they hadn't faced enemy fire, they were still unseasoned as far as I was concerned, was a sure way to start a lot of unnecessary shit; a lesson I had learned in the police department.

There were two main candidates for the position, SSG Marshall and SSG Thomas Payne. From conversations with Captain Hargett, I had heard of Marshall before. He was a Special Forces sergeant who served with an Army Reserve SF HALO Team in El Paso, TX. He was a lawyer in his civilian occupation, and he was a nice guy. I didn't know Payne, formerly from the 2nd Ranger Battalion and 5th Special Forces Group but I could see he was from G Company; wearing a tower of power on his left shoulder over his T-patch and master parachutist wings on his chest with an Expert Infantryman's Badge above them.

Some of the blues told me later about how the board went. When they asked the question, "How will you incorporate the Blues Platoon people who are not airborne qualified into the new unit, with the other incoming people who are?" SSG Payne told them that he would smoke them like a cheap cigar until they got airborne qualified or they left the detachment. SSG Marshall told them that he would work with all the people in the unit and help them get ready for airborne school. Payne became a team leader for a surveillance team and Marshall was made the Detachment Sergeant. He and Captain Hargett were involved with coordination meetings with the 124th Cav and when he came back from one of them he told me, "I don't know if they understood what we were saying or if they didn't think we knew what we were talking about."

I was assigned to the Commo section because the commander's driver was a privates' position. Several drills later it was announced that SSG Marshall would be leaving the Detachment because he was involved in a legal case which involved the 49th Armored Division Commander and there was a conflict of interest. SSG Payne became the Detachment Sergeant and then the fun began.

He confronted me in the hallway of the Armory and asked me what I was doing in the unit. I said I was there because Captain Hargett asked me to join. Before he could reply, I told him that I had something to offer as far as training went and he said, "Oh yeah, like what?" I could tell that he wasn't impressed with me being there, but I forged ahead and told him that the two things that I considered very important were night operations, especially the ability to navigate cross-country at night, and communications. Without communications, I told him, this unit doesn't have a mission. He looked at me for a moment and grunted something and walked off.

Selection of Team Leaders

The table of organization called for the team leader to be Ranger qualified. Initially, most of the ranger qualified sergeants were in G Company. When the word got out though, some transferred. Young ranger sergeants coming of active duty were steered this way if they were going to attend the University of Texas.

Once an LRS team had left the aircraft and was assembling on the drop zone after falling through the night sky and being slowed by their parachute to have an 80% chance of landing without a disabling injury, everything was now on a sergeant and the other five members of his team. There was no supervision by his superiors in the rear.

The team leader had to be an independent thinker and many of the Ranger veterans were not able to solve the many problems that occur in these small units. These sergeants would prove to be the best and worst of the leaders, but the difference was determined by their brains and their character.

1988: Single Channel Radio Operator MOSQ

The unit's first priority was for everyone to be to be MOS qualified. Both the cavalry scouts who had to reclassify as infantry and the commo section would reclassify as 31C, Single Channel Radio Operators. Both sections would have classes every drill, taught by instructors from the Army Reserve, and a two-week training course in September. We showed up for our first class and were issued a number of self-paced lessons. The lessons were on equipment and techniques that we would not be using in our unit.

When the Army decided to have Long Range Surveillance Units, they overlooked the communication plan. The LRS concept owed much of its doctrine to the Vietnam War and to units like the Long Range Reconnaissance Patrols (LRP/LRRPs), the 5th Special Forces Group Special Projects such as Project Delta, and the MACV Special Operations Group (SOG). Most of these operations were run using Frequency Modulation (FM) radios and radio relay sites or aircraft to relay their communications.

In the event of a Soviet invasion of western Europe the LRS doctrine called for stay-behind teams go underground (literally), and let the

Warsaw Pact forces pass over them. They would use High Frequency (HF) radio and data burst techniques for long distance communications with their headquarters. This type of communications was only used and taught by Special Forces. Their commo course was not available for other soldiers so until the Ranger Department developed a commo training course, the LRS units had to either find a Special Forces commo man to teach them or they had to teach themselves, using manuals and talking to people.

Our instructor, the Army Reserve sergeant, didn't show up on Sunday, the second day of our course. It wasn't until the February drill that we found out that he had quit the reserves and there wasn't anyone else in the Austin Army Reserve unit that could teach our course. We had been told that unless we were MOS qualified, the unit could not deploy to perform its wartime mission and we couldn't go to other Army schools or get promoted.

We informed our higher headquarters, the 124th Cavalry, in Waco, but they were an armored unit and most of them knew nothing about airborne ranger LRS operations. As far as the Cav was concerned, we were a pain in the ass and they did the minimum to support us.

I was put in charge of the commo section and was reading everything I could find on High Frequency communications while we taught the self-paced courses to ourselves during drill, the blind leading the blind, so to speak. I knew a guy named Carl from my time in G Company. He had worked in the commo section and was now working for EMS as a Paramedic. I went to lunch with him and picked his brain on how to form the commo section and train them so we could communicate. He gave me some good principals to work with and I started making the training schedule.

There was very little equipment at first but eventually we received some AN/PRC-74B HF Radios. We found a manual for them and some batteries and took them out to the woods near our armory during drill. I had a little experience with this radio eight years before in G Company and one of the other guys said he knew something so we assembled two of the radios and tried to make voice communications using the whip antenna that was part of each radio's equipment. We tried all the radios but none of them would talk to each other. I asked the full-time supply sergeant if he could get the radios checked out at the commo repair shop at Camp Mabry. He had me fill out the forms and agreed to take them over during the week.

I went to the commo shop on one of my days off to check on the radios and met a young sergeant named Rick. I told him who I was and asked him if he knew anything about our radios. He did, and told me he had checked them out. They needed to be re-calibrated and he had done so. "Do you know about using these radios?" I asked him.

He assured me that he did so I asked him if he would like to come over to our unit and take charge of our commo section. "There are two SSG E-6 slots and an SFC E-7 Commo Chief slot open." That got his attention and within a month he was running our commo section. He didn't have a 31C MOS so he went to MOS training with the rest of us.

We continued training as best we could until September of that year when the whole commo section was sent to Camp Swift for our two weeks of AT. There we were taken in hand by a black staff sergeant who knew his stuff and he taught us how to be single channel radio operators. He was an old soldier and he was good with radios and he was good with troops. Before our training was over, we were begging him to come to our unit and be our commo chief. I think he was tempted, but he lived in San Antonio. He wore jump wings, but he told us that he was too old to start jumping again. We all managed to pass the final examinations. We got our certificates and now all we had to do was go to airborne school and we would be LRS qualified.

A Rumor of War

Around this time, I was looking at the newspaper one day and read that the US Congress had voted to expand the role of the military in the war on drugs. By 1988, drug abuse and the collateral damage had become part of the culture. We had a stream of drug overdoses every night at the ER, as usual. There were law enforcement bulletins about a new version of cocaine called 'crack' that was fueling gang warfare in Los Angeles and creeping into Texas. The newspaper article specifically mentioned that the National Guard would be used along the US/Mexico border. I had a feeling that the LRSD would be a likely candidate for a border surveillance mission.

1988: Operations Sergeant-LRSD

SSG Payne was keeping an eye on me during the months before AT. In October he called me into his office and asked if I wanted to help out with training schedules. If there was one job that I wanted this was it. Now I could use my experience to help the CO and SSG Payne develop realistic training for our unit. At first I was just typing the training schedules which were sent to our higher headquarters and posted on the bulletin board. This was clerk work, but I attended all the training meetings and eventually my suggestions were taken more often than not, that's how I became the operations sergeant of the LRSD.

The 'New Army' Training Plan

The rebuilding of the regular army from its degraded state after Vietnam had been dramatically successful. It was a 'new' army and the whole atmosphere had changed. The level of professionalism had greatly increased. Individual and collective training standards had been carefully spelled out in manuals that were made available in quantity at the lowest unit level.

Training the individual soldier was and is sergeant's business. The training of the individual soldier was covered by a Soldier's Training Publication Manual for every Military Occupational Specialty (MOS). Task, Conditions, and Standards were defined with step by step procedures for everything a soldier was expected to not only know but be able to demonstrate in a hands-on performance evaluation. These soldiers' manuals were for all the ranks holding a particular MOS.

An infantry private had many tasks to master. The squad leaders and platoon sergeants had even more required skills and were expected not only to have mastered the privates' skills and their sergeant skills, but they were also responsible for training the soldiers under them. Of course, those soldiers who aspired to leadership and those who were ambitious could read ahead and learn the higher level skills in anticipation of using them when they were needed.

Unit collective training is in most units, officer business. Collective training was covered in a series of Field Manuals that explained the army's training philosophy, Battle Focus Training. In these publications were the methods used to develop a Mission Essential Task List

(METL), training cycles and many other aspects of the modern Army training plan. Finally, there was a manual for each type of Army unit to be used to evaluate the unit's readiness for war, the training evaluation program or ARTEP. This was the commander's tool to measure his unit's combat readiness against strict standards.

There were a lot more manuals that expounded on leadership and the tactics, techniques, and procedures that were, at that time, considered the best possible answer to any threat on the modern battlefield. Getting the leaders in the Guard to read these manuals with comprehension would be the weak link in this system for some time to come.

The 'New Army' Promotion System

The 'new' army enlisted promotion system had evolved, and soldiers competed for promotion by accumulating points for various achievements. Schools, decorations, time in service, education, etc. were given a numerical value and lists were published periodically that showed how many points a soldier needed to promote to the next rank in a particular MOS.

When soldiers are promoted to Sergeant, they become Non-Commissioned Officers (NCOs). Sergeant promotions required education and with the new training system came a new non-commissioned officer education system (NCOES). For a long time, the officers had had a complete education system and now the NCOs would have the same.

In 1973, the Army started the Sergeants Major Academy. This institution was eventually given responsibility for the entire enlisted education system. There had been some NCO academies here and there in the past but now, in the regular army, if a soldier wished to be promoted he had to attend a course that was supposed to prepare him for greater responsibility commiserate with his new rank.

The Non-Commissioned Officer Education System

The first level of NCO school was called PLDC or the Primary Leadership Development Course. This was for all soldiers that were to be promoted to Sergeant. The next was the Basic Noncommissioned Officers Course or BNCOC. This course had two phases: common core subjects that were supposed to be known and practiced by all sergeants everywhere and an MOS specific phase where, for example, an infantry sergeant staff sergeant learned the skills necessary to lead an infantry squad. The next level was Advanced Noncommissioned Course or ANCOC. The training was in two phases and trained platoon sergeants.

The Sergeants Major Academy also taught a First Sergeants course, a staff NCO course, and finally the Sergeants Major course that prepared a man to be a sergeant major on the staff or a Command Sergeant Major, the senior enlisted man in a battalion or higher unit. While this NCO education system was in place in the regular army, the Guard was still promoting people without this education.

The young sergeants coming out of the regular army were living testimonies to the efficacy of this system. As we used to say, they knew their shit. When they got to the Guard, many of them were appalled. The Guard was still way behind the Regulars in establishing and maintaining standards. Many of these NCOs had tried to maintain their 'new' army standards but the Guard usually wore them down pretty quickly. It took a strong character, real skills, and good personal traits to uphold standards and not alienate the rest of the unit.

All the improvements in the army's training and education system were very good at training the soldiers on what to think but were not very good at training them how to think. For most soldiers in the larger combat formations this was not a big problem. They had always been told what to do and they were supervised closely by the higher echelons of command that may not have been there next to them but were at the end of the cable that connected the handset to the radio. The only places where this type of leadership was not practical as far as existing technology went was in the small special mission type units where a sergeant would be expected to lead a small team of men for days at a time without any supervision from his higher chain of command.

This new unit, the LRSD, presented a unique opportunity for the true believers of the 'new' army system to put its tenants into practice. The NCOs would have the critical role in training this unit because each team would be led by a sergeant and the officers would seldom see them. The commanders in the rear had no input to this small unit leader in his daily decision making. This required a special kind of man who was still able to think independently when required and one who was able to remain focused on the mission, doing what was necessary for the accomplishment of the mission and the survival and welfare of his team.

CPT James (Mike) Hargett

Airborne!

LRSD-Learning HF Commo with the AN/PRC 74b and Long Wire Antenna

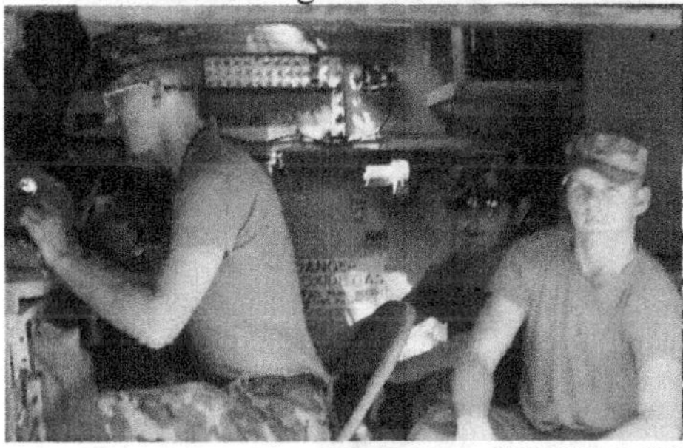

LRSD- MOSQ 31C

LRSD-Some of the originals

LRSD-Airborne Operations-SFC Payne Jumpmaster! SGT 'Bobby' Garcia (Rigger)

LRSD: CWs Mike Mullenix Sam Warren (Pilots)

LRSD-Methods of Entry

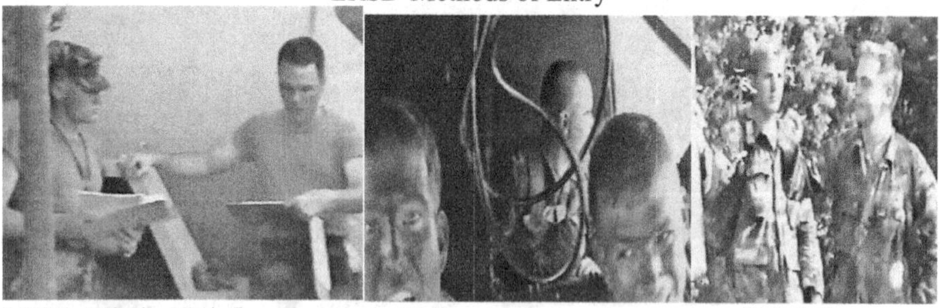

LRSD: Mission Briefing Helicopter Extraction, Debriefing

Fight the good fight of the faith…
1 Timothy 6:12

Chapter 3- October 1988

Operation Gatorade

In October, G Company conducted the first low-visibility border surveillance operation for the Texas National Guard. Code named Operation Gatorade; it was an inter-agency operation with the Border Patrol (BP) in the Big Bend National Park. It was the opening shot of the Texas National Guard's participation in the War on Drugs. It was not classified as such, but everyone involved signed a security statement and there were no press releases. There were a few stories in the newspapers. One was the result of a leak and another was the result of all the activity around Marfa.

May 1989: 'Wild Bil' Enney

One day Alan Waters came to see me at the hospital and told me that there was a guy named Bil over at Camp Mabry who wanted to meet me. I guess the Lord decided it was time for me to have a career change. Alan had brought me to the church and shepherded me for 10 years and he had helped me get my last three jobs. When I met him, he was a parachute rigger for G Company and now he was a police officer with Austin Police Department.

Bil Enney was a retired Special Forces master sergeant from Wisconsin who heard Barry Sadler's siren call, 'The Ballad of the Green Berets' in 1966 and signed up. He had fought in Southeast Asia with MACVSOG, running recon in Laos and Cambodia and working with Thailand Special Forces on some special projects.

Bil had also served in Panama where he helped run a jump school for Central and South American soldiers. LTC 'Bo' Gritz was his Battalion Commander, and it was clear that he hadn't like 'Bo.'

"That 'so and so' would give you the choice of an Article 15 or a sparring session with him while he practiced his Tae Kwon Do." Bil must have chosen a sparring session with him and he had neither forgiven nor forgotten 'Bo' Gritz.

Bil served with the 10th Special Forces Group in Germany, where he was the first medic to be promoted to Team Sergeant. When he was stationed with the 1st Group on Okinawa, he had married a local woman. They had two children, a son and a daughter, and he loved them very much. His last assignment before he retired was as a senior regular army advisor to the Texas National Guard. When he was told to prepare for a tour in Germany he decided to retire and stay in Austin so his kids could finish high school without disruption.

In 1988 he was working as a clerk in the personnel department of the Texas Army National Guard. When the Pentagon notified the Guard that there was a mandate from congress and funding for the military to support law enforcement narcotics investigations, the staff at Texas National Guard Headquarters got nervous, especially when they found out that one of the missions being considered was sending small teams to the border to watch for drug smugglers along the Rio Grande River.

The Chief of Staff at that time, Colonel Fred Jones, didn't shy away. He went to a retired army colonel named Kurt Hurst who was working at Camp Mabry. Hurst had served as the first operations officer for Delta Force when it was being stood up under Colonel Charlie Beckwith who had also retired in Austin.

When Colonel Jones asked for help with the planning for the first border surveillance mission, Colonel Hurst asked Bil to help him. Bil had served with Colonel Hurst in the 10th Group in Germany. The two of them planned the operation and wrote the OPLAN for Gatorade on Colonel Hurst's daughter's computer in the back of a flower shop being run by Colonel Hurst's mother. This was the first operation conducted with the Border Patrol in the Big Bend Park during the fall of 1988.

They worked with the full-time staff of G Company to plan and conduct the first mission. There were no major glitches, so the plans were made for a second mission. Colonel Hurst had some other matters that needed his attention, so he recommended that Bil be given the job of running the next mission.

Bil ran the next mission in the spring of 1989 and once again managed to win the approval of Operation Alliance, the Border Patrol, the senior leadership of the National Guard Bureau, and the Texas National Guard. There was no funding for a civilian so Bil continued to work for the Texas National Guard at GS-5 wages while becoming the de facto head of the Counterdrug Program.

Alan had been visiting his friend Captain Monte Akers and had met Bil. They talked about the Guard becoming involved in the War on Drugs and G Company's secret missions in Big Bend Park with the Border Patrol. Bil was looking for some other people to conduct these missions. Alan had told him that he knew just the guy for that kind of work and that was how I found myself sitting in front of Bil and answering his questions.

He was not a big man, but he was a presence. He had a mischievous demeanor, ready to see the humor in anything. We made some small talk and I mentioned the name of an SF sergeant who had been the senior Army advisor a few years before. He let out a few explicatives and let me know that he wasn't very impressed with the sergeant's professionalism. "Neither were we," I said. "He was always too busy working on his correspondence courses to be of much help to us."

Bil now knew that we had the same view of military service, so he allowed that he was involved with some operations being conducted in support of law enforcement. Since I knew about the recent G Company operation through an old friend who would get in trouble if Bil found out that he told me, I had to play it cool while Bil slowly unraveled his purpose in talking to me. He asked me what my job was. I told him I was the operations sergeant for the Long Range Surveillance Detachment. He wanted to know how we were doing on commo. I told him that we were just getting started, that we had the old AN/PRC-74 radios and that we were moving at a slow place toward commo proficiency.

Then he asked if I had heard anything about the recent operations. There had been an article in the Houston newspaper about G Company doing a mission on the border and I mentioned it. Now he could lay a few of his cards on the table. He asked if the LRSD had some people who could conduct real world operations. I said we could, but the commo was a problem.

Then I went for it and told him, that in my opinion, even if we were commo proficient; high frequency communications was not the way to go for operations supporting law enforcement. It would be better to get Satellite Communications (SATCOM) radios, I told him; the troops could talk directly to a cop without having to send their messages through long wire antennas using one-time pads and digital burst devices. "Cops talk on the radio all the time," I said, "for them to have to sit there and wait for longer than one minute for a reply to their question would drive them crazy."

I found out later that Bil liked all my answers and especially what I said about commo. They had a lot of commo problems with G Company and during the second operation some regular army commo people from Fort Bragg came to Texas where they were infiltrated into Big Bend and linked up with several teams. They had SATCOM radios and they saved the day. We talked about other training and personnel issues and he thanked me for coming by. I told him that if he was looking for someone to help him with these operations, then I would like him to meet SFC Payne, our detachment sergeant, who was a student at UT at the time. Bill agreed to talk to him and I left.

1989: SFC Tom Payne

SFC Tom Payne, the detachment sergeant and I had a rocky start when I first met him at a LRSD drill. Since then he had seen that I could do training schedules and other operations type work that nobody else knew how to do or wanted to do. He had taken me out of commo and made me the operations sergeant. Since I was a 'leg,' I had to endure a constant stream of airborne insults but the troops accepted my work, so I felt like a member of the unit.

Tom was a Californian like me. He became interested in military things at an early age. His interest was in tanks and other armored vehicles. When he was a teenager, he was saving money to buy a surplus army half-track to drive on the streets of his hometown. When he was old enough to join the military, he told the recruiters that he wanted to be in an Army Reserve armor unit. The recruiters told him that there weren't any armor units in Northern California where he lived. Then the recruiter asked him, "Do you like to go camping?" Tom said yes and that's how he found himself in Special Forces.

He completed basic and infantry training, jump school and the SF qualifications course. After a few years of serving in the reserve SF unit, Tom decided to go on active duty and enlisted for service with the Rangers. He was assigned to the 2nd Ranger Battalion and it was there that he saw a vision of military perfection. He completed Ranger School and was not happy when he received orders reassigning him to a Special Forces Group.

The 1st Special Forces Group had been disbanded after Vietnam and was now being stood up again. Every SF qualified soldier in the army that wasn't already in SF was transferred back into SF to fill the new unit. Tom chose to go to the 5th SF Group at Ft Bragg and several weeks later the 2nd Ranger Battalion was jumping onto the Point Salinas Airport on the island of Grenada. Tom missed the only combat jump that was made by U.S. forces since Vietnam and he was not happy.

He served out his enlistment with the 5th Group, where he did special reconnaissance and served with a Special Projects group that evaluated the security of strategic facilities around the world. After his discharge he moved to Austin and enrolled at the University of Texas where he studied Persian and was hoping to get hired by the CIA when he completed his education.

I contacted Tom after my first meeting with Bil and told him that Bil wanted to meet him. He agreed to go with me. A few days later I made an appointment and Tom and I went back to Building One to talk to Bil. Bil asked him about his military background and about his current duties as detachment sergeant of LRS. At the end of the meeting, I said to Bil, "Why don't you hire him?" Tom glared at me while Bil told him to call him the next day. We left and outside, Tom said, "I can handle my own job interviews," I just shrugged, and we parted company. Tom called me the next day and told me that Bil had hired him.

I told Tom that I would like to work with him if any openings came up. I was tired of the problems at the Police department and I knew that I would never get promoted there because of the photo incident. Tom was non-committal and I found out later that he was hoping to hire a Ranger buddy from the 2nd Ranger Battalion. When he asked, the guy told him that he was being hired by the BATF.

Two weeks later he called me and asked me if I want to be a clerk. I did want out of the PD but not to be a clerk so I turned him down. Two weeks later he asked me if I wanted to come on active duty and be his assistant, planning and coordinating the surveillance missions on the border. "Hell yes," I replied. "Show up at Building One the day after tomorrow at 0700 to process in." When I told my boss, Sergeant Sam, that I was taking a military leave of absence, he looked angry but didn't say anything because he was a pilot in the National Guard himself and knew that he had to let me go. He did however attempt to get me fired but that came later. It was now May 30, 1989.

June 1989: The Texas National Guard

In 1986, President Reagan had declared that Drug Trafficking was a threat to National Security, so the Department of Defense had begun to increase their attention to this 'Drug War.' The National Guard had been using their helicopters to spot marijuana that was grown in places like Hawaii, California, and Texas since the 1970s. In June of 1989 the Texas Guard's Counterdrug (CD) program was just getting started. The program was being run out the Guard's Headquarters, in Austin.

OTM-D (The Counterdrug Program)

The Guard had benefited from the military buildup during the Reagan years. There was a lot activity in the headquarters when Counterdrug showed up. With the expanding staff there was need for more room, so we were given part of a large room on the second floor at the end of the building.

In order to fit us into the existing system, we came under the Plans, Operations, and Training Officer, Colonel Richard Brito. He was a busy man with an office down the hall and I had only spoken to him once. I misunderstood his name when I heard it and called him Colonel Burrito when I was talking to Tom. He warned me not to call him burrito, "It pisses him off," he said. Since I was the lowest ranking guy on the program, I took his word seriously. Colonel Brito was a controversial figure in the counterdrug arena. We only heard bits and pieces because we were so enthusiastically busy with the border surveillance project. I was blissfully unaware of most of the shit-storm that was going on in the offices twenty feet away.

Under Colonel Brito was the Plans, Operations and Military Support Officer, Lieutenant Colonel B. Under LTC B was Operations and Training Management-Drugs or OTM-D, that was us. LTC B or "Uncle Bennie" as he was called by some of the staff, was a heavy-set man with graying hair and a nice smile. He didn't seem to be overly worried about the CD program and I got the impression that he was coordinating with Bil when it came to decisions. For his part, he knew the Texas National Guard system.

Personal Computers were rare, and the internet was a mystery to most of us. There was a small room in the corner of our area with a teletype machine that was constantly printing classified messages from the Pentagon. Some secretaries would come in every so often and tear off the long pages of messages. They looked at us with critical, if not disdainful eyes as they walked past our area. We were called the 'druggies' and I don't know if they thought we were all using drugs or just selling the stuff, but they were never nice to us.

The rest of the people in the headquarters for the most part ignored us except for someone who made up the duty roster for the staff duty officer and NCO, a nightly duty position requiring an officer and a sergeant to sit in the headquarters all night answering phones and making sure that the alarm system for the weapons storage vaults was working. This person was happy to see us because we could be added to the pool of available people for this onerous position. A lot of people liked it because you got the next day off, but we were 'doing things' and would stay at work the next day because we didn't have any time to waste.

In 1988, there were two or three people working on the CD project. By the end of 1989 there were at least 20 people working in the headquarters and a number of others working in the field.

When I showed up for work, I was informed that I was on orders for 39 days. That would take me to annual training (AT). After that, assuming that there was money I would be brought back on orders until the end of the fiscal year, September 30, 1989. After that it was anyone's guess as to what would happen. Since I was on a military leave of absence from the Police Department, I didn't give it much thought and managed to be free from anxiety about my employment future. The Lord had brought me here and here I would remain until he sent Alan around to tell me what was next.

The whole CD program and its budget were the subject of much discussion in the headquarters, in the Pentagon and in the Congress. Some of the staff members made it the main subject of conversation and questions about what we had heard lately about the budget became a source of irritation to those of us who were busy with what we were being paid to do. One of the first things that we were told when we came on was that it was not a retirement program. Anyone with 17 years of active duty would be laid off before they reached the magic 18 year mark. Then you enter into sanctuary at 18 years, and it would take a crime like treason to discharge you before you reached twenty years.

When someone asked about the possibility of the CD Program members becoming AGR, Active Guard and Reserve, the status that many of the full time Guardsmen where in, Bil said that they decided against it because then we would be what was known as Title 10, US Code. That would add a whole new level of bureaucracy and many more rules which would limit our support to law enforcement.

In the beginning, all of the officers were Army but soon we began to see bluesuiters showing up and visiting Uncle Bennie and Bil. Before long there were several Air Guard officers and enlisted personnel working in the office with us. The Air Guard is very different than the Army Guard. I used to say that the Air Guard was kind of like a military organization, but not quite.

Their work ethic was different than ours and you could always tell when it was 5 o'clock by the stream of blue flowing out the door. The Air Guard would allow their people to retire from the program which caused some animosity between us. Not all of it was bad; one of the flyboy officers that showed up early on would eventually become our Task Force Commander and a fine one he was.

The CD headquarters had two functional areas, the staff which handled personnel and pay issues and the Special Projects elements which included Tom and me. The staff was thrown together and in typical Guard fashion, people who had no idea about how to do what they were assigned to do were given great responsibility for staff work.

One LT, a pilot, was told to do travel vouchers and since he didn't know what to do with them, or didn't care for the task, being a pilot, he started storing them in a desk drawer until people started complaining about not receiving their travel and per diem payments. The first I knew of the problem was when Uncle Bennie came into the work area and pulled out a stack of 1351s and began working through them like it really mattered. The LT disappeared shortly after that, like a lot of officers who wandered across the CD stage during the early years.

The first project, beside the surveillance operations with the Border Patrol, was sending Military Police to the border to assist the U.S. Customs Service with cargo inspection at the ports of entry. Every day, thousands of trucks cross the border from Mexico into the U.S. with tons of products made in Mexico. Since the North American Free Trade Agreement this traffic has greatly increased. Customs did not have the resources to search every truck and train coming into the U.S. and so the Guard personnel were greatly appreciated. The cargo inspectors became a mainstay of the CD program, catching more drugs directly, than any other project. There were plans in the works to provide Intelligence support to law enforcement and eventually Intel would become the largest CD project, but that would come later.

June 1989: Spartans in an Athenian Camp

Our day would start at 7AM and go until we were done or until we were too sleepy to do any more that night. Bil educated us with his 20 years of Special Forces experience. He gave us a list of sayings he had written, but he never really elaborated on some of them. This caused us to talk about a particular saying for several days at a time during a break or at lunch where we would each mention some small insight into what we thought he was trying to say. Because I was a low-ranking sergeant, one of my favorites, for obvious reasons was; "It's not your fault if you are outranked, but it is your fault if you are outsmarted."

Bil would have a meeting each morning and use an easel with butcher block paper to illustrate the lessons for that day. Every day Bil educated the CD staff which included a Lieutenant Colonel, some Majors, Captains and some senior NCOs. One of the best was entitled, 'Spartans in an Athenian Camp.' Here he showed us that while we were very excited and motivated toward this new mission, the rest of Camp Mabry, which included the headquarters for both the Texas Army and Air National Guard, not only didn't know what we were doing, but they didn't care and that it would be hard to get anything from them unless we had rapport and followed the proper procedures.

Rapport, rapport, rapport is what Bil preached. This included rapport with the rest of the National Guard. Bil knew that even though there were regulations and procedures for everything in the military, it was a lot easier to get something or get something done if the other person liked you or was afraid of you. Bil wanted us to cultivate rapport and he would provide the scare factor if there was no other way to accomplish a critical objective.

He had a gravelly voice and some problem with his neck so that he had to turn his whole body to turn his head. He had put on weight since he had retired but he was still a bit of a rooster. He showed me his barroom betting gimmick where he lay face down on the floor, put his hands out in front of him like he was diving, and did a push-up. He didn't drink all the time but when he did, he got real drunk.

Since I did the same thing back when I was drinking, I was able to understand Bil or at least accept him for what he did for the program rather than looking at the trouble he caused when he got drunk. He and I were the only two Vietnam vets in the program headquarters and he seemed to consider me as a trustworthy military acquaintance.

MSG (Ret) William 'Bil" Enney
Counterdrug Coordinator-Texas

Texas National Guard Headquarters
Bldg 1, Camp Mabry, Austin, Texas

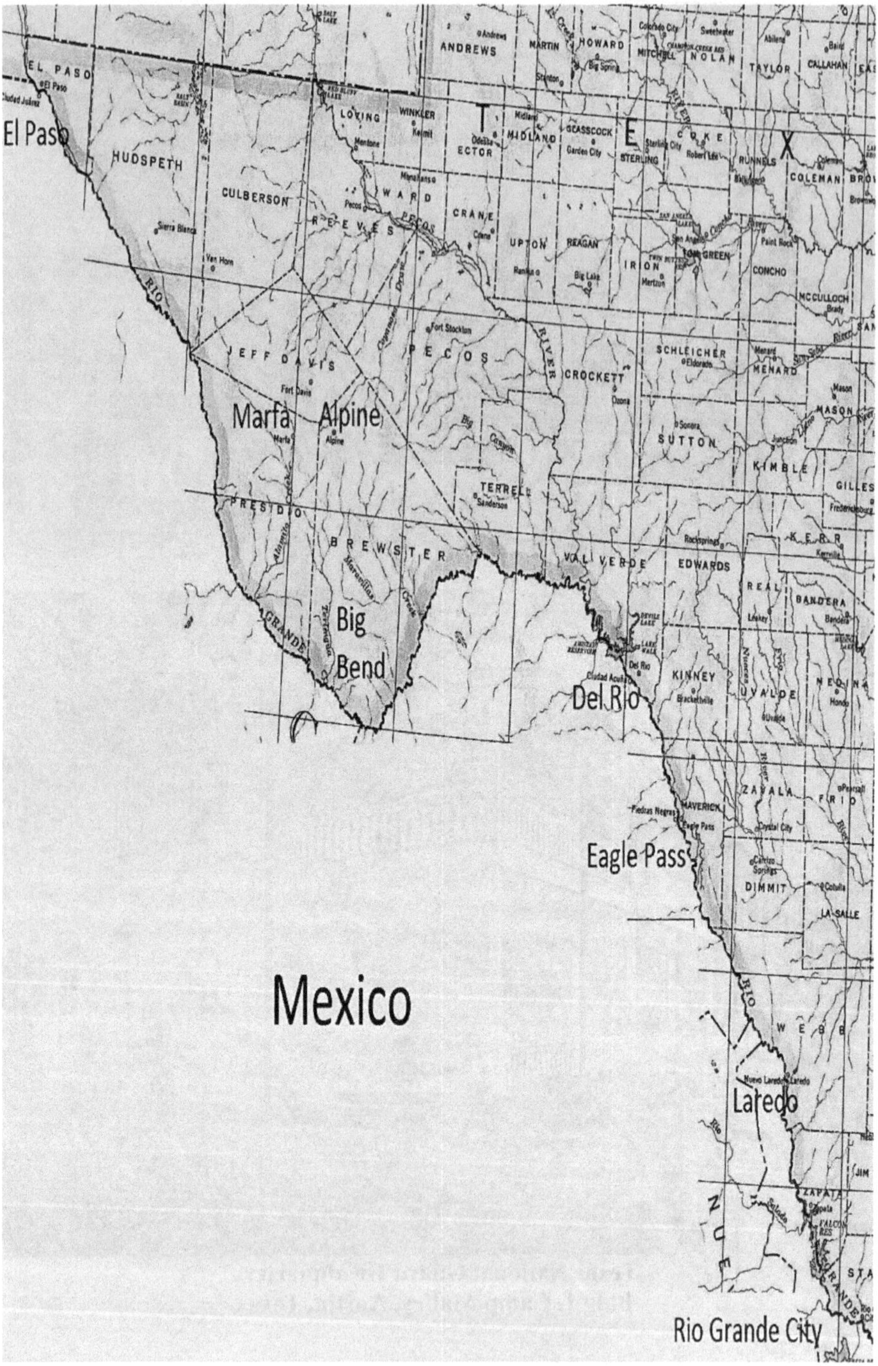

West Texas Border Region

Proclaim this among the nations: Prepare the battle; Rouse the mighty men; Let all the men of battle Draw near; let them go up.
Joel 3:9

Chapter 4-June 1989

Warning Order: Operation Unity 3-4 Del Rio

Tom and I worked on various tasks during my first week on the program. We studied the first two operations and tried to work on an SOP based on the lessons learned. One day Bil called Tom to his office. When he returned, Tom told me that we had gotten a mission request from the Border Patrol (BP) and that we would be doing site selection in the Del Rio area for the next operation. He explained that we had to meet the BP Sector Chief Patrol Agent, establish rapport with him; then plan and coordinate an inter-agency surveillance operation that was being run by Operation Alliance in El Paso.

We began to gather what information we could about Del Rio. I went down to the camp's maintenance office and got a complete set of 1:24,000 U.S.G.S. Topographical Maps of the Del Rio area. These are excellent maps and we used them extensively until GPS and digital maps became available 10 years later.

Operation Alliance

Operation Alliance was a multi-agency task force that was charged with planning and coordinating drug interdiction efforts on the Southwest Border from Texas to California. Modeled after a similar drug interdiction effort in the Caribbean by the South Florida Task Force, Operation Alliance was formed in 1986 in an attempt to get every law enforcement agency on the border on to the same sheet of music; but up to now there had been more cacophony than music.

U.S. Customs, the Drug Enforcement Administration (DEA), and the Border Patrol along with all the other Federal law enforcement agencies were caught up in a perpetual turf battle which had its roots in the budget battle that took place every year in congress. The Federal Government was unable to cut through this bureaucratic nightmare and establish a strong central authority, or unity of command. They tried to establish a unity of effort but institutional self-defense and the animosity inherent in the members of these agencies was often stronger than the higher goal of coordinated drug interdiction.

One key factor in the request for funding was the measure of success established by the body count of the drug war, seizures and arrests. It was hard for an agent of one organization to give vital information about a drug smuggling event to another organization because when the bust was made there was no allowance to quantify the information that led to the seizure so the agency that made the bust got all the credit.

In an attempt to mitigate this conflict, each agency involved in an investigation was allowed to claim the seizures and arrests of a single bust. These statistics were tallied at the end of the year and if five agencies were involved in a one hundred pound cocaine bust, then five hundred pounds had been seized.

The Border Patrol

I would work with all these organizations eventually but for the next two years most of our work would be with the Border Patrol. It was established in 1924 and the first agents were Texas Rangers, sheriff's deputies, local police and some cowboys. They were rugged individualists who were used to working alone on horseback in the desert and mountainous terrain of the border with Mexico.

This tradition had been passed down and there was still that flavor when you met a BP agent, but current BP policies had far more agents working at vehicle checkpoints on the major highways than cutting sign on the river. The BP had become unionized and so even though they wore military insignia they were far from a military organization.

The Border Patrol had five sectors in Texas: El Paso, Marfa, Del Rio, Laredo, and McAllen. Each sector was commanded by a BP agent with one star on his uniform. Each sector had a number of stations commanded by a BP agent with a silver oak leaf.

The Border Patrol morale was generally pretty low at that time, although morale is easier to discern in a military unit than it is in a law enforcement agency. Law enforcement work is basically an individual effort and getting cops to work as a team is difficult. Getting different agencies to work together is almost impossible. The reasons for low morale in the Border Patrol were many, depending on who you were talking to at the time. Some of the agents I had contact with over the next few years were line agents who drove me to remote sites on the river and we shared the front seat of a Suburban for hours if not days at a time. I became a kind of sympathetic listener to many a tale of woe from the agents who felt like the red headed stepchildren of Federal Law Enforcement. If you can't complain you might get depressed so many agents would take the opportunity to unload on me because they had already discussed these things with their fellow agents to the point where they never talked to each other. A new ear like mine was a welcome relief to these guys who had spent many a lonely night on the border considering their plight.

One of their complaints was the fact that the entry level pay grade for the Border Patrol was GS-5 while the DEA and FBI would come in as GS-7 or even GS-9. BP agents were required to become literate in the Spanish language. One year after they completed the academy, they were tested on their Spanish and if they failed the test they were fired. All the other Federal agencies paid extra money if you spoke another language; the BP didn't. Another complaint was that some the agents had been assigned to what were called hardship posts for a lot longer than the BP policy required.

Others were unhappy with the lack of urgency on the part of the government to stop what they saw as a flood of illegal aliens from all over the world crossing the Southwest border from Mexico into the U.S. every night. The amnesty law that had been signed by President Reagan in 1986 had caused some in the Border Patrol to believe that their work was part of a Darwinian process of only catching the less intelligent, less motivated individuals and letting the ones who were able to run the gauntlet get a free pass after staying below the radar for a few years in the U.S.

Border Patrol culture was a lot like most cop culture with some amusing variations. Spanish phrases were used a lot more among typical white men than you would hear in most places, even in Texas. 'De donde en chingamos estamos?' was one of my favorites.

Bil had made arrangements with the Border Patrol for us to go down to the border for site selection. BP agents would take us to various places they suspected were being used for drug smuggling and we would identify the location on our map. Bil set things up and a week later Tom and I were flying to El Paso for meetings with Operation

Alliance and the Border Patrol's special mission unit. Bil met us in El Paso and we entered a whirlwind of meetings involving senior members of the Border Patrol who were assigned to Operation Alliance.

Operation Alliance had coordinating authority over us when we started the border surveillance operations. We dealt with the BP agents assigned to Alliance, but later, because drug interdiction was not their primary mission they set up a new entity called the Border Patrol Special Coordination Center.

June 1989: The Border Patrol Special Missions Unit

After some meet and greets and long listening sessions, we were sent to meet the BP agents we would be working with. The decision had been made in Washington that military support to law enforcement would require that a law enforcement agent be in 'close proximity' to the military personnel providing the support. Close proximity had been defined as within 15 minutes travel time.

The legal ramifications of military involvement with domestic law enforcement were many. Soldiers enforcing civilian laws have been prohibited by the Posse Comitatus Act of 1878 which was enacted after the Civil War to prevent soldiers from enforcing the law in the South. The Act had gone through some revisions over the years and actually, the Guard did not come under this prohibition, but the National Guard Bureau decided that it would comply with it as policy.

Because of the distances involved in the border environment it had been decided that a law enforcement agent would accompany the military surveillance teams we sent out. The BP agents going out with our teams were members of the Border Patrol's Special Missions Unit (SMU). These were the high-speed, special mission agents of the BP and they were our introduction to the BP culture.

The Border Patrol's Special Mission Unit was founded in 1984 and their first selection and training class was run by the U.S. Marshals Service-Special Operations Group. The SMU held the line at some prison riots and later, they developed their own selection and training program. The concept was to select and train Border Patrol agents to SMU standards and use them as a special reaction team when needed.

The SMU commander and several supervisors were stationed at Headquarters in El Paso. The actual SMU agents were stationed all over the US in an On-Call status. While on Operation Snowcap, SMU agents worked in places like Bolivia where they destroyed cocaine labs and blew up clandestine airfields. These were the agents who were happy, 'living the dream,' as we say; down on the border; tracking illegal aliens and drug smugglers, out by themselves in the middle of the night, watching and waiting for a chance to do something interesting. These were the type of men we had in LRSD, men that gravitated to SMU and the Emergency Response/Special Operations teams within the Border Patrol sectors.

Site Selection-Del Rio

We were issued a Chevy Blazer from the Border Patrol at Operation Alliance. It was an undercover vehicle with no agency markings or government license plates. It had a Border Patrol radio, cop lights and a siren hidden in the engine compartment. This was part of a deal that Bil made with his buddies at Operation Alliance and it was greatly appreciated because our own system had not yet figured out how to get vehicles for us. We were told to drive to Del Rio and meet Bob, an SMU agent, with whom we would conduct site selection.

It's a long way from El Paso to Del Rio and it took us most of the day to get there. Tom had a phone number for Bob and when we got to Del Rio, he called him and arranged a meeting at a bar in a shopping center near the motel. When we got to the bar Tom spotted Bob by the clothing description and after exchanging IDs we sat down and began the rapport building process Bil trained us on.

Bob was a little younger than me with a big mustache and the tan of a man who spent a lot of time outdoors. He didn't initiate new subjects but answered our questions. As we talked, his sense of humor became obvious and we had a pleasant time getting to know each other. I let him know that I was a cop on a leave of absence from my department and that gave us a connection from then on. Tom's Special Forces background tended to bring a more competitive vibe to meetings like this because it is endemic in the special world that everyone must be better and have bigger balls than everyone else. After Tom and Bob had a few drinks they began some one-upmanship talk that threatened to get ugly, but they kept it under control. We agreed to meet for breakfast before we drove to the Del Rio Border Patrol Sector Headquarters.

At breakfast, Bob suggested we all drive together in his suburban which had four-wheel drive and a Border Patrol radio. The suburban had a strange black and white paint job and Bob called it Shamu, after a killer whale that put on shows at Marineland. We drove through Del Rio to the old part of town and sector Headquarters.

When we arrived at Sector Headquarters and were introduced to the Sector Chief, the chief had already been briefed by Operation Alliance in El Paso, so he had been expecting us. Bob explained that we would be doing site selection for an up-coming border surveillance operation using SMU agents and members of the Texas National Guard. The chief told us to cover the Northern half of his sector during the operation which would be in two phases, each lasting 10 days.

He said we should start at Comstock which was upriver from Del Rio and work our way down river. He also told us that he wanted to put a boat with one of our teams on Lake Amistad to watch for smugglers using the lake. He gave us the name of an agent in Comstock and a phone number and wished us luck. As we left, Bob laughed and told us we were in for some long days. "There's more than 200 miles of river in this sector and we'll have to drive 20 miles from the highway to the river in some places; yeah, we're going to have some long days."

It us almost an hour to get to Comstock; on the way, Bob told us how he envisioned the operation. He said that there would be two agents with each team. We could put five teams out for each iteration. Tom was writing everything down and asking questions about equipment, infiltration vehicles, etc.

Bob was speculating about support for the operation, but he said not to worry about it because they were used to improvising in the Border Patrol and that everything would work out. I was comfortable with that because cops always do things that way; but Tom knew that we would have to have a solid plan to present to Bil.

We found the Comstock BP Station on the side of the highway at the edge of town. The Station Chief greeted us and introduced us to one of his agents who would take us around the station area. He showed us on the map where he thought we should put teams and asked if we had any questions. We had a million questions; but we followed Bob's lead as he thanked the station chief and we left.

The agent assigned to us was happy to let Bob drive. He got into the front seat of Shamu and Tom and I sat in the back. He asked us where we wanted to go and Bob told him we would like to start at the furthest upriver site and work our way back to Comstock. Off we went, Bob and the agent doing some BP gossip while Tom shuffled through the maps and I got my camera and new video camcorder ready.

I had recently bought the camcorder to video tape our training at LRSD, with the thought that someday I would make a documentary showing the building of an Airborne Ranger type unit from scratch. When I found out that we were going to do site selection, I decided to take a video tape of the sites and use the tape for briefing the teams when they were preparing for the mission. Bil was strict about all aspects of our work so I didn't mention using video for site selection. Tom didn't object so I brought it along.

Before we drove upriver on Hwy 90, we asked Bob to zero the odometer so we could use the mileage to determine where we turned off the paved road to follow a dirt road to the river. When we got to the first possible surveillance site, the Comstock agent got out and walked up to the gate. He lifted a chain that looked like it had ten padlocks on it and began inspecting the locks until he found one that he recognized.

He took a large key ring and went through the keys until he found one that matched the lock and opened it. He pushed the gate open and Bob drove through and waited until the agent locked the gate behind us. The agent got back in and Bob drove down the road.

Tom had the map of the area and had found the road we were on. He marked the road entrance with a pencil and made some notes. I know some military organizations don't want you mark your map in case the enemy gets a hold of it and discoverers some sensitive information; but we would have to take that chance because we were getting way to much information to memorize.

The first site was about 15 miles down the dirt road. When we drove through a rancher's front yard, the agent asked Bob to stop and he went up to the door and talked to someone for a moment before returning and getting back in. "Go ahead," he told Bob, "they know me and I'm in uniform so there's no problem. The chief likes to keep good relations with the local ranchers and landowners so I usually stop and let them know what's going on."

I asked him if the locals ever refused permission for the BP agents to cross their land and he told us that they had legal access to any land within 25 miles of the border. "We can cut the locks on their gates if we have to," he said, "but we usually don't have a problem with keeping one of our locks on each road that goes to the river."

We finally got to the river. The agent showed us where the crossing was and as he talked, I stood next to him and videotaped. This way I was able to make a tape that would be seen by the surveillance team when they were preparing for their mission. Since it would have a local agent narrating, it would add to the credibility of the information we gave them for their mission planning.

While I was videotaping, Tom was on the other side of the agent taking compass azimuths of the various points of interest as the agent described them and making notes on the map. Then Bob and Tom looked around for potential hide sites and he noted them on the map.

All of this information would be 'packaged' into what we called a target folder that contained the information that a team would need to plan a surveillance mission at this place. The team would get a laminated map along with the target folder. A route sheet described the route to the site from the highway, the drop-off point and the location of a hide site and observation post (OP). There would be recommendations for special equipment and camouflage items that would be useful at this particular location. The target folder also contained an intelligence section with light data, weather patterns, known smuggling activity, and information about the locals. Finally, there would be an Escape and Evasion plan that would assist the team in planning emergency recovery procedures.

When we had collected all the information about the site and I had taken a few still photographs of the area that would be included in the target folder, we packed our stuff in Shamu and headed back to the highway. When we stopped at the ranch house on the way through the barnyard, the agent had a short conversation with the rancher. After he got in and we started up the road, Bob asked him what they were talking about. "I just told him that we were looking for places to put agents to watch for smugglers," he said.

Tom and I looked at each other in shock but didn't say anything while Bob continued to drive. When we got to the highway, the agent suggested that we go to Langtry for lunch. Bob was good with that and Tom and I agreed so we went to the place where Judge Roy Bean, 'the Law West of the Pecos' used to preside.

We ate at a small cafe and the agent told us stories about the area. He had been stationed there for 20 years and was fascinated by it. He was an avid collector of Indian arrowheads and showed us several that he had in his pocket. The next place we were going had been an Indian camp in the past and that he wanted to look around a little while we were there. It was a long drive from the highway and he hadn't been there in a long time. I watched Bob's face while the agent was talking but he never changed his expression.

After lunch we took off down the road and went to two more places before the agent said we would have to call it a day. We agreed to meet him the next morning at the Comstock station and finish up the Comstock sites. As we drove home Tom sat in front with Bob and asked him about some of things that came up, especially the agent's conversation with the rancher at the first site we visited.

Bob was a little pissed off about that, but he told Tom that the local BP agents had to maintain good relations with the locals. We were concerned about operations security (OPSEC) and Bob was concerned about the same issues, but he was an experienced BP agent who knew much more than we did about BP policies and politics. I asked him if the local agents would tell the local people about the operation when it was actually in progress and he said that he would get the higher-ups at Operation Alliance to talk to the sector chief about OPSEC for the upcoming operation.

These kinds of problems would irritate us from time to time because there was no way to determine if our surveillance efforts were effective if we believed that the operation was compromised. The military tries very hard to safeguard its information but when it's an inter-agency operation there's not much you can do about your counterparts' failures to protect the information regarding the operation. This was our first experience with inter-agency cooperation, and we didn't know yet how big the OPSEC problem really was along the river.

The next day we went back to Comstock where we met the local BP agent. He took us to several more sites and we finished about 4PM. At the last site I walked with him off the road a short distance to an 'Indian Mound' where he had found some arrowheads in the past. We spent about ten minutes there, talking, while he sifted through the sandy soil looking for arrowheads.

At one point he called my attention to a dried-up plant and asked me if I knew what it was. I didn't so he told me that it was a 'resurrection' plant. I was interested in anything to do with resurrection, so I asked him why it was called that, and he told me to go back to the Suburban and get some water and bring it to him. Bob and Tom were discussing something and ignored me as I grabbed a water bottle and ran back to the agent.

"Watch this," he said. He poured some water around the edge of the plant then he sat there waiting. I sat next to him and after a minute it began to turn green and opened up. "That's resurrection," he said. I told him that I had read in the Bible that in creation God was able to manifest his power. He agreed and we walked back to the vehicle and joined Bob and Tom. We took the agent back to the Comstock station and said goodbye. I never saw him again, he retired soon after our time together, but I always remembered his showing me the 'Resurrection' plant.

We returned to Del Rio, ate dinner together, and agreed to meet the next morning when we would go back to Sector Headquarters and meet with the Del Rio Station Chief for a briefing on the sites he wanted covered. After dinner, Tom and I organized our notes and filled out our Target Folder Checklists for each site we visited.

The next morning, we met Bob for breakfast and drove back to Sector headquarters and walked to the Del Rio Station office. There we met Simon Garza, the Station Chief. Bob briefed Chief Garza on the purpose of our visit, and he listened attentively but didn't ask any questions until Bob was finished. Chief Garza had been a Border Patrol agent in Del Rio for a long time. Before he joined the Border Patrol he had been in the Army in Vietnam, which gave us an instant bond.

He wanted to discuss our methods based on the previous missions in Big Bend. While those missions were supposed to be 'secret,' we soon found that there were no secrets on the border. When we told him that we wanted to select one site for each team, he asked if he would have any say on which sites we used for the operation. We told him that there would be five teams and that their site would be chosen by us during this trip.

He didn't like this and said that there were hundreds of smuggling sites in his station's area and that sometimes one was hot and other times another was hot. "Why can't I tell you what sites I want you to work when you get here" he wanted to know? This was getting interesting. He had informants and other means, such as sensors, of determining where the best places were to put a team and he wasn't happy with having no say-so in where we were going to operate. Bob looked a little anxious because this was Border Patrol business and he didn't want to get cross-ways with Simon Garza.

I opened my mouth and asked what he would recommend. He said that we should make twice as many Target Folders as we had teams and when the time came for the operation he would let us know which ones he wanted to use. "Sounds good to me," said Tom who looked at Bob. He nodded in agreement; we had made the first alteration to Bil's model for conducting these operations. Simon assigned us an agent and we went to look at Lake Amistad and some sites below the dam.

The Lake Amistad site was unusual because it would be conducted from a floating surveillance platform. The Air Force Base in Del Rio had a recreation area at the lake, and they had some large pontoon party boats for rent. We made arrangements for one to be used by a surveillance team for the operation. Bob drove out to the lake and we took some photos and some video. The lake was big, and the map showed lots of inlets on both sides that could be used for smuggling. Bob said there were agents who used boats in Florida and that he would try and get one of them for the next operation.

Mexico across the river

After we left the lake, we went down river and looked at several sites. We were able to get the information we needed a little faster and so we finished all the sites upriver from Del Rio by the end of the day. After we dropped the agent off at the station, Bob suggested we cross the river and go to Ciudad Acuna for dinner. "There's this great Mexican restaurant called O'Brien's and I always eat there when I come to Del Rio." "All right!" I said, "I've been wanting to buy a guitar from Paracho and I'm sure there will be plenty of them across the river." Tom didn't look very happy. "I don't think we're supposed to go to Mexico," he said, which brought a smirk to Bob's face, which Tom saw. "Don't worry about it," I said, "They'll never know."

Tom wasn't comforted as we pulled up to the motel. Just as we stopped and I was opening my door, Bil Enney stepped out from behind a wall. "Hi Bil," I blurted, "we're just getting ready to go to dinner at

O'Brien's, want to come?" "What the blank is O'Brien's," he asked. "It's a great Mexican restaurant in Acuna. Bob's been there a lot and says it's a great place." "Where's Acuna," asked Bil. "Across the river," I told him brightly, "I'm going to get a guitar there after we eat."

Tom wasn't saying anything and Bil started looking around to see if anyone was listening. I realized that things had turned serious as Bil considered the situation. We had been invited by a senior Border Patrol agent to eat dinner in Mexico. He had hammered rapport building into us, so he was really weighing all the issues. Finally, he smiled and said, "all right, so and so, but don't tell anyone about this because we're not supposed to cross the border."

We got cleaned up and met Bob a half hour later. He would drive us to the bridge in Shamu and then take a taxi across the river to Acuna. Tom and I left our pistols in the hotel room;

June 1989: The Armed Forces

The weapons issue was a touchy one and we were not authorized to be armed for our coordination visit. I told Tom that I was carrying whether it was authorized or not. "I'm not going down to the border and hang out with law enforcement officers without a gun," I said, "That would be bad manners." "What do you mean, bad manners," Tom was baffled by the logic behind my statement.

When we asked to be armed back at headquarters, they told us that the Border Patrol was armed and that they would protect us. When I explained to Bil and the colonel that we would be going to remote areas and that it was wrong to expect a law enforcement officer to have to change the way he operated, putting himself at risk, to protect us. "Besides, this is a 'big boy' game. How's it going to look for you to send your two surveillance experts down to the river unarmed? They won't respect us; law enforcement is an armed society."

The arming of the troops was one of the issues that came up early on. The Chief of Staff of the Army, General Carl Vuono, told the Adjutant General, Major General James Dennis, that he didn't want the troops to carry weapons on these missions. This was typical post-Vietnam thinking by the senior leadership in the Army. They were afraid of an incident that might hurt their career. This kind of thinking continues to this day among many professional soldiers, which makes you wonder why they call it the 'Armed Forces.'

General Dennis told him that the border was a dangerous place and if his troops couldn't defend themselves against deadly force then he wasn't going to send them in harm's way. He fought in the Korean War with an airborne unit and he considered us members of the 'Armed Forces.' Operations were one thing, what Tom and I were doing was another. All my appeals fell on deaf ears. It wasn't, I don't think, so much an anti-gun thing; it was more of a logistical problem. The headquarters didn't have any guns and no procedures had been set up to get any, so I used my own and so did Tom.

Acuna, Mexico

We crossed into Mexico and I was excited. I had spent a lot of time there in the past and always had a good time with the Mexican people. My Spanish was sufficient for simple conversation and that impressed Bob. When we got to the restaurant, I translated the menu for Bil and Tom and they were also impressed. Bil had a little Spanish from his time in Panama but Tom was a 'no comprende' guy.

Bil sat next to Bob and asked him about how things were going. Bob gave him a good report and told him that he was really impressed with our using video for site selection images. Bil gave me a mean look and I figured I was in trouble again. Then he smiled and told Bob that yes, video was the way to go, "No more black and white, slightly out of focus pictures, we use color video for our operations." We ate a delicious meal and Bob, Tom, and Bil had a few beers. I was worried that Tequila might enter into the night's activities but I was spared this time.

After dinner, we paid up and left the restaurant. There were some shops on the street, so I walked over and found one with a guitar in the window. Bil wanted to get back across the river before we got into trouble but I wanted to get a guitar, so they all waited outside while I went into the shop and negotiated for a guitar. "Hecho en Paracho?" I asked the shop keeper. "Si, como no," he said, showing me the label inside the sound hole of the guitar. Sure enough, it said 'Paracho, Michoacán.'

I had been to Paracho years before. It is a small town dedicated to making guitars. The wood they use comes from trees growing in the mountains around the village. I was going to return to Paracho and have an acoustic bass guitar made with frets, but I never got back there so buying this guitar would have to do.

I came out with my guitar and we took a cab back to Del Rio. Everybody but me breathed a sigh of relief as we crossed the river. I was sorry we couldn't have stayed longer. We passed through the customs inspectors and went back to the motel. When we got there Bil told us that he wanted to go with us to talk with Simon the next morning. Tom and I went to our room and discussed the day's events. He was a little worried about Bil's sudden appearance and I told him that everything was ok. "He went to Mexico with us, which means he likes us."

The next morning, after breakfast, we went to Simon Garza's office at Sector headquarters. Bil sat down in the chair closest to Simon's desk and began telling Simon about the National Guard drug program. He was using a lot of military acronyms that was making Simon wince. This went on for about 15 minutes. Simon managed to get a few words in at the end and told Bil that we had agreed to do twice as many Target Folders and that he would let us know which targets he wanted us to work just before the operation started. Bil sputtered at this for a moment and then agreed. When Bil was finished he got up and told us that he was going to visit the Sector Chief about some other business and said he'd see us back in Austin in a few days. Simon assigned another agent to go with us and show us the down- river sites.

Mitchell's Car Wash

The places were close to town so we could see quite a few that day. We got into the Agent's Suburban and he told us that the first place we were going to was 'Mitchell's Car Wash.' Tom and Bob and I exchanged glances and waited while he drove along the river bottom until we got there before we asked any questions. There were no paved roads to the place he showed us on the map so we couldn't understand why there would be a car wash there.

We had a U.S. Customs photomap in our map case and I pulled it out to see what was there. The photomap was an aerial photograph that had been gridded and the place names and some topographical information had been added. We found the place we were going and sure enough, there was the name, 'Mitchell's Car Wash,' but there was nothing there! When we got there, we got out of the Suburban and looked around. We were on the bank of the Rio Grande. Mexico was about 40 yards on the other side of the river. There was a high bluff behind us. We had driven here on a small dirt road and there was no car wash.

The agent noticed us whispering among ourselves and asked what was wrong. "Where's the car wash?" I asked. The agent started laughing, "You're standing on it," he said. "You mean people drive down here and get their cars washed," I asked. He laughed again, "No, this is Mitchell's Car Wash," he said and he proceeded to tell us the story.

There was an agent named Mitchell who had worked at the Del Rio Station some years ago. He liked to fish and when things were slow on his shift he would drive from the station to this spot and go fishing. One day, while he was fishing, his Border Patrol vehicle rolled into the river and was almost completely submerged. Mitchell had to walk back to the station and tell the Station Chief that his car had rolled into the river. The chief knew that Mitchell was probably fishing and now he was going to catch him. He called for a wrecker and had it meet them at the station and then Mitchell, the chief, several other agents who came for the show, and the wrecker, drove out to the place where Mitchell's vehicle was sitting submerged in the river.

When they got there, the chief watched the wrecker driver wade into the river to the place where the tip of the roof was breaking the surface of the river. He ducked underwater and attached his tow chain to the bumper hitch on the rear of the vehicle. Then he waded back out and was walking to the wrecker's cab, when the chief told everyone that he didn't want anyone touching the vehicle when it came out of the water until he said they could.

The wrecker driver pulled the vehicle out of the river and everyone started moving toward it. The chief yelled for everyone to stop, and he walked up to the driver's side door and opened it. River water poured out all over his pants and shoes. That didn't improve his humor at all. He waited until most of the water drained out and then he leaned inside to make sure that the parking break was not set, but it was. He looked again; the parking break was completely set. He was furious.

By now the others had opened the passenger door and were all watching the chief as he sat on the wet seat and grabbed the hand break and checked to see if it was engaged, it was. He got out sputtering about how there was no way that that vehicle could have rolled into the river with the hand break set. He looked at Mitchell, who shrugged and said nothing. The chief told the wrecker driver to take the vehicle back to sector headquarters and give it to the maintenance supervisor.

"The chief told Mitchell he would get to the bottom of this but he never did," said the agent. "You see, as soon as Mitchell saw his vehicle roll into the river, he took off all his clothes and swam down to the drivers compartment and set the handbrake. He got out of the river, dried off, dressed, and walked to the station to inform the chief."

Mitchell retired from the Border Patrol and lived in Del Rio. "You can go to his house and he will tell you the story himself," the agent said. Soon, the place became known as Mitchell's Car Wash, When the Customs people were putting their photo-map together, they came to the Border Patrol to find out the place names along the river. One of the Border Patrol agents told them with a straight face that this particular place was known as Mitchell's Car Wash.

Flying the River

After Mitchell's Car Wash, we went to eight more sites in the next two days and we were done with site selection. We had one more task and that was to get aerial photographs of the sites. The aerial photos would be used by the surveillance teams to help them with their route and mission planning. Since I had a 35mm camera and the video camera I was elected to do the air mission. I was somewhat panicked because we had covered hundreds of miles of river and I wasn't sure that I could find the places again from the air. I decided to mark all the sites on a Joint Operations Graphic (JOG) map because there was no way I was going to follow the river with the smaller scale U.S.G.S. maps. The JOG is a military map at the scale of 1:250,000, which means that for every inch on the map there are 250,000 inches on the ground.

Tom and Bob took me out to the Del Rio airport and we went into the Border Patrol Air Operations hanger. Bob introduced me to the chief pilot and told him what we needed. He decided to fly me himself and told them that we would be back in an hour or so. I had my map and my cameras, a bottle of water and my gun in a small bag that I carried with me as I followed the pilot outside to a single engine high wing plane. It was painted white and light green and it was small. "Is that a Piper Cub," I asked. "No, it's a Christen Husky."

They had to shoehorn me into the back seat and I found a place to store my small equipment bag on the floor by my feet. The pilot gave me a set of earphones with a microphone and showed me where to plug them in. He completed his pre-flight inspection, climbed in and started the engine. After speaking on the radio, he taxied to the end of the runway, turned around and took off. I saw Tom and Bob watching by the door of flight ops and waved to them as we soared into the wild blue yonder. This was great. I had flown in a few small airplanes before, but this was real seat of the pants flying, the windows were open and there was a small latch holding the door closed next to my knee.

I did a commo test with the pilot over the intercom and found that it was easy to communicate. We had decided to fly to the upriver sites first, then fly down river to the last site we looked at on the ground. I decided while we were flying to photograph the sites as we flew upriver and video the same site as we flew down river. Now all I had to do was find the sites. Fortunately, the pilot had been flying in Del Rio for many years and he was very familiar with all the sites I mentioned to him when we were back at flight ops.

Soon after we took off, he told me were approaching the first upriver site. I looked over his shoulder to the front and then to the left and right out of the windows. When I saw the river, I began looking to the left and right and up and down. I didn't want to miss a site or look stupid in front of the Border Patrol pilot, so I prayed the Special Ops prayer, "Oh Lord, please don't let me screw up." I was starting to panic when suddenly everything changed, and I recognized the place we had been a few days before. I started snapping photographs and asked the pilot to steer left or right until I got the shots that I needed. I tried to get as much of the road approaching the site and the site itself as I could. The same thing happened at the next site and after that I was sure that I could find the sites that I had been to on the ground.

We continued flying upriver past Comstock to the first place we checked few days before. After I took the photographs, I let the 35mm camera dangle from the neck strap and grabbed my new video camera. This was one of the early SVHS camcorders. It was big and heavy. Power was a concern because it used a lot of electricity and I only had two batteries. In order to see clearly in the view finder, I had to take my glasses off, and block out any outside light. When I took off my glasses I couldn't see anything far away so I was moving my glasses on and off while trying focus the picture in the view finder, trying to figure out where we were and talking to the pilot at the same time.

As we circled around, I realized it would be better if I could shoot from the river toward the highway. I mentioned this to the pilot, and he obliged me by flying down the center of the river and gradually he moved over until we were flying on the Mexico side. Eventually we were flying a good ways into Mexico as the river turned sharper than we did. I didn't say anything, figuring that he knew what he was doing. In the many flights I had with the Border Patrol in the next few years I noticed that the boundary line on the ground was much more zealously acknowledged than the one in the air. Later, when I showed the video to the teams in isolation, they complained that it looked like I was flying in Mexico. I told them that this was an illusion created by the use of a 'double refractive lens.'

We finished all the sites and flew back to the airport. I thanked the pilot for the great flight, and he complemented me on knowing where the targets were. It was then that I realized that the map was still folded in the bottom of my bag. I was pretty happy about my first aerial photo mission with the Border Patrol and hoped that the pictures I had taken turned out ok. we said goodbye to Bob in the parking lot of the motel and drove back to Austin.

SFC Payne-Site Selection

Typical River Smuggling Sites

Unity 3-4: Site Selection and Aerial Photos
1st 'Selfie'-Photo Recon, Flying the river

Unity 3-4: Mitchell's Carwash and Ciudad Acuna

Author and SFC Payne

Let every person be subject to the authorities over him, for there is no authority except from God, and those which exist are ordained by God.
Romans 13:1

Chapter 5-June 1989

Operation Unity 3-4: Mission Planning

When we got back to Austin, there were a few new faces in the headquarters. One was Major K, a small, red-faced artillery officer who Bil introduced to us as the officer who would oversee the Del Rio operation. Later Bil told us that while Major K was a figurehead, we should keep him happy while we ran the operation. Tom tasked me with the operations order and the many supporting annexes and appendices.

Authority and Leadership

This didn't sit well with me and I told Tom if he wanted something written, he should do it himself. That didn't sit well with him and he told me to meet him outside behind the headquarters for a 'discussion.' When we were outside, he moved so he was standing in front of me and said, "Sergeant Murry, come to the position of attention, now!" Not knowing what else to do, I did, and stood there waiting to see what was next. "I am in charge of this operation," he said "and when I tell you to do something, I expect you to do it. If there is a problem, I expect you to let me know but don't ever tell me to do something that I have just ordered you to do, understand!"

I was totally caught in my rebellious nature when I first responded to him and now the Lord had put me in a situation where I could either accept the authority he put over me or I could stay in my rebellion and this man would deal decisively with me. At that moment my future was at a crossroads. After a moment, I looked him in the eye and said, "Yes sergeant, I understand."

From that time on I had a good relationship with Tom. He never minded my strenuous objections and suggestions because he knew that once he had made a decision, he could give me an order and I would do my best to carry it out. I learned a great lesson that day and from then on when I was in a position of responsibility and I had to confront someone about their failure to follow instructions or their direct challenge to my leadership role, I would remember Tom's 'in your face' style of dealing directly with the problem and use it myself.

Leadership in a group of young men in never easy and most of the time it is not pleasant. If an Army sergeant is afraid to deal with these kinds of problems, then he shouldn't be a sergeant. It comes down to placing yourself in a position where the other guy might be enough of a hothead to take a swing at you and if that terrifies you then you should take your leadership to more pleasant climes like logistics.

Operations and Plans

So I began to write. When an operation is planned, the product of that planning is called an Operations Plan or OPLAN. This is what is presented to the decision makers for their approval before an operation starts. It is the basis for a series of briefings and in our case, it would culminate in a command briefing to the Adjutant General and senior representatives of the Border Patrol.

I used my Macintosh computer because we didn't have a computer assigned to us. I was not a staff trained NCO, so I found a copy of Field Manual 101-5, the staff officer's bible and went to work. There is the English language and there is the Army language using English words. Every excruciating nuance of descriptive writing is covered in this manual; for example, when you describe the points of interest at river crossing points you do so from left to right, facing the river.

These procedures are designed to minimize confusion in combat operations and if everybody read these manuals and could remember every rule listed in the 250 pages of this mind-numbing tome on military operational paperwork, it might cut down on casualties. The written message that resulted in the Charge of the Light Brigade was a lesson in point. Bil knew these things pretty well so I had to learn them if our work was going to pass his scrutiny.

Bil wanted everything perfect and was the final approval authority before the command briefing. He and Uncle Bennie would hold a 'murder board,' something that he had done in Special Forces. We would brief the OPLAN to him and they would ask 'what if' questions to see if we had made contingency plans for every possible problem that might arise during the conduct of the mission.

Our planning included determining the number of volunteers from the LRSD to serve as surveillance team members and support staff, transportation for them to El Paso for pre-mission training with the BP agents, food and shelter in El Paso, transportation by air to Del Rio, ground transportation for insertion and extraction, transportation back to Austin after the mission, communications equipment needs and procedures, emergency medical evacuation, logistical requirements for field rations and batteries for the radios, pay, etc., etc., etc.

While Tom and I were working through all these issues, I was also writing the intelligence annex with appendices for each site. This required me to write up all the information given to us by the local BP agents when we visited these places and whatever other information I had gotten from the Del Rio Sector Intelligence officer. I had spent several hours with him while we were there, and he had given me a lot of historical data regarding past smuggling activities at these locations and this information was added to each appendix as a Tab. As the information became more detailed, I had to search the manual for further subdivisions of the Intelligence Annex.

The Military/Industrial Complex-Part One

Bil sent Tom up to Fort Worth to General Dynamics, a large aerospace company that made the F-16 Fighter plane and other big-ticket items for the Department of Defense. They had a small plane they were touting as the drug busting answer we were all looking for. It was a single engine high wing turboprop with a Forward Looking Infrared (FLIR) sensor. They told Tom that it was invisible at night and silent. Then they asked if they could participate in one of our operations so that they could demonstrate its usefulness to us and the Border Patrol. Tom was ok with it and told them to call Bil.

This was big business and it was our first glimpse of the Military-Industrial Complex at work, but it wouldn't be our last. Tom was laughing when he came back from talking to the General Dynamics reps. He said, "They call it a Low Intensity Conflict Aviation Support or LICAS aircraft, the Lick-Ass." Tom always had a way with acronyms.

Mission Planning

Late at night when Tom and I finished work, I drove home with my head a-buzzin with ideas, worries, things that I had forgotten, things to do the next day, on and on. My wife would have some dinner for me and an occasional stink-eye about my absences but I was on a mission as they say. I told her that we were real busy right now but things would slow down soon, which they did after I retired 15 years later.

The next morning, we met for the morning staff meeting and then we were be back at it. Tom kept Bil up to date with our progress and I sat in our tiny cubicle which contained two desks, pounding away with two fingers at my Macintosh keyboard. When Tom and I were both in there, sitting back to back, our chairs collided when one of us got up to do something. The Macintosh didn't have a hard drive so I had to keep everything on floppy disks which included the Word program. There were quite a few disks to keep track of as we wrote the plan.

Briefing Charts

As we got close to completion of the plan, Bil directed us to prepare large posters that would be used in the command briefing. He sent us to the print shop at the camp and meet with a certain person there that he knew. This person would provide us with large pieces of poster board and colored letters of various sizes that would be used for the points Bil wanted displayed during the briefing. Tom and I had to make very light lines with exact distances between them and then use the letters to make each chart. This was a giant pain in the ass but the more charts we made the faster we got and when Bil told us that they looked good I felt like we had really gotten a hard thing done.

Remember, this was before PowerPoint and all graphic presentations were hand made. Shortly after this, we were introduced to an IBM program that helped you make graphic presentations that could then be transferred to 35mm slides which would then be projected onto a screen. This was a big advance and the staff officers loved it because they could make their presentations look good which made them look good, regardless of the quality of performance that would come after the briefing was over. I was a little disappointed that we started using the computer-generated charts on the next operation, but I got over it.

PowerPoint changed everything a few years later. The art of the briefing became the most important aspect of the staff officer's skills. By the time I went to Afghanistan in 2003, the PowerPoint Ranger was the number one point of ridicule and the envy of all staff officers.

The Murder Board

As the time approached for the command briefing, Bil told us to get ready for the 'murder board.' Major K began to come around and look at the OPLAN drafts and seemed to be happy with our work. He wanted to have a complete review before the 'murder board' so we agreed to walk through the whole plan two days before the board so that we would have time to make any changes before we presented the whole package to Bil and Uncle Bennie.

Tom and I got with Major K in a vacant office at the agreed upon time and we went over the entire plan. At the end he told me, "Greg, that is the best OPLAN I have ever seen," which made me feel good but maybe I should have wondered how many other OPLANs like this he had seen. Tom and I noticed a few holes and after Major K went about his happy way, we sat down and filled them.

On the day of the 'Murder Board', I was nervous but Tom assured me that, "it ain't nothin but a thang," and we marched in to face our inquisitors. Bil said "Begin" and Tom took him through the operation from start to finish. Bil interrupted from time to time and ask about our contingency plans for this or that possible train wreck.

We had to show him an answer in writing in the plan for each contingency or we were going to have to make corrections. It was a real education from a man who had done staff work in Special Forces that included flying units from the states down to Panama where they parachuted by night into a remote drop zone to conduct jungle training operations. Bil was a planning master and I was grateful for this education even though it was painful. At the end of the briefing, he told us that this plan was "a piece of shit" and we better fix it fast or he would find some other people who could. I was dumbfounded. Major Ken surprised me when he stood up; him being a kind of timid guy. I figured since he had told me that this was the best OPLAN he had ever seen that he was going to put Bil in his place. "Bil, I want you to know that I told Greg that this plan was a piece of shit!"

BAM! I was 'gob smacked' as the Brits say. All I could do was sit there in silence while Tom told Bil that we would fix the OPLAN as soon as we left his office. "Make sure you do," were his final words as we filed back to our cubicle. Major K suddenly remembered that he had an important meeting with someone elsewhere and disappeared. Tom and I went out on the porch where we talked.

Tom started laughing as I began to sputter about Major K and his knife in the back technique that he had displayed a few minutes before. "Where's the duty, honor, country that these officers are supposed to represent," I asked him. I was in shock and Tom thought it was funny! He assured me that this was standard behavior for most of the officers he had known and that Bil knew it and now I did too.

"You can't trust em," he said, "that's why I hate all officers," a quote that I would remember because I had heard it before in two of my favorite movies; 'From Here to Eternity' and 'The Cross of Iron.' When we went over Tom's notes, we realized that there were only a few things that needed fixing. I made the corrections and before I went home, I hit 'Print' and we were ready for the next step.

The next morning the Border Patrol reps from Operation Alliance and SMU, along with Air and Army National Guard aviation representatives, showed up for the rehearsals for the command briefing and command post exercise that would take place the next day. The command post exercise was a series of 'what if' scenarios that helped both the law enforcement element and the National Guard element visualize possible problems: legal, logistical, and operational, that might arise during the upcoming operation. This is where our contingency planning would be tested by our law enforcement counterparts.

Bil told us that he would handle the entire military portion of the briefing. He wanted us to pay attention to his methods, he said, because he expected us to do most of the military part on the next operation, if there was one. After a rehearsal of the command brief where each briefer stood up on the stage we were finished. We used the next day to go through Bil's part of the briefing using the poster board graphics that Tom and I had labored on a few days before. We said farewell to the BP guys and returned to the office for a review of the day's events with Bil. During the briefing some problems arose with the charts and Tom and I spent another late night making last minute corrections. I asked Bil if I could videotape the briefing and he grudgingly OK'd my proposal. This was all I needed to pursue my current hobby of videotaping everything that moved at our LRS drills and at Counterdrug. I secretly hoped to make a surfing video someday, but that goal seemed to be a little remote in Central Texas and on the Rio Grande. As a result of my hobby I accumulated more than 400 hours of video covering the next ten years and I'm still trying to come up with some reasonable editing scheme to document all the sweat and tears that went into our efforts. There was a little blood also, but that was self-inflicted.

Command Briefing

The next morning we showed up in Class B uniform: dress trousers, dress shirt without a tie, low quarter shoes, spit shined to a glass-like appearance for me; and for Tom, spit shined Cochran Jump Boots with his dress trousers tucked into his boot tops. Bil told me to wear my CIB and Tom wore his Master Blaster wings, EIB, and Ranger and Special Forces Tabs on his shirt pocket.

The conference room began to fill with a lot of brass including a Brigadier General, the assistant Adjutant General for the Army Guard, and the Assistant Chief of Staff, Colonel Danny Kohler who expressed some surprise when he saw me. We had served together in D Troop in the 70s. He was a major then and the troop commander. I always remembered the way he flew out to Lake Belton and rappelled off the cliff with us during one summer camp, leading from the front.

The number of people coming to what I thought was a compartmentalized and classified briefing was way more than we thought was right. We went to Bil and took him aside; "Who are these people and what the heck are they doing in our business?" Tom asked with exasperation. Bil told him 'at ease' and not worry about it. This was, he explained, typical in a headquarters populated by a staff of chair-borne commandos. For some, coming to a briefing like this was a kind of vicarious participation in an actual military operation, made even more 'sexy' by the fact that the troops participating in the operation would be armed and carrying live ammunition. For others, being seen in the briefing was a kind of power statement in itself and if the military is nothing else, it is a power projection statement. Some people who should have been there were absent, not wanting to be associated with something that might prove to be a disaster and a career killer. Some people were there that didn't need to be there, but they were the risk takers, wanting to be seen as players in the biggest game in town. Headquarters politics was something I was spared from during my entire career. I knew it was going on but since I didn't know how to nor was I inclined to learn, I never had to participate in that kind of bloodless war-fighting that goes on in every chain of command.

I set up my video camera on a low tripod near the front, focused the camera, and made sure I had a spare battery in case the first one didn't last. I took a seat next to Tom and looked up at the stage and around the room, checking for any last minute problems that might have escaped our attention. Then it was show time. We used this term when discussing moments like this. Our whole operation had become a 'Cecil B de Mille production' and it seemed like we had entered into show business. This was not the straight leg infantry anymore.

Bil took the stage. He had his friendly face on and he was wearing a dark blue blazer with a Special Forces Decade Club insignia on the left pocket. He stood at center stage and started the briefing. Following a strict protocol, he thanked the guests, starting with the General, for coming. He then introduced the BP agents from Operation Alliance and SMU and at the end of the introductions he introduced Tom and me as members of his staff who would be assisting him in conducting these types of operations. We stood and faced the audience and I saw Colonel Kohler giving me a smile and thumbs up.

Bil went on to give a brief history of the first two Unity operations in the Big Bend and how they had proved that the National Guard and the Border Patrol could work together in the 'War on Drugs.' He explained that a number of lessons had been learned and that this operation would demonstrate improved interoperability between the two agencies. With that he introduced a Border Patrol agent, Lee Justice, from Operation Alliance who gave the drug smuggling situation in the area where we were operating and covered the Border Patrol use of force and primacy in any law enforcement situation.

Rules of Engagement?

This was a sticky point. Most of the Guard leadership had never carried a weapon with live ammunition except for qualification purposes in the ultra-controlled environment of the rifle and pistol range.

Rules of Engagement was a term that was in the manuals but outside of most of their understanding, much less their experience. Now the Border Patrol was going to have the soldiers follow the legal requirements of the federal government for the use of force, to include deadly force which was all the Guardsmen had to offer if a violent confrontation was presented to them. Lee covered the use of force policy which he would also explain to the Guardsmen when they were conducting pre-deployment training just prior to going to the river.

According to National Guard Bureau rules, the Guardsman could load their weapons only on the command of the team leader. The lawyers had been looking for some doctrinal guidance regarding the arming of troops during peacetime operations and had found what they were looking for in the Civil Disturbance field manual. This manual had been written after all the fiascos of the Vietnam era when the military was called in for riot control in places like Detroit, Michigan in 1968 and Kent State University, Kent, Ohio in 1970.

In Detroit, the Michigan National Guard had fired all their ammunition, killing at least eleven people and eventually requiring more than 4000 paratroopers from the 82nd Airborne Division to quell the rioting. The crowning piece of work that was a driving force in the writing of the Civil Disturbance doctrine was the incident at Kent State where four kids were killed and nine wounded by the Ohio National Guard.

We, however, were not conducting civil disturbance operations. Expecting the troops to carry unloaded weapons in the dark along the Rio Grande until the team leader determined that there was a threat and ordered them to load was just the kind of legalism that had permeated our military since Vietnam. This kind of thinking was amply demonstrated in Beirut, Lebanon in 1983, when two marine sentries, guarding the entrance gate to the Marine Barracks were still trying to load their weapons as a truck drove through their checkpoint and into the front door of the barracks, where the driver detonated the equivalent of 12,000 pounds of TNT, killing 241 Marines and other US servicemen.

The Border Patrol, like all law enforcement officers, carried loaded weapons on and off duty all the time. As far as our soldiers carrying weapons went, Tom and I agreed that the team leader would tell them to load their weapons as soon as they got out of the vehicle that dropped them off at their insertion point.

The Border Patrol agents would have primacy in any law enforcement situation. This meant that should the team come in contact with anyone while they were conducting their mission, the BP agent would take charge of the team and direct them while he talked with, arrested, or shot whoever they had encountered. Bil said that everything would be OK politically if the first bullet in a bad guy belonged to the BP.

Lee covered the smuggling situation and the legal/BP primacy issues in a very smooth and almost soothing way. Lee had been in the BP for a long time and had served in some of the hottest parts of the Border, including serving as the station chief in Rio Grande City, Starr County, Texas, arguably the hottest point on the whole border. He had a good presentation and when he finished, he introduced Bil, who proceeded to brief the military portion of the plan.

As Bil went through the charts that Tom and I had worked so hard on, I began to see how a good briefer would pause for a few moments to allow the audience to read the bullet statements on the chart, not reading them aloud himself, but after a moment, adding some further details to what was written on the chart. Bil was a great briefer and I could see that he had put a lot of preparation into his brief, knowing that the whole program's success rested more on these briefings than anything that occurred in the field.

When he got to the end, he concluded with assurances that we had developed a good, joint plan and were prepared to execute it on order. He asked the General for approval of the plan and the General said he approved it. Then he asked if there we any comments or questions. The General spoke first and made some vague comments that seemed to be positive and then he got up and walked out.

The next to speak was Colonel Kohler. He wanted us to get volunteers from G Company, men who had gone on one of the first two operations because their experience would be helpful. This caused Tom and me to see red because we had considered that this would be an LRSD mission, G Company having had their chance and failed, as evidenced by the fact that Bil had hired us, from LRSD, to run these missions. We looked at Bil who ignored us and assured Colonel Kohler that that is exactly what he was planning to do. Our jaws dropped for a moment and Tom elbowed me in the ribs before I could stand up and shoot my mouth off. The next thing Colonel Kohler asked about was 'bulletproof' vests. He wanted to know if the BP could provide them for the soldiers. Bil looked over at Bob, who stood up and told him that BP didn't wear them on the river, it being 120 degrees in the daytime. The colonel nodded and asked a few more questions which Bil fielded quickly. Bil then asked if there were any other questions and since there were none Bil declared that we had an approved plan and that this concluded the command briefing for Operation Unity 3 and 4.

Tom and I began gathering all the charts and papers. I turned off my video camera and as I packed it into my travel bag, I felt someone tap me on the shoulder. I stood up, it was Colonel Kohler, smiling as we shook hands, "How are you doing Greg, I didn't know you were back in the Guard, much less involved in this, what have you been doing since I last saw you?"

The last time I saw him was in 1980 when I was leaving the Guard but still working as a ground's keeper at Camp Mabry. He was doing PT out on the running track and we had talked for a moment. I told him I was getting out and he told me that if a war started I could go to war with him. That meant a lot to me then and I was glad to see him being successful as the deputy Chief of Staff, the highest-ranking full-time position in the Texas National Guard at that time.

I told him that I was a police officer on a leave of absence to support this operation. I think he was surprised that I was in law enforcement, because of some of my actions and statements when we served together in D Troop would have led one to believe that I spent most of my time on the opposite side of the law. He shook my hand again and told me to come by and see him sometime.

Bil, Tom, and I went to lunch with all of the Border Patrol agents at a local Mexican Restaurant, no doubt staffed in the kitchen with illegal aliens. We had a 'breaking of the bread' meal with our counterparts, continuing the rapport building process that Bil had mastered in faraway places with exotic peoples under much more stressful conditions. After lunch, the command post exercise went smoothly, and we received several complements from the Border Patrol jefes. Bil was in fine form throughout the day; he was in his element and loved it. When it was over, we said farewell to Bob and Lee Justice from Operation Alliance, pledging undying friendship and fraternal feelings as we shook their hands and watched them drive away.

I was really learning a lot, but it was like drinking from a fire hose. Tom and I realized we were processing an enormous amount of information, much of which would be useful as we conducted more of these operations. We kept talking about how we had to write this stuff down into an SOP, but we were too busy just trying to keep up with the tasks that Bil had given us. We put our materials away in our cubicle that night and went home for a few hours rest.

The next week was filled with logistic and personnel preparation for the upcoming operation. Tom said that we would ask for volunteers from LRSD and G Company during the next drill. Fortunately, we had the help of George C, an old paratrooper officer. George had lost his commission somehow, and was now a Sergeant First Class (SFC) and would soon be a warrant officer. He was the S4 or logistic guy for Counterdrug and he really moved mountains to get us what we needed. He was a member of the airborne mafia at Camp Mabry and they helped each other when they could.

"Amateurs study tactics and professionals study logistics," was an old saying that I had heard and read for years and now I was beginning to see why it kept its currency. Everything we needed for the operation, from rations to weapons required a unique requisition process that Tom and I were ignorant of. George took us by the hand and schooled us on what he needed from us before he went out and got what we needed.

He broke new ground every day in logistical procurement operations at Camp Mabry. "You know what I hear up at USPFO," the central supply point for the Texas National Guard, he said. "We can't do that," they would say, "and when I asked them why, they would say, because we have never done that before. It's kind of like asking someone for directions and being told, you can't get there from here," he would say with a laugh. George really got into supporting our operation along with all the other projects and was a key member of the Counterdrug staff and greatly responsible for our success, beside which he managed to keep us all out of jail which is an occupational hazard in the world of logistics.

Volunteers of America

When drill came there was an undertone of excitement in the unit. The word had gotten around that Tom and I were working up at Guard Headquarters on secret things. There were already rumors about G Company's secret missions in the Big Bend, thanks to the ex-G

Company members of LRSD and a few news reports. Tom and I had developed training scenarios that would require the teams and commo to perform all the tasks that would be required on a Counterdrug mission, without telling the unit that we were training for a counterdrug mission, OPSEC being an important factor in what we were doing.

As the drill went along guys were coming up to me and Tom at various times and trying to elicit information about the 'big' secret that was hanging in the air. Finally, at the noon formation on Sunday, Tom announced a special meeting in one of the classrooms after lunch. When the time came, all the members of the unit were in the classroom, chattering like magpies. The tension in the air was thick enough to be cut with a Randall or Gerber knife, the kind that you see taped upside down on the web gear of some death dealer from the movies or in real life. A lot of those knives got lost that way and I often wondered about how many ended up stuck in the foot of the wearer.

Tom came into the classroom with LT Bruce, the XO, who was currently in charge and hoping to get command of the unit in the future. The troops were called to attention, and then given "at ease", as Tom strode to the front of the classroom. He directed all the full time Guard members like the readiness NCO, unit clerk, pay clerk, and supply sergeant to leave the room; the supply sergeant looked very unhappy because he wanted to do cool things. The rest were insulted, but they didn't show their feelings. Being full time they thought they were above the rest of us and most of us didn't like them anyway, because they were REMFs, (rear echelon military forces) while we were the fighters.

Then Tom read our OPSEC statement that warned everyone that what we were about to discuss, while not classified, was highly sensitive in nature and would not be discussed with anyone not associated with our Task Force. He then asked for volunteers for two 2-week periods in the middle of the summer. "We will still have our annual training, (Summer Camp), but these periods will be in addition. Everyone who can or may be able to come should keep their seats and the rest of you go outside and do something useful until this meeting is finished," Tom said, putting one of the sergeants in charge.

Tom then told the remaining troops, about half the unit, to put their name on one or both rosters that would be coming around. "This doesn't mean that you are going, because we still have a lot of things to work out, but we will let you know as soon as possible. There will be orders issued for you to take to your employer if needed when we have made final determinations, any questions?"

Every hand shot up and Tom began answering them, shouting "at ease" every few minutes to quiet the excited troops. I was in a corner videotaping the whole thing, already planning to make a historical record of these exciting times. "What's SGT Murry doing videotaping this? He shouldn't be doing that!" said SGT Albert, a former submariner turned army paratrooper.

I didn't know it at the time, but the submarine fleet was in a wartime status, playing tag with the Russians, during the Cold War. The reason I, and most of the rest of the country didn't know about it was because the Navy really practiced OPSEC better than the rest of the services.

Tom told SGT Albert not to worry about it and continued answering questions for about five more minutes before dismissing the troops and directing them to report to their section leaders to continue with the tasks at hand until the end of drill. Tom and I joined the LT in his office where we briefed him on a few more details regarding the operation. He was very eager to be part of it and told us so. Tom promised him that he would talk to the people running the program and try and get the LT onto the operation.

After final formation, at about 5:50 in the evening, I walked out to my car which was parked near the front gate. The gate guard, a retired National Guard sergeant called me over, "Why are you guys keeping these soldiers so late on Sunday night?" he asked. "We got a lot of things to do," I told him flippantly, thinking that he was teasing me.

"Well, let me tell you," he said, "I'm going to complain to headquarters about how you are abusing these troops!" I didn't know what to say but I remembered Bil's class, 'Spartans in an Athenian Camp.' Then I remembered rapport building and so instead of giving him a smart ass reply I told him that I would relay his concerns to my chain of command and that seemed to mollify him enough so that I could get out of there and go home.

Monday morning, we were back at it, working our asses off for the next week. We had our two-week annual training on the 18th of June. Tom and I were both deeply involved in training plans for this event, so we worked even later, and our weekends were completely taken up with these activities.

LRSD-Annual Training

During the last drill before AT, Captain Hargett gave me a piece of yellow legal paper with the numbers 1 to 14 penciled in in the margin. On number 8 and 9 he had written, 'Rubber Boat Training.' "Give this to Sergeant Payne and tell him to fill out the rest of this. It is our training schedule for AT." "Sir," I said, "there's a training meeting this afternoon after we dismiss the troops, could you give it to him then in case he has any questions?"

He put on the face of a typical officer in a hurry who doesn't have the time or inclination to explain anything to a sergeant and told me to, "Just give it to him, he'll know what to do" as he grabbed his hat and limped out the door. He had been injured in a parachute jump during a drill that spring and wouldn't be going to AT with us.

I met Tom before the training meeting and told him that Captain Hargett wouldn't be there but that he had given me his training guidance for AT as I handed him the paper. Tom looked at it for a moment and looked up at me, "this is his guidance, rubber boats on day 8 and 9, and that's it? Looks like we can do what I've wanted to do after all." We went into the meeting and began to hammer out what we would do for two weeks at Fort Hood, TX

We devoted the first week to classes on individual and team tasks like land navigation (day/night), communications training, tactical movement formations, immediate action drills and an airborne fun-fest with multiple jumps. Then we would have our rubber boat training which would include drown-proofing, where the soldiers would learn to make water wings out of their trousers, and poncho rafts, an improvised tactical water crossing technique.

The main event would be the use of large inflatable rubber boats that can be used for tactical water crossings. We were being supported by an active duty LRS unit at Fort Hood and they would supply the rubber boats and accessories. We envisioned black Zodiac commando boats with silent motors but what we got was a little less high speed but good for entry level commandos such as ourselves.

The final event would be a surveillance mission that included an isolation phase where the team would plan the mission and prepare their equipment, culminating in a briefback where they described their plan to the commander and made any changes based on his guidance.

Then they would be inserted by vehicle into a simulated drop zone, navigate to a specific location, and surveil their target, sending back radio messages describing enemy activity. At the end of the mission they were extracted and debriefed. All of the doctrinal elements of an LRS mission would be used on our border missions that we would be conducting shortly after we finished AT.

As our involvement with CD went on, the LRSD would continue faithfully to prepare for a war that would never happen, that is, the counterattack in Western Europe against the Russian hordes pouring through the Fulda Gap. In training for this scenario, the soldiers would also become proficient as surveillance experts against a skillful enemy right here on the border and later, inside Texas.

We spent the week before AT working on the Del Rio operation until the CD staff went home and then we worked on AT issues until we were exhausted. When we went to Fort Hood for AT, we had a staff meeting the first night. LT Bruce, the XO, would ask a question and if it were Tom or I that was required to answer he would have someone wake us up and he would ask the question again. After one of us would answer we would go back to sleep until the next question. The next morning, after getting some decent sleep on the hard ground of North Fort Hood, we dove into the task at hand.

Our unit was combined with the LRS Detachment of the 1st Cavalry Division and they gave us some training and logistical support. The first week's training went well. The airborne operation went as planned. The riggers showed up with a truck load of parachutes. Two UH-1 Hueys, flown by pilots I knew from my old Air Cav days, dropped stick after stick and the men got at least two jumps and some got three.

The water operation was the hit of AT. Tom taught the unit drownproofing and poncho rafts. After the rubber boat training from the active duty LRS troops, he directed our unit to have a boat race across the lake wearing full combat equipment with their rifles. The active duty LRS guys were incredulous and told us that they would never be allowed to carry their rifles in the boats for fear that someone would drop one overboard and lose it in the lake. The teams had a great race and the morale in the unit was sky high.

One night, the senior full timer, who was despised by all, disappeared and Tom had the whole detachment in formation at midnight answering a roll call to document his absence which was used to get rid of him. After his name was called three times with no answer, the troops broke into a spontaneous rendition of the chorus of 'Kiss him goodbye' "Na na na na, na na na na, hey hey, goodbye."

After that though, things went downhill. One of the Intel guys at CD, SFC D, decided to task us with some surveillance operations. He worked in the G2 and convinced them that he could put together a training mission for LRSD. When he came over and told me that I should expect a mission alert notification, I asked him to cancel his project because we weren't ready to put teams out on a G2 training mission as we were still in the crawl phase of our communications training.

He told me that it was too late and that we would be putting the teams out whether they were ready or not. We got the tasking notice and gave missions to five of our teams. This was the start of three days of chaos which culminated in one team flagging down the division commander's helicopter for a water resupply and then disappearing before he returned with some water.

This led to the G2 standing at attention in front of the General wondering why he ever listened to SFC D's great idea. We took some heat from the G2 people but when I blurted out about how we had told SFC D we weren't ready, they left for more compliant company.

On one of the last nights we had a detachment fun time where we set off some excess pyrotechnics, had some chicken fights, administered some pink belly treatments, and had a unit songfest. At the end of the two weeks we headed back to Austin and our 'real' war.

Operation Unity 3-4: Del Rio (Part 1)

The troops assembled at the armory on 51st Street with all their field gear. They were allowed one rucksack and one duffel bag. We kept them in the hanger where the helicopters were worked on until the C-130 taxied onto the tarmac and lowered its ramp. We filed out of the hanger as quickly as possible and walked up the ramp into the airplane. The loadmasters stood on the ramp directing us to drop our rucksacks and duffel bags in the center space and then to move forward and take a seat. This was an ERO or 'engine running on-load' and we were ready to go in about ten minutes.

While most of the troops were very familiar flying in a C-130, most had never landed in one but that didn't seem to bother them as they all went to sleep shortly after take-off. The trip was uneventful and some of the guys didn't wake up until the aircraft bounced the first time when we landed at Biggs Army Airfield, in El Paso, TX.

We filed off the plane, grabbing a ruck or duffel bag as we passed the pile on the ramp and carried them to where Tom was standing near the terminal building. He had a big smile on his face, and I could tell that he was as jacked up as me. "Load your gear on the back of that truck," he said, pointing at a flatbed truck nearby, "and get on the bus" which was parked next to the truck. We were driven to a mess hall nearby where we would eat lunch. Tom gave a quick word about OPSEC and told us to, "eat up and get out here in 20 minutes." After lunch we got back on the bus and were driven a short distance to the Border Patrol's Special Mission Unit's headquarters where we would conduct our mission planning and run the command post.

Isolation

"Leave your gear on the truck and file into the building," Tom told us. He led the way to a large classroom and directed everyone to take a seat. Bil walked in and stood at the podium. He took the introduction paper that I had written and introduced himself and Mr. W the SMU commander. In the introduction I had written that Mr. W had graciously offered us the use of his headquarters. "Graciously huh", he said with a mean laugh, "graciously, Mr. Gracious Mr. W."

I wasn't sure why Bil was being so cynical about this whole gracious thing until I visited Mr. W's office during the operation. There on his wall was a plaque indicating that Mr. W had been involved with MACVSOG, the secret special operations unit of the Vietnam War. Bil had run recon with SOG and since Mr. W had been in the Marines in Vietnam, there must have been some doubt in Bil's mind regarding the veracity of that plaque. He never said why, but several times during the operations when we were alone together, he would inquire, "and how's the gracious Mr. W today?"

Mr. W welcomed us and left the room. Tom and Bob went to the front. Bob called the names of the BP agents and gave them their team number, then Tom called the soldier's names and assigned them to a team. The men all shook hands and each team sat together for the rest of the briefing. Tom told the teams where the sleeping areas and the latrines were and the rules for activities in the building. He gave the meal schedule and told everyone to keep a low profile at the mess hall that we used. Everyone was dismissed to get their gear and I moved my things to a small room near the operations room.

The next event was a medical briefing given by Mike T an SMU agent who had been an SF medic in a former life. He gave a practical trauma management class and first aid procedures for desert operations, to include treatment for cactus thorn's and ant bites. The teams were dismissed to have some personal time before dinner.

I was in my room relaxing when I heard some voices raised above the dull roar. I looked down the long hall and saw an agent reach across and grab one of the soldiers by the shirt with one hand while he formed his other hand into a fist and drew it back by his ear. I ran down and jumped between them, knocking the agent's hand off the soldier's shirt with my shoulder. "Don't you think it's a little early for that, you hardly know him," I said with a smile. He wasn't smiling but my remark seemed to break the tension.

I turned and told the soldier to report to Tom in operations. "What's the problem," I asked the agent, still smiling. "You need to tell that little smart ass to keep his mouth shut about how he's so much better than us because he's a ranger and he has jumped out of airplanes," he said angrily. So much for the rapport building talk we had given to the troops before we left Austin.

I asked him if I could talk to him and after a moment, he followed me outside where I proceeded to do everything but roll over on my back and beg for mercy. I told him that I was a cop on a leave of absence and that I knew just how he felt. "These guys have only gone to some training and you guys have been out doing the real deal for years. Their just jealous of you guys and want someone to pay attention to them. If we don't kill them, they will grow up some day." This seemed to mollify him. I assured him that we would find someone else for his team and we went back inside. SSG N, the offending soldier, was down in the ops room with Tom, telling him the BP were a bunch of legs and blah, blah, blah. Tom told him to shut up and that if he created one more problem, he would be on a plane back to Austin and he would never work for CD again. That got his attention.

Tom told me to go get Sergeant X and tell him to report to operations. Tom told him that he was being reassigned to a different team.

He explained what had happened and told him to be careful with his BP agents. Sergeant X agreed to keep his airborne mouth shut and I introduced him to the pissed off agent who by then had cooled off. Sergeant X did a good job and the next day I saw him and his agent sitting and laughing together before the mission briefing. Tom put SSG N on Sergeant X's team and everything was smooth until the next crisis.

I set up the Command Post (CP) while Tom visited the soldiers and made sure everybody was happy. We were told to move our supplies into the SMU's supply room which increased tension because no organization likes to co-mingle supplies; midnight requisition habits are hard to break and there would be several accusations of theft by both organizations before the operation was over.

Satellite Communications

We started unpacking the satellite radios that had been procured from active duty units by someone's behind the scenes deal making. I had never seen a satellite radio before, so I was quite interested in seeing how they worked. There was a technical manual that came with the radios, so I was skimming through it and when I got to the part about batteries; I read it several times but I couldn't remember us ordering any of the batteries mentioned in this manual.

As we reviewed the supply situation it suddenly dawned on us that we didn't have the batteries needed for the satellite radios we were going to use for the operation. This was a 'showstopper,' more of that show business lingo that I was learning from the 'special' army. I let Tom know and we asked Bob if he could make inquiries with his people, but the SMU didn't have any batteries either.

We finally reported the problem to Bil and he was speechless. He started getting red and I thought we were going to witness one of his famous explosions which didn't happen that often but when they did they were worth watching if the blast area didn't include you but I was afraid that Tom and I were going to be at ground zero on this one.

The radios had come from two places, the 18th Airborne Corps out of Ft Bragg, North Carolina and from some element of the U.S. Air Force. The Air Force radios were not configured for use in the field and had rechargeable batteries for their power supplies. We were all scratching our heads until LT Kevin, who had been looking at the equipment then the manual then back to the equipment spoke up. He was a commo officer from G Company who had just joined the CD program. He told Bil that he knew some people at Ft Bliss and if he could get a vehicle he would see if they could help. Bil said ok and LT Kevin disappeared out the door.

After dinner the teams had classes in counter-tracking, taught by a senior member of the SMU, known for his skill as a combat tracker. After that there was a commo class taught by one of the guys from Fort Bragg. He didn't say anything about the battery crisis and gave a good simple class on the use of the satellite radio and the encryption device that was attached to it. The satellite radio introduced us to a whole new world of communication.

There were a number of satellites in geo-synchronous orbit around the earth. This meant that they followed the earth as it turned on its axis and stayed in the same location in the sky. In order to send a radio wave to the satellite you had to aim the antenna, using a compass, in a certain direction and then you had to elevate the antenna a certain number of degrees above the horizon. This directed a fairly low powered signal into space where it was received by the satellite and retransmitted back to earth on a different frequency to the ground station.

It was very cool for us who had been struggling to communicate using high frequency radios in LRS operations. No more one-time pads, long wire antennas, hand cranked generators, and batteries that went dead in the middle of a transmission. We could send an encrypted transmission by voice and receive an answer within seconds. We were in the high-speed world of modern communications.

There was a downside however, DOD, which owned the satellites, had designated one up-link and one down-link frequency to all counterdrug activity. Customs was on this channel along with DEA and all the other CD activities in the Western Hemisphere. That meant that we would have very small windows of time when we could use the channel for routine traffic. We scheduled a Blind Transmission Broadcast every hour with a time hack and weather updates and commo check/situation reports with the teams twice a day. All other radio traffic would be initiated by the teams for spot-reports of smuggling activities or emergency medical evacuation requests. After the teams did a commo check, they drew their weapons from the supply room and went to the range where they zeroed their rifles and conducted tactical training.

The Mess Hall

There were continual problem-solving opportunities well into the night. Tom and I went over our checklists to take care of the problems we had anticipated, and we met with Bil for the ones that seemed to be sprouting up like weeds in the springtime. We finished at about 1 AM and went to sleep. My alarm woke me at 5 AM and I was ready to go.

The troops were directed to walk to the mess hall by teams. I joined one of the teams and listened to their bitching about having to wear their soft caps rather than their berets. We were trying to maintain a low profile at this facility but it would prove impossible. Besides not wearing berets they were wearing sterile uniforms, with no patches or name tapes. We thought that this would lessen the chance that someone would see that we were from the National Guard and 'doing' something unusual. Actually, the sterile uniforms greatly increased the amount of interest of those eating there when we walked in.

We had to show our IDs to the headcount to get our food. I looked around the room and saw quite a few men dressed in blue work shirts and pants eating their food and watching us intently. The mess sergeant came over to me and said hello. I had met him when we were here preparing for the mission. "Who are all these guys in the blue clothes?" I asked him. "Oh, those guys are the federal prisoners," he said. " The prison has a contract with us and they eat all their meals here." "Any idea what kind of crimes they're in for?" I asked. "Mainly drug trafficking." "Oh." I ate as fast as I could and returned to the CP to prepare for the mission briefing that would start at 8 AM. Tom, Bil and Lee Justice would give the mission briefing to all the teams and I would give site specific information later, one team at a time. The teams started filing in about 10 minutes before the start time and there was quite a bit of nervous chatter while everyone waited to get started.

Mission Briefing

Tom walked in with Lee, Bil, and Gracious Mr. W. After introductions were made, Lee took the podium and began with a short history of the Border Patrol and the SMU. He segued into an explanation of the use of force rules that we would be required to follow. This was very important for the soldiers because they had been trained to follow what the military calls 'Rules of Engagement' and this type of operation required a completely different mindset that was based on a complex legal theory that was built around the primacy of self-defense.

The military has rules on when soldiers can 'engage' or open fire on their enemy. Law enforcement never opens fire on a criminal except in extreme circumstances. Law enforcement officers are given firearms to defend themselves in the line of duty, not for killing a person for committing a crime.

Our use of force policy followed the Texas Penal Code almost to the letter. We were allowed to use force to defend ourselves from someone who was trying to hurt us while we were performing our duties. If that person was threatening to or using deadly force, we would be able to use deadly force to defend ourselves.

I had received a lot of training on this subject in the police academy and it was a subject of refresher training at some of the police schools I had attended in the past. Use of force and search and seizure laws are the meat and potatoes of police work. Cops discuss and debate these concepts as vigorously as some attorneys would in court. I listened carefully to Lee's explanation and watched the faces of the soldiers to see if there was any sign of comprehension. When he finished, he asked for questions and we had to listen to a number of 'what if' questions, that, while they seemed to make sense to the questioner, they didn't with the rest of the room. After a few minutes Tom put a stop to it and told the soldiers that if anyone had more questions they could talk to the agents on their team.

With that, Tom handed out the mission packets to each team leader; within these specially made bags were the target folder and some maps of the team's site. He had them inventory their bags and after determining that each team had a complete mission packet, he turned the briefing over to Bob who gave the overall intelligence briefing for the Del Rio Sector. He covered the light data, a weather forecast and a general overview of what they might see on the river.

Tom gave the mission statement and how the mission would be conducted. There is an enormous amount of detail in one of these mission orders and the team members were taking notes as fast as they could write. Tom covered how they would get there, what they would do, and how they would get back.

He covered supplies and communications which included emergency resupply based on no radio contact, and MEDEVAC by helicopter in the event of a medical emergency. Last he went over the final contingency which was what the team was to do if we couldn't get them out—their escape and evasion plan. Once the briefing was completed the teams were given the rest of the day to look at their target folders and prepare for their mission.

During this period, I gave each team their individual site briefings and showed them the video of their site. They had the aerial photographs in their target folder, so they used these and the video to determine if the place we had picked for their hide site was the best location. After I showed them the video I tried to answer their questions.

It is much better to have the team leader do the recon so that he can see what he needs to see. Because the team leaders were traditional Guardsmen, we were unable to bring them on site selection. This required the teams to trust us for information on where to hide once they got to the target area. I've never been in a unit where the fighters completely trusted anyone in the headquarters. Trust is earned by shared experiences and we wouldn't be going out with these teams, so we had to allow their intense scrutiny of our plans.

I told each team that the hide site we had chosen was for them to use as soon as they got out of the vehicle the night of their insertion. "After you have moved your gear to this location, set up security, made your initial entry report and have eyes on your target area, you can look for better place if you want," I told them and this seemed to mollify all but the most anti-headquarters elements on each team. Because Bob had been with us and had helped select these locations, the agents were ok with our plan and that helped in eliminating counterproductive activities during the teams' planning.

The Range

After dinner, the teams were transported back to the range. Bob and another agent gave a quick class on the tactical procedures to be followed if the agents had to make an arrest. Our agreement with the BP stated that if a law enforcement situation was unavoidable, that is if someone stumbled onto the team, the agents would take charge and direct the team to cover them while they dealt with the situation.

This was very different from our standard response to discovery while on a wartime mission. An LRS team was trained to run if they could or kill whoever it was who discovered them and then run. It's easy to discuss a major change in the mindset of a person but difficult to predict how that person would react under pressure. All we could do was try and reinforce the principles whenever we heard a soldier talking about 'taking someone out.'

This wasn't an idle threat. Eight years later, a Marine corporal shot and killed an eighteen year old boy in Redford, Texas, who, ironically, had dreamed of being a Marine. Most people considered murder but I blamed JTF-6 for putting their people out there without major retraining and better supervision.

After the demonstration phase the teams did "talk through-walk through" rehearsals under the watchful eye of Bob and his amigos. Once he was satisfied that they had grasped the concept, he called a halt to the training and announced a lunch break. Bad sack lunches from the mess hall were given to everyone. We took our food and sitting in the sand of the northern Chihuahuan desert, we ate and complained.

After running through two live fire scenarios per team, the agents brought out some of their non-standard firearms like the HK-53 and the Steyr AUG along with assorted shotguns. After some good-natured challenges from the soldiers they put on a shooting demonstration which impressed the troops and this led to the agents letting the soldiers fire the weapons until they ran out of ammunition. After chow at the mess hall, the teams went back to the range for night firing. I heard them coming when they returned to the headquarters and there was a lot of loud talk and laughter, so I figured that the morale had gone up with the trip to the range. They then had the rest of the night to conduct mission planning which consisted of their dividing the duties within the team and memorizing the plan.

I got with Tom and Bil and we reviewed our plans. The big problem that remained was the battery situation for the SATCOM radios. Bil had been calling people all over the country trying to get different radios but so far, he was getting nowhere. He always displayed optimism under stressful conditions, and he told us to go to bed and that tomorrow would take care of itself.

Mess Hall-Again

We had our next problem the next morning. Tom and one of the officers had to go down to the mess hall to talk to the mess sergeant. One of the soldiers had gone back to the serving line for seconds before the stated time. The cook wouldn't give him more food and an argument ensued resulting in the cook shouting at the top of his lungs, "I know all about you people and your secret drug mission, and I don't give a shit, you're not getting seconds until its time."

I'm sure that the federal prisoners found that exchange both amusing and enlightening. One of the agents told us that the prisoners had telephone privileges and that news of our secret mission would be everywhere by tomorrow. We discussed this with Lee Justice and the older BP agents who told us not to worry about it, "It's a big border and we will be operating hundreds of miles from here."

Tom took the offending soldier aside and warned him that any more problems with him and he was out of there. The soldier tried to defend himself, but Tom reminded him that the mess hall procedures had been briefed and that there was a sign on the wall at the serving line stating the policy for seconds. "It's your fault that the incident occurred, don't let it happen again."

Batteries and Briefbacks

LT Kevin had come back to HQ while I was sleeping. He had gone from unit to unit over at Ft Bliss asking for radio batteries. The LT had noticed that the battery terminals on the satellite radios were the same as the batteries for the AN/PRC-77 radios which were standard issue in most of the army. He found a unit that was getting ready for a big inspection and they had an excess of the batteries we needed.

He was already in deep shit with Bil for making a deal with some other agency that would obligate the CD program to spend money that we didn't have, a serious logistical offense, and his job was on the line, but when he showed up with the solution to our communications train wreck, Bil was all over him, as much as Bil could be all over any officer, with brotherly love and affection. LT Kevin continues to work for CD to this day as an LTC.

There was little time to rejoice over the battery find. Bil got a phone call from Austin, telling him that the C-130 mission to Del Rio that was scheduled for tomorrow was canceled. There was a hurricane in the Gulf of Mexico and the plane was needed for disaster recovery work. We had considered these problems during our planning and it had been decided that we would use an INS bus to drive the troops to Del Rio. Bil got with Lee Justice who contacted Operation Alliance and arranged for an Immigration bus, normally used to transport illegal aliens, to take the teams to Del Rio.

The teams stayed in the area preparing for the briefbacks that would take place that afternoon. Tom had decided that I would remain in the command post for the first mission while he went to Del Rio to command the Forward Support Element. He was responsible for resupply activities and liaison with the sector agents during the operation. He took two satellite radios in case one of the teams had a broken radio.

Because of the long bus ride to Del Rio, the briefback times were pushed up so that they could all be completed that night. When the troops got the news about the change in schedule and the bus ride there was a good deal of bitching but after they had released their tension, they went back to preparing their briefbacks.

After lunch that afternoon we started the briefbacks. The whole command group was sitting in the classroom and each team filed in and took a seat in front of them. The team leaders were the agents, the only time during my 15 years in CD that National Guard soldiers would be under the direct command of a civilian law enforcement agent. Each team talked their way through the mission. At the end the command group asked a few questions and given a 'go.'

That afternoon, the teams made commo checks with their satellite radios, using the new batteries, and they all worked. The last thing they did was to test fire their weapons. They wouldn't clean them again until they came out. Bil had structured the isolation schedule to include the things he had done 20 years ago when he climbed into a helicopter and flew across the border into Laos. Now he was sending troops down to the border with Mexico.

The next morning the teams finished their packing. They stored their duffel bags with supply and waited for the bus. Tom left with Mike W, an LRSD man and the sergeant from the Air Force who had brought the radios and all the problems because of the batteries. The Air Force sergeant was a little standoffish with us and seemed a bit disdainful of the operation. The Air Force in those days thought about war in terms of megatons and throw-weight and our little operation seemed to him like a waste of his time. His attitude was to cause us some big problems with the Border Patrol before the operation was over.

Before he left, Tom told me that I was in charge of the CP. "I know there are some officers but Bil told them that you and I are running this operation, so run it." I assured him that I would do my duty. Even though I was nervous, I gave myself to the Lord and his Grace and went into the radio room and worked on the radio operator's roster.

When the INS bus pulled up, the troops were ready to go. The bus was pretty small, so the troops had a tight squeeze getting their rucks and themselves into a seat. This bus was used for short trips from a detention facility to the border and this trip was going to take nine hours with one stop in Marfa. I was introduced to the driver by Bob who called him 'Speedy' due to his propensity for speed while he drove. Speedy assured me that he would get them to Del Rio, "No problema." I went back inside and instructed the first shift to monitor the radio because if the bus had any problems, the teams would notify us by radio. The operation was on.

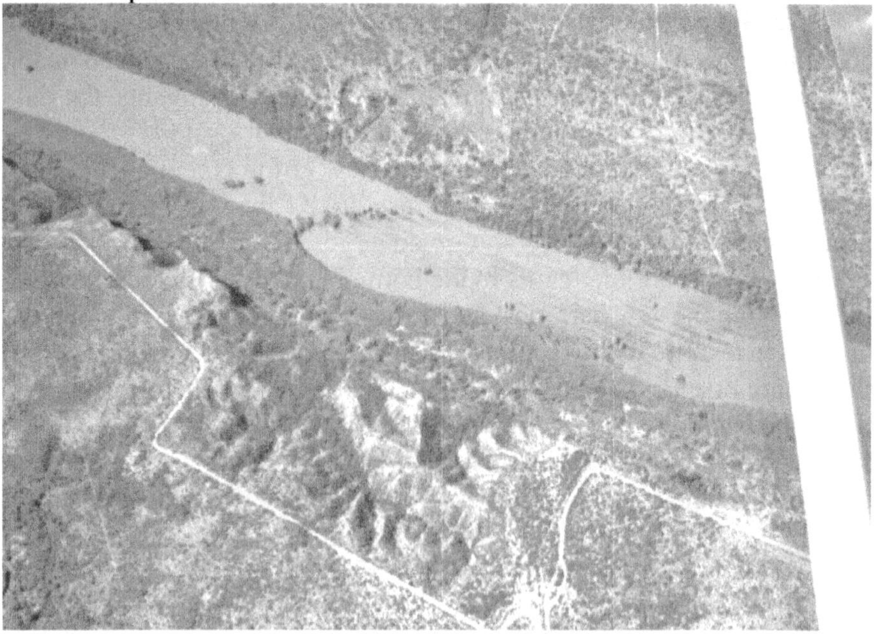

Rio Grande River-Del Rio, TX

Deliver me, O Jehovah, from evil men; From violent men preserve me Who devise evil schemes in their hearts and continually stir up wars.
Psalm 140:1-2

Chapter 6-July 1989

Operation Unity 3-4: Execution

The bus took the teams to Laughlin Air Force base in Del Rio. They arrived at 7PM and Tom met them there. They loaded into Border Patrol vehicles for their insertion into the target areas. Tom made sure all was well before waving goodbye as they drove into the sunset for the first night of Operation Unity 3. Meanwhile, back at the ranch, I was settling into what would be part of my routine for the next 10 years, running a command post. Bil would be there much of the time with Lee Justice and the tactical decisions would be theirs but the daily activities in the CP were directed by me.

The next event for us was the initial entry reports by the teams when they had reached their target areas. These reports started coming in around 9PM from the teams that were close to Del Rio and it was 4AM before we heard from the team that was farthest from Del Rio. I reviewed the CP commo procedures with the shift leaders and went on a run before lunch.

Bil met me at the door when I got back and told me to take it easy because we were a mile higher than Austin. He knew I was getting ready to go to airborne school, but he told me several times that he thought it was a bad idea. He was always trying to get me to leave LRSD and get into a unit where I could get promoted but I was determined to be a paratrooper. I think he was worried that because of my age I would get hurt and he knew about these things but I was caught up in the airborne slipstream and wanted to prove to myself and my comrades that I could jump out of a perfectly good airplane in flight.

The radio operators had a hard time transcribing the messages they were hearing onto the paper message forms we were using. The teams were talking fast, they were being stepped on by other radios on that frequency, there was a lot of static that made it difficult to hear words clearly, and this made it difficult for me to leave the CP and get any rest until we worked this out. One thing I did was to buy a tape recorder at the PX and they used it to record the messages so they could play it back when they were trying to write what they had heard on the radio.

As a police officer I was used to doing police work on a radio and writing down what was needed, but these soldiers were just getting into voice communication so it would take a while and there would be some mistakes before they felt comfortable in this new environment.

The SMU supply man had an abundance of large military camouflage nets and he gave me several. I kept myself busy and out of the radio room by cutting them up and making individual nets. I detailed a couple of soldiers to watch what I was doing and then do the same thing because I wanted to make 50 nets. These two soldiers were from G Company and felt that this work was beneath them, so my leadership challenge was to persuade them to do what I told them.

This I accomplished by entering into a two-hour debate with them on why they should or should not do what I told them to do. I didn't want to fall back on the LT or Bil so I went round and round with them as they threw obstacle, after reason, after excuse as to why they couldn't perform this task. Unfortunately for them my debating skills had been honed to a sharp edge through losing some pretty strenuous arguments with my wife so in the end they were forced to concede to my logic and help me make the nets.

On the Border

For the next few days, the teams in the field called in the things they saw. When they saw something, they would run it by their agent who might have been sleeping, and after he looked the situation over and decided it looked suspicious, they would report it. Usually it involved either a vehicle driving through their area, a boat passing their location on the river, or an airplane flying over them. The radio operators tried to get the details like license plate numbers, vehicle descriptions, people descriptions, and activity. This took a lot longer than I was used to as a police officer and I had to make myself stay out of the radio room while they went back and forth on the radio; "Say again all after…"

If the report seemed important to Lee, he would call the Sector HQ in Del Rio and they called the station where the activity was reported. The station dispatched an agent to wait on the highway where the road came out to try to intercept the vehicle and search it.

LT Bruce, our commander of LRSD had a great LT idea and told a joke during the mandatory listening time when the teams made their commo checks. Bil happened to be walking by the radio room while the LT was giving the punch line, speaking into the SATCOM handset, and realized that the LT was broadcasting his joke to all the counterdrug operations in the Western Hemisphere. He went 'Bil'istic.

Bil came into my room where I was resting and began hollering at me to get the LT on the next thing smoking at the El Paso airport. I let him go off for a while until he slowed down and then I asked, "Well, was it a good joke?" He stopped for a moment and then smiled before allowing that it was an OK joke for an LT. I asked Bil if he would spare the LT and give him one more chance because Tom and I liked him and thought he was doing a good job running LRSD. Bil grumbled a bit and then told me to go and tell the LT that he had one more chance and that he had better watch his step.

Tom was down in Del Rio doing resupply. He was coordinating with the Border Patrol for vehicles and drivers to take food, water, and radio batteries to the teams. The teams had prepared bundles with these items, to be delivered to them about halfway through the mission by what we called automatic resupply. If something came up, then we conducted on-call resupply. Both of these took place during the night.

His end of the operation went smoothly except for a big problem caused by the Air Force sergeant with the Air Force SATCOM radios. He was made to feel that the problem with the radio batteries was his fault which it wasn't. But since he was there, he had to listen to all the bad mouthing of the Air Force and those who wore the blue uniform, so he wasn't happy. In his angst, it seems that the good sergeant told a BP agent that this operation and the BP was FUBAR from A to Z.

The agent told his supervisor who told his supervisor until the message got to the Chief Border Patrol agent in Washington, DC. From there it went to the Justice Department and from there it went through the Defense Department to the National Guard Bureau who sent it to the Texas Adjutant Generals Department's Chief of Staff who notified Uncle Bennie who called Bil in El Paso. When the secondary explosions finally subsided in Del Rio, Washington, DC and Austin, Bil was left to add another saying to his list of 'Enneyisms,' 'The 15 Minute Rule,' which stated, "Bonehead actions and statements will surely be in your counterparts headquarters in 15 minutes!"

On the last day we gave the teams their extraction times and one by one, they showed up at the debriefing location at Laughlin Air Force Base looking tired and grungy, some of them with a good start on a beard, where they were given a snack and then they were seated in front of Bil and Lee and debriefed. Bil had devised the debriefing procedure. The format he used was similar to the ones used by MACVSOG. Every message was gone over and the team was given an opportunity to clarify or correct each one. Any updates to the maps they were using was noted and then the team was dismissed to begin recovery of their equipment and preparation for movement.

No drugs were seized as a result of the first operation. Some illegal aliens were caught but we weren't there for immigration law enforcement. One of the teams had some walk very close to where they were hiding but managed to stay hidden. The boat team caught a lot of fish and generally had a great time floating around Lake Amistad. The radio operator on the boat team showed some smarts when he realized that the coax cable for his satellite antenna was broken. He used a paper clip to make the connection and was able to request a coax cable resupply.

What Happened on the River-Lessons Learned

We learned a lot about conducting these types of missions. Our soldiers were a mixed bag with regard to LRS standards. Some of them did well and others were not able to handle the inactivity of surveillance. Lying on your stomach under a mesquite tree in 120 degree heat and looking through a pair of binoculars for hours, ten days at a time, is a difficult kind of soldiering and not many people can do it well. I don't think we gave much consideration to this or we would have put the unit sleeping on the ground with no tents for the two weeks of AT.

Training operations can be deceiving for the soldier and the leader because it is difficult to replicate combat conditions in a peacetime environment. During AT the Cav officers wanted us to truck hot food out to our teams on an overnight training mission. Even units like Rangers and Special Forces are forced to curtail the realism in their training due to safety regulations and the housekeeping policies of the bases where they train. For example, most training areas will not allow you to dig fighting positions, so most infantry soldiers deploy to combat never having dug one. It takes time to mentally and physically adjust to the daily demands on the body when living in an outdoor environment. I realized that the airborne mentality was working against us because airborne was all about aggressive movement and surveillance was all about passive observation. As this was an airborne unit, we would have to try and focus that aggression toward skillful field-craft.

The biggest revelation was the fact that some of our sergeants were not suited for leadership of small teams doing this kind of work. This information came to us not at the debriefing but from individual soldiers who told us about some pretty bad events that took place on the river during the operation.

On one team, we were told; one of our guys put his underwear on his head and danced around on the rocks naked for the amusement of the team. This seemed to create a kind of nonchalance with the rest of that team. Their agent decided to add to the party by walking to a nearby store on the highway and ordering a large pizza that was delivered to the store where he picked it up and paid for it wearing his tactical uniform. We were glad that Bil never heard about this because the explosion that ensued would have ended our relationship with the SMU; but that would have to wait for another year.

The Border Patrol agents had a saying, 'What happens on the river, stays on the river' which I found out later was what people said who went to Las Vegas. The Bible says, 'that which is said in secret shall be shouted from the roof tops.' I'll go with the Bible because nothing that happened on the river stayed secret for long.

The tales we heard were from soldiers who were either dismayed by unprofessional behavior or getting revenge for some problem they had with another soldier. As these stories came trickling in, we checked on them as best we could and made our own determinations on who we would choose for leadership responsibilities in the future. Most of the army is run by a strict chain of command and no one does anything that is not observed by at least two levels of supervision at any time.

These missions required leaders that could work without supervision for long periods and could enforce strict standards of behavior on their soldiers without any backup. We began mulling over the idea of a full-time cadre of team leaders that could better control the part time soldiers that would be used to man these operations in the future.

The Lick-Ass Debriefing

After we debriefed the teams, we debriefed the air crew from General Dynamics that was flying the 'Lick-Ass.' They had been up and down the river each night and the teams had talked to them on the radio but were unable to signal the airplane with their infrared lights to show where they were because the FLIR could only see heat. The teams had all complained about the noise that this 'silent' aircraft made as it passed over their position. The agents were pissed because they were sure that the reason we didn't catch any dope was because the smugglers heard the plane and stayed home.

The air crew had spotted some illegal aliens near the town of Del Rio and had tracked them with their FLIR; the heat from the friction of the illegals' shoes left a signature on the ground for some time after the people had gone. The BP was very interested in this technology but they weren't interested in the aircraft. Bil listened to all this and he still invited the General Dynamics guys to come to the next operation which proved to be a mistake.

I was disappointed that we hadn't caught any dope and so was the BP. Bil wanted to emphasize that the mission was successful because of the 'intelligence' we had produced but he did that as a face-saving gesture to the people up the chain. 'Intelligence' at its best is information that can be acted on to produce a positive result. The results of our efforts in Del Rio told the BP what they already knew, that drug trafficking probably didn't occur at the locations we surveilled but might have occurred at all the other locations along the river. I told this to Bil many times and I'm sure he understood but he had the big picture in view which was to enlarge the program by not failing. This was the same in bureaucracies as success. He was counting on Tom and I to refine the way we did things until we could catch dope.

I told Tom that law enforcement people knew that you didn't always catch the crooks every time, but you had to keep going out. We determined that we would measure our success or failure on the basis of whether our planning was sound and the execution of the plan was done in a professional manner. Finally, we would measure our success on our relationships with the law enforcement people we were supporting. If they invited us back it was a success because if they didn't, we wouldn't get another chance to make them successful with our support. Catching dope as a measure of success we would leave to the cops.

When we returned to Austin, we had to turn in all the borrowed equipment and once again George C was a great help. I spent some long days writing up the after-action reports and the intelligence summary. Bob showed up a few days later to arrange for one of our helicopters to come out to El Paso to support the next SMU selection and training session. We spent some time with him laughing and talking about the last operation and speculating about what might be next.

Bob was admiring Tom's Ranger Tab and saying how he wished that the SMU had something like it. Tom grabbed a piece of paper and sketched out a Tab with the SMU name on it and showed it to him. Bob thought it was great and that's how the first SMU Tab was designed, by a Ranger who had a high regard for these law enforcement agents who were a cut above the rest. I'm sure that most of them now days have no idea how that tab came into being nor do they know how their leadership screwed its designer.

August 1989: Back at the Athenian Camp

Soon after we returned from Del Rio, we had to move. The Counterdrug Program was growing at such a rate that there was no more room in the Guard headquarters. George C found a temporary building designated T-6D or Tango Six Delta. Tom and I moved our things over there and settled in. We were glad to get away from the flagpole. Now we could wear BDUs to work instead of the Class B uniform we wore in the headquarters. We set up a small area with two desks and continued to recover from the Del Rio operation.

The Operation Guardian project was moved into the same building a few days after we got there. This was our support to U.S. Customs where Guardsmen helped them inspect cargo trucks coming out of Mexico. There was a young Cav LT in charge of Guardian named Smith. The cargo inspectors were coming from out of state for two-week tours. They were MPs from other states' National Guard.

There was no furniture for the Guardian people, but I never gave it a thought as I was working 12 hour days. Eventually I noticed SFC D eyeballing our office when he came in and went out. He was the G-2 sergeant who now wished he had never heard of LRS after the fiasco he was responsible for at the end of AT. I heard him and Tom talking behind me one morning and he ordered Tom to give him one of our desks. Astounded to hear Tom agree, I was directed to help them move Tom's desk out of our area and across the room to the Guardian area. I was infuriated with this kind of military bullshit and when lunch came, I went outside with Tom and asked him why he gave up the desk. He just shrugged and said, "Because he outranked me." Tom was right but it never would have occurred to me to use my rank to do something like that and so I always kept a weather eye on SFC D after that.

George C took care of us when he showed up afternoon with a flat-bed truck and told us to follow him to nearby warehouse where there were hundreds of desks, chairs, and other office furniture gathering dust. We found several nice desks, some chairs and a table for map work that we brought back to T6D. I saw SFC D and the other Guardian people looking at our new furniture and told them that George was outside and he had a lot more of these, hoping to preempt another rank raid on our resources. It worked and we went back to work.

Tango Six Delta was the only building on the camp that had a 'Dutch' door. Bil had security signs placed on the door which declared that this was a secure area, and no one was permitted to enter unless their name was on the access roster. I was sitting at my desk near the front door when I heard a knock and upon opening the top half, I saw a full bird colonel standing there. "Good morning, sergeant, I'm Colonel Jones and I'm coming in." This was the chief of staff for the Texas National Guard. "Yes sir", I said, "Let me check your name on the roster and I'll let you in." I scanned the roster which didn't have many names on it and didn't see his.

I was getting nervous now; this was the highest-ranking active duty officer in the Adjutant General's Department. He wanted in, his name wasn't on the roster and I was a buck sergeant. He was starting to breath hard as he waited for me to open the bottom half of the door that barred him from entering. I looked at the roster one more time, front and back, and told him, "Sorry, sir, your name isn't on the roster so I can't let you in." He didn't like this answer. His face got a red and he stood up on his toes and said, "Sergeant Murry do you know who I am?" "Yes, sir," I answered, "you're Colonel Jones, the chief of staff." "That's right" he said, "and I'm ordering you to let me in this building right now!"

Now I knew I was deep doo doo. There was no way I was going to let him in, but I was disappointed because I knew I would be fired and I was really getting into the whole drug war thing. "Sorry about that, sir" I told him. "You'll have to talk to Mr. Enney if you want to get in here." He gave me his meanest stare and said, "We'll see about this" as he turned on his heel and stalked off. I went back to work and waited for the hammer to fall. Tom came back and I told him what happened. He said he would talk to Bil and I continued working on my projects. Bil came over later and asked me what was going on and when I told him about my confrontation with the colonel he just laughed and said, "He knows better than that, don't worry about it."

Bil never trusted the full-time National Guard staff that worked at Camp Mabry. He learned about National Guard politics when he was a regular army advisor. Every organization has its rules and its politics. The Guard politics were blatantly obvious to someone who had grown up in the much more subtle atmosphere of the Special Forces where there was a fairly strict adherence to the ethical code of the soldier.

Bil believed that trust was earned, and he had a good relationship with Colonel Jones which stemmed from the first surveillance operation in 1988. Was this one of Bil's devious loyalty tests? I never found out and Colonel Jones never came back to T6D.

CDTF Intel Operations

The Counterdrug program expanded by leaps and bounds. Operation Plus-Up was in high gear, sending soldiers to work with the Border Patrol doing administrative tasks in the sector headquarters that were being done by agents. Plus-Up was designed to get these agents out of the offices and back onto the river. The BP loved it and couldn't get enough of the free labor we were providing.

The latest 'big deal' was intelligence support for drug law enforcement agencies and was designated Plus-Up II. Captain L, SSG R, and SFC D who had hijacked Tom's desk, all members of the G2 section of the 49th Armored Division, had pitched the idea that they could send out soldiers to work with the narcs and be intelligence analysts. This concept had already been approved at the National Guard Bureau, so it wasn't a hard sell. Bil was looking for some staff to manage it and when these three showed up, they were it.

The DLEAs had been told about this at a recent conference and the requests were coming in. The idea was to have school trained military intelligence analysts supporting the cops. Doing what, no one seemed to know. There was no one in CD familiar with law enforcement work but me and I had been a patrolman. I told them what I knew but they weren't really interested in talking to us 'knuckle-draggers' as we were known in the Intel world. SSG R had been a team leader in LRSD and SFC D wore a Special Forces Tab but now they were in the 'Secret World' and were slightly disdainful of our operations.

Captain L and his two sergeants were school trained Intel people and were very apprehensive about sending people who knew nothing about Intel to work with the cops. Bil was in a hurry to fulfill the requests he had for Intel support, so he directed his new staff to hire anyone with a pulse and send them out. The first guy was a tank mechanic who was looking for work and heard about Counterdrug. They sent him out to some small town near Texarkana to do Intel.

They thought this would lead to disaster, but they failed to recognize the fact that the National Guard soldier has some capacity for adaptation that the regular army soldier doesn't have. They also failed to understand cops. A free body working for them was just fine; they would figure out what the person was to do when he arrived. We sent a lot of people out that summer and fall. Sure, there were some train wrecks along the way but for the most part it was a great success and continues to be so to this day. Some of the school trained Intel people ended up washing cars for the cops and a logistic clerk ended up playing a major role in the successful FBI investigation of a Mexican Drug Cartel. There are a lot of interesting stories about what happened in the Intel program, but I'll leave it to one of them to tell the story.

September 1989: Military Industrial Complex II-E-Systems

Tom, after a meeting with Bil, told me that we would be going on a road trip with CPT L to Greenville, Texas to meet with some civilians who worked for E-Systems, a company that did Government work. Tom said that they would participate in the next surveillance operation along with the Lick-Ass aircraft from General Dynamics. CPT L was going because their support involved Intelligence products and Bil wanted the Intel project to mesh with our surveillance project. We all met at T6D at 5AM the next morning and got into Tom's BMW. He liked driving and he wanted to pick up some mileage money.

The two of us talked with CPT L on the four-hour trip. He was a Signal Intelligence officer who had left the regular army, joined the Guard, and was looking for an active duty opportunity and it looked like he had found one. He was an empire builder and with his two ambitious sergeants he would try and take over the program, but that would come later. Right now, he was getting the lay of the land and asking Tom and I a lot of questions about our project. I could tell that Tom was holding back a little, so I kept my answers as short as possible without being impolite. We passed the four-hour drive pleasantly. In Greenville we turned off the Interstate and drove south on a two-lane road for several miles, until we came to the turn off for E-Systems.

There was a guarded gate and beyond were some enormous buildings. The guard checked our ID cards and called someone on his radio. He told us to park and wait for our escorts. A small van came out from behind one of the buildings down the street and sped toward us. Two civilians who I will call Dick and Harry jumped out of the van and walked quickly over to our vehicle. We got out as they walked up and they introduced themselves, shaking our hands, and welcoming us to E-Systems. They asked us to go with them in the van and we set off into the heartland of the Military Industrial Complex.

As we drove through a maze of buildings, I could see several large aircraft in enormous hangers, so I figured that there was an airfield nearby. I didn't know the half of it. We finally got to a small building and we followed Dick and Harry inside. There were several civilians standing there and they welcomed us to E-Systems. They offered us coffee and other refreshment and let us use the rest room before asking us to follow them to a small conference room that had place settings for each of us to include cardboard name plates with our names on them.

There was a small color brochure at each place and when I picked mine, I saw that they had printed our names on the back. The brochure welcomed us to E-Systems and had an hour by hour agenda with the names of the people who would be talking to us. This was way beyond what I was expecting. I had no idea what was going on, so I decided to try and keep my mouth shut while I figured it out.

The senior man welcomed us again when we were all seated and proceeded to spin his web. E-Systems, it seems, was troubled by the reports of rampant drug trafficking on our Southwest Border and since the Cold War was getting colder all the time, they had decided to lend their expertise in war fighting to the Drug War which was starting to look like the only game in town, though he didn't say that.

What the Border Patrol needed was an integrated system of sensors and intelligence personnel to create an electronic wall along the border. E-Systems had been developing such a wall and they wanted to demonstrate it during our next operation. With that, the briefer turned the meeting over to Dick and Harry who would fill in the details.

Dick unrolled a large map of the border and began to explain how their system worked. They had a complete system of ground sensors that they would place alone the entire border from the Pacific Ocean to the Gulf of Mexico. These sensors would all feed to a central point called the 'Fusion Center' where the sensor 'hits' were compiled in a large computer. The center would be filled with analysts who would analyze each event and determine whether the BP should be dispatched to investigate. The computer, they said, would analyze all the activity that was detected by the sensors and generate reports that would help the BP allocate their personnel resources in a more efficient manner.

What these guys had come up with was a border sensor system. B.I.A.S., which is what they called it. Brilliant—what these defense contractors could do for us. They asked us if we had any questions or comments. This was a bad thing to ask me because I hadn't just ridden in the back of Tom's car for four hours to hear about a sensor system that the Border Patrol already had. I bit my tongue however and let CPT L display his stuff to these guys who had obviously served in the military intelligence field. They discussed computer systems and software. "Your using Oracle," said CPT L with a knowing smile. This brought on a long discussion about computers which I tuned out because I was reading the handout that they gave us regarding B.I.A.S.

The handout mentioned an aircraft that would act as a relay for the more remote sensors, so when a break occurred in the one-upmanship of mutual knowledge displays that had been going on with CPT L and the E-Systems guys, I asked about the plane. They were delighted and insisted on showing it to us. We left the building and got back in the van and were driven to a large hanger nearby. We followed Dick and Harry inside and there in the middle of a large room was a small strange looking, single engine prop plane.

Blast from the Past

Wait a minute! I'd seen that plane before. I looked it over while Dick and Harry gushed on about what a great piece of technology this plane was. Then I remembered where I'd seen it before. It was in a small pictorial history of the McNamara Line from the Vietnam War. This was the exact same plane that was in that book. Then everything came into focus. Somehow, E-Systems had inherited all the left-over equipment and plans for the McNamara Line. This was a 'great idea' operation of the Vietnam War where the Secretary of Defense, Robert Strange McNamara had tried to incorporate western technology as a way to defeat Asian willpower. He envisioned an electronic cordon across the DMZ and along the borders of Laos and Cambodia with South Vietnam that would stop the North Vietnamese infiltration of troops and supplies using the Ho Chi Minh Trail.

One of the main elements of this cordon was a sensor system that was monitored at an intelligence center located on a U.S. base in Thailand. These sensors would alert U.S. Air Force units to the presence of personnel in the area of the sensors and they would bomb and strafe the area day and night until there was no more activity. A lot of money was spent developing this system, but it was never fully employed due to North Vietnamese artillery which decimated the Marine engineer units that had been tasked with constructing the line along the DMZ. The system was in place around the Marine Combat Base at Khe Sanh and worked very well according to some who were there. A number of other areas along the borders with Laos and Cambodia were sown with sensors dropped from aircraft and these provided some targeting information to the Air Force who was trying to stop the traffic on the trail.

The whole project was abandoned after McNamara was fired by LBJ and its remnants ended up in some hangers and warehouses in Texas where they gathered dust until some bright guys at E-Systems found them and got the OK to pitch a 'new' idea to the Border Patrol. This all came to me in a flash, but I kept my mouth shut as they showed us the airplane I had seen in the book.

We went to lunch after we left the hanger. I waited until we were away from Dick and Harry before I told Tom and CPT L about what I thought was going on. Since Vietnam was before their time and neither had heard of the McNamara Line, they were pretty skeptical about my ranting denunciation of the whole morning's presentations.

I argued that this was a business deal between E-Systems and the U.S. Government that we were being finessed into endorsing because we were actually working in this new theater of war so our endorsement would have some credibility. "The BP already has a sensor system, probably using the same sensors left over from the McNamara Line. They have it tied into a computer system at each Sector Headquarters which gives them reports that are easily analyzed at the Sector and used to put agents in hot spots when patterns are detected."

I was on a roll. "You know that these Sectors are like kingdoms and no one is going to like the idea of a central intelligence office that can take agents out of one sector and move them to another based on computer reports of sensor hits that no one has verified." Tom agreed with me but CPT L, being an officer, was being cautious. He wanted to hear the rest of the presentations, so I agreed to keep my suspicions to myself for the time being.

We returned from lunch and spent the afternoon being briefed by E-Systems scientists and Intel people. I was really getting an inside look at how a huge company like E-Systems was able to make billions of dollars doing everything from maintaining Air Force One, the fleet of large airplanes used to fly the President of the United States and the airborne command posts that would be used by the National Command Authority in times of nuclear war. They also invented 'new' solutions to any kind of threat to their right to continue to feed off the tax payers of this country. As the afternoon went on, I determined that Dick and Harry were the driving forces behind this project. I guessed that within a company like E-Systems, entrepreneurial initiatives by the employees were encouraged and possibly rewarded based on the financial success of the project they developed. Dick and Harry had a lot riding on this, and they were doing everything they could to make us happy.

After the last presentation, one of the senior E-Systems people asked us for our comments and suggestions. CPT L gave me a 'keep your mouth shut' look and proceeded to give them some glowing feedback on the whole system. Maybe he was trying to get a job there in case the drug program went under later or perhaps he really thought that this project was a good idea.

Tom was asked what he thought and while he remained non-committal, he agreed that the next operation would be a good test of their system. He turned to me and asked me what I thought. CPT L was looking at me and the guy saw this and said, "Look, we really want your input." I looked at CPT L and said to the guy, "Are you sure you want my two bits", and he said, "Absolutely."

"Well," I said, "I can see that you fellas put a lot of work into this project, but I think you should consider the lessons of history before you go too much farther. This project is like the Phoenix Bird, rising from the ashes of the Vietnam War." I proceeded to give them the story of the McNamara Line, and from the looks on some of their faces this was news to them. How soon we forget, everything!

"The problem I have with your project is one, that the BP already has a sensor system that does everything yours does. Two, your project would centralizes their system above the sector which not only greatly increases the cost but takes away the sector chief's authority over the use of his personnel. With the current system, the sensor alerts go to a computer in the sector headquarters where the intelligence people give the reports to the sector and station chiefs for the most efficient use of their limited manpower. What your project will do is start a war within the BP and Operation Alliance."

The room was pretty quiet when I finished talking. Then the E-Systems guy said, "You bring up some interesting points. We'll have to consider them as we proceed, thanks." That was it for the presentations, but Harry had one final stroke to give us and this was a BBQ at his house that evening. We went back to the motel and on the way, CPT L remonstrated with me about my candor with the project manager. He wasn't really angry, but he made it clear that he would have preferred to be the only spokesman for our group, being an officer and all.

I didn't see it that way because Tom and I were running the surveillance operation and we were NCOs. As long as Bil was happy with our work we didn't need to trouble ourselves with worrying about what some officer was thinking. Dangerous business in the army but what the hell,' I want to be an airborne ranger, I want to live a life of danger.' We were living the life and enjoying it.

When we got to Harry's house, we were greeted by his wife and invited into the back yard. Harry was grilling some fajitas and offered us a beer. Dick was there with his wife and everyone seemed pleasant enough. I turned down the beer but joined the banter around the bbq pit while Harry cooked the meat. There was a nice view of the sunset and I stepped away from the conversation to watch it. I heard someone moving to stand next to me and saw out of the corner of my eye that it was Dick. He had had a few beers and I could tell that he hadn't been happy with me since I gave my opinion of his project that afternoon. He leaned in close to me and said, with a slightly menacing voice, "Nobody appreciates a military historian," and then he walked away.

We did get one tangible benefit from E-Systems. They gave us several sets of small seismic sensors that would prove to be useful in the years to come. They were buried in the ground near a trail or other likely avenue for foot or vehicle traffic. Each sensor had its own ID number and when it alerted, the number was displayed on the receiver, telling you where the activity was.

Operation Unity 3-4-It's a Wrap

When we returned to Austin, CPT L and Tom told Bil about the trip and I continued to wrap up the Intelligence Summary for the Del Rio operation. I wrote it on my Macintosh following a format that I found in one of the Field Manuals that covered intelligence operations. Since we didn't catch any dope, all we could do was state the obvious which was that each place we watched had the potential for drug trafficking even though we didn't actually see a smuggling operation.

This was hardly news to the BP which received a copy of this document. Other copies went to NGB and who knows where else and it's possible that a copy of it is still sitting in a file drawer somewhere. I've still got a copy and it always amazes me to look at something that I put a lot of work into while other people were prodding me to get done because it was so very important; but now it was a vapor trail in the sky that had dissipated and finally disappeared.

The After Action Report (AAR) was different. This document went to the National Guard Bureau and was our way of telling the congress that we were successful and needed more money. I used the format from the AAR that was done after the second operation in the Big Bend and wrote up the information for the Del Rio operation. Daily, Bil asked for a draft and red inked some things and demanded more detail for other subjects.

Needless to say, any mistakes on our part were left out of the AAR in order to keep the document short and more readable for the busy people in Washington. Actually, no mistakes by anyone were mentioned. The purpose of this kind of AAR was to promote good relations between all the agencies involved and to make known any problems that were caused by a lack of funding from Washington. This was also a wish list where we told NGB that if we just had more night vision equipment, satellite radios, etc. we would be more successful next time.

Tom and I were happy with overall results of the operation. Bil told us we had done a good job and that he wanted us to get ready for the next one. We conducted a very real, no BS internal AAR and used the lessons learned for the SOP we were writing for these operations.

Unity Briefing Slide

Operation Alliance Blazer (Loaner)

Rio Grande, Del Rio, TX

JTF-6 LOGO

Border Patrol Christen Husky

Unity Site Selection

BP Agent

Lone BP Agent

SMU Agents

For I also am a man under authority, having soldiers under me. And I say to this one, Go, and he goes; and to another, Come, and he comes; and to my slave, Do this, and he does it.
Matthew 8:9

Chapter 7-October 1989

Operation Unity 5

The next operation would be in the Rio Grande Valley. It is in south Texas and was well known as the heart of drug smuggling on the border. I drove down in the Alliance Chevy Blazer to the Border Patrol's McAllen Sector Headquarters to conduct initial coordination. When I got there, I contacted the Deputy Patrol Agent in Charge named Steve. We hit it off right away. He was a good guy with a great sense of humor. After I went over the basic format for the operation we had conducted in Del Rio, he gave his approval and told me to meet with Harry who was in charge of the sector's ERT or Emergency Response Team.

I drove east to a town where Harry was the station chief. He was a nice guy and we got along immediately. He asked me to tell him about the Del Rio operation, although he probably knew all about it through the drums along the river that were beating all the time. When I finished giving him an overview of the operation, he let me know that things were a little different in the McAllen Sector. "We don't use the SMU in the McAllen Sector," he said. "Our ERT guys are much better at working the river in McAllen than those guys who have never been here before." That made sense to me, but I asked him if Operation Alliance would have a problem and he said that Chief Reyes ran this sector, and nobody told him what to do here. Harry told me to meet him at Sector headquarters in the morning and we would go over some details with Steve. I met Harry at Sector the next morning and we went into Steve's office. Steve left the office for a minute and when he came back, he said that the Sector Chief, Silvestre Reyes, wanted to talk to me. Chief Reyes was the equivalent of a Brigadier General and I was a Sergeant E-5. Nowhere else in the military would I have had an opportunity to even see a man of Chief Reyes' rank, let alone sit down and talk with him in an inter-agency coordination meeting, but there I was. Harry and Steve took me to the chief's office.

Chief Reyes invited us to sit down. I saw a picture on the wall of a young man sitting behind an M-60 machine gun in the door gunners' seat of a Huey helicopter. I asked him if the picture was of him. He said the picture had been taken of him when he was a door gunner in Vietnam. "I was there in 66 and 67 with the 1st Infantry Division," I told him, playing the veteran card and this seemed to relax the situation. "When were you there?" I asked him and he told me he was there around the same time and where his base camp was. We talked a little about Vietnam and then he changed the subject.

"Harry told me that the National Guard was coming down here to do our job for us, how can you do our job?" he said in a slightly sharper tone of voice. There was silence in the room for a moment as he sat there looking at me. I was under the impression that all the higher levels of approval had been taken care of, but now it looked like the sector chief was questioning our legitimacy to even be here. He was staring at me and I realized that if I didn't say something, especially the right something, we were not going get a chance to operate in the biggest dope smuggling area in Texas.

I quietly called on the Lord, and then I started speaking. "Sir," I began, "the Texas National Guard has conducted surveillance support operations for the Border Patrol in the Marfa and Del Rio Sectors without incident. We have learned valuable lessons during each mission, and we have incorporated these lessons into our pre-mission training. Our soldiers have learned a lot from your agents, and I hope we can continue to do so in your sector. We are willing to do whatever we can to help you accomplish your mission in this sector. We are committed to give you and your agents a hundred percent of our energy and skills toward whatever task you assign us."

He nodded and I sensed that he wanted me to continue. "We are from Texas" I said. "A number of our soldiers are Spanish speakers. This is our home, and these are our people. Our soldiers are well trained in surveillance skills. All the men involved in these missions are from the Long Range Surveillance Units of the Texas National Guard. These units are very similar to the LRRPs in Vietnam. They are airborne qualified. Many of them have active duty experience with the Rangers and some are Special Forces soldiers. The missions that we have been performing for the other Border Patrol Sectors are almost exactly the same as our wartime mission."

I stopped for a moment and once again it seemed that he wanted me to continue. "Finally, sir, there is no way that we could come down here and do your job. We have neither the training nor the statutory authority to confront and arrest drug traffickers or illegal aliens. What we can do for you, however, is to hide in the vegetation along the river for a week or longer and give your agents timely and accurate information regarding suspicious activity in their area which would allow your people to seize contraband and arrest violators."

I stopped and watched Chief Reyes. After what seemed like a long pause, he smiled and said, "I just wanted to hear what you would say. Steve and I talked about how we were going to use you ever since Operation Alliance told us that you were coming. We thought about putting your men out near the Sarita Checkpoint near Kingsville to watch for smugglers walking around the checkpoint but, by the way, last week, did you know?" He looked at me, "a National Guard truck was inspected at the Kingsville checkpoint and my agents found 500 pounds of marijuana in the back." My heart sank. "Great," I thought. "Some dirt bag has hurt our reputation in this sector before we had a chance to show them our stuff".

Chief Reyes saw the look on my face and chuckled. "Don't worry about it. After hearing the qualifications of your troops, I have decided to put your guys on the river in Starr County. This is the heaviest drug trafficking area on the border. If you can hide there without being spotted, then you can work anywhere on the border." I breathed a sigh of relief and thanked the chief for giving us a chance. He stood up and walked over to me and shook my hand. "You did good," he said. "I thought you were going to fold but you surprised me. You made your case and I appreciate your candor, good luck."

We left his office and Steve and Harry took me to lunch at their favorite Mexican Restaurant. When we had eaten and were talking, I asked Harry why he told Chief Reyes that we wanted to come down and do their jobs and Harry laughed and said that he didn't, "The Chief just wanted to see how you would react."

During my years on the border I would hear a lot of talk about Silvestre Reyes. In some of the other sectors, I heard Border Patrol agents say some bad things about their sector chief, but everyone in his Sector respected him and always spoke highly of him. After working with Harry and getting to know him, I asked why this was and he told me it was because Chief Reyes really took care of his people. He required them to do a good job and at the same time he was getting them the things that they needed to do their jobs. Chief Reyes is a U.S. Congressman now and a key voice in the ongoing concerns about our border with Mexico. He is a straight shooter and I'm sure he is doing his best for that difficult situation.

After talking to the chief; Steve, Harry, and I agreed to meet again in a month and that's when we would do site selection. I asked about getting an aerial over flight of the sites and Steve said he would take care of it before I came back.

Operation Unity 5-Planning and Briefings

Tom and I went together to the Rio Grande City BP Station where we met an agent who knew Harry and was trusted by him to pick the best sites in the area. We got into his Suburban to visit some sites on the river and a few minutes after we left the station, we heard another agent run Tom's license plate number on the radio. Why he did it was anyone's guess, but he wanted to know who owned the BMW that was parked outside the station.

The agent filled us in on life in Starr County. "One of our agents at the Falfurrias Checkpoint searched a truck and found a load of dope. The defense lawyer accused him of illegal search and seizure. The judge asked the agent what his probable cause was. The agent immediately replied, 'Well, your honor; the truck was from Starr County.' The judge ruled he had PC and the driver was convicted."

This was the home of the Choo-choo, a Chevy pickup truck tricked out in Chattanooga, Tennessee, with colorful murals on the tailgates celebrating Mexican revolutionaries like Pancho Villa. The players drove them around the valley, letting everyone know that they were in the trade. The local law enforcement couldn't be trusted as far as he was concerned. Many county sheriffs had been arrested for involvement with the traffickers.

He took us to a partly completed home overlooking the river and showed us a colorful mural on the living room floor depicting some border voodoo demon that the dopers were invoking for assistance in getting their stuff across the river and past the checkpoints. On top of that, by the agent's estimation, most of the people in Starr County were either actively involved in drug smuggling or profiting from it.

The terrain along the river was flat and there was a thick belt of what the agent called Carrizo Cane along the banks of the river. He tried to find places on the river where a team would be able to see some distance either way but working here would pretty much limit the team to one crossing. There were so many crossing sites that it was impossible to sensor them all and to make a new one was easy. This would be a hit or miss operation like the ones in Del Rio but the chance of the team being discovered was much greater because of the high population density near the river.

After site selection in Starr County, Tom and I went back to El Paso for a joint planning session with Harry of the McAllen Sector, Lee Justice from Operation Alliance, and Bil. The main topics were the transportation plan to get everyone from El Paso to McAllen and then to the field. Harry, Tom and I were talking to Lee about the problem of having five BP Suburbans driving up the highway from McAllen into Starr County around the same time. Lee had worked in the Rio Grande Station in Starr County and agreed that the word would get out that something was going on. He also told me on the side that one or more of his agents was working for the bad guys when he worked there. The local agent said that there were many watchers and a sophisticated communications interception operation that monitored all the law enforcement radio traffic. According to the agent, this group could detect changes in the normal pattern of radio traffic, so that even if the communications were encrypted, they still knew when something was going on.

Operation Alliance had spent a lot of money on a repeater system that would allow an agent to talk on his hand-held radio from almost anywhere along the river. The system was able to use encryption so that no one could listen in, but the agents would lose their radio's encryption when they changed their batteries so after a while most radio traffic was in the clear. Harry, Tom, Lee and I began 'what-iffing' and came up with a possible infil plan that would avoid using the highway into Starr County. It would start with a C-130 flight to Corpus Christi Naval Air Station where we would be met by an INS bus. The bus would take us to a remote area north of Starr County where we would be met by the BP Suburbans keeping radio silence. The teams would then leave at 30-minute intervals and be inserted into their surveillance areas.

We were pretty happy with this idea and when Lee presented it to the planning group, it was approved. After the meeting, Tom and I were talking about our great idea when Tom suddenly started laughing about our 'Desert One' debacle in the making, referring, to the failed hostage rescue attempt in Iran ten years before. "One of the Suburbans is going to crash into the bus and the fire will be seen from Laredo to Brownsville," he laughed. "We have to get everyone off the bus when we get there and have them hide in the bushes on the side of the road or they'll all be killed," which, of course, when the time came, is what we did.

October 1989: Joint Task Force 6

We also had meetings at JTF-6, the new DOD entity that was providing support to Operation Alliance. It was headquartered at the old stockade on Biggs Army Airfield near Fort Bliss. Commanded now by BG Sherman Williford who had recently commanded Delta Force, he was quickly removed because of a turf war between some Generals.

They must have had a battalion of staff officers and the plans and policies coming out of there required a brigade of lawyers. Each 'great' idea generated by one of the staff officers had to be scrutinized by a squad of lawyers and was the subject of many an incredulous conversation between the various entities involved in the 'War on Drugs.'

There was still some question regarding command and control of military forces which, when it was finally straightened out, directed that the National Guard would have the right of first refusal when the cops requested support from the military. If the National Guard couldn't provide the support, then JTF-6 would try and help. This didn't set well with the 'regulars' who always looked down on the 'Weekend Warriors' of the National Guard. They were as excited as we were a few months earlier but now that we had a few missions under our belt we looked at them as newcomers who were going to screw up everything.

This command arrangement required the exchange of Liaison Officers between JTF-6 and our headquarters in Austin. The first guy we got was an Army major who spent the next two years in the longest Microsoft Solitaire tournament in history. The next one was a Marine Corps major and helicopter pilot who stayed busy. When the Marine Corps birthday came around, he organized a birthday party in our headquarters conference room and we all had to stand at attention while he read a long version of the entire history of the Marine Corps to a bunch of soldiers; he had style. The next one was an Army major who was getting ready to retire but he got us into one of our most successful missions because he saw that we would be better at what the cops wanted than what JTF-6 could provide.

November 1989: Operation Unity 5-Command Briefing

When we finished all the preparation for Unity 5, we had another Command Briefing. This time, someone in the headquarters created the briefing graphics on a computer and exported them to 35mm slides. Tom and I produced the operational information and Captain L and his boys came up with the Intel. Bil paid close attention to this process and made sure that each slide said what he wanted said.

On the big day, we showed up in our Class B uniforms. Bil told me to wear my CIB. He had decided to have Tom, Captain L, and I take a part of the briefing along with Major K, my OPLAN writing buddy. Bil gave the introduction and introduced us to the audience, which was pretty much the same since the last time. Major K went first with an overview of the operation and told everyone that the teams would operate in 'goose eggs' along the river. Tom and I looked at each other with alarm. "What were goose eggs and what did they have to do with this operation?" Since he was an artillery officer, we figured later that this was some kind of artillery talk that he fell into under the pressure of briefing the General and his staff.

He finished his part and turned the briefing over to Harry. Harry gave an overview of the drug trafficking situation in the McAllen Sector and the ERT unit that he commanded. Captain L followed him and proceeded to tell us everything there was to know about Starr County except where the smugglers would cross the river and when. SSG R assisted him, pointing to places on the large map hung on the wall. It was a typical Intel brief which showed that a lot of effort had gone into its preparation but in the end had very little substance.

After Captain L and SSG R finished their part, Tom gave a quick description of our isolation training. Then he covered the infiltration, insertion, mission conduct, exfiltration and debriefing phases of the operation using the slides that were projected on the screen and changed by me when he nodded.

At the end of Tom's part, he introduced me, and I gave the communications part which was fairly short. When I was finished, I turned the briefing over to George C who covered logistics. Bil followed George and covered command and control He closed the briefing by asking for approval from the General who gave it and we were done.

Operation Unity 5-Execution

We had problems with the orders for the volunteers from G Company and LRSD. These people needed to give them to their employers a few weeks before the operation, but they weren't ready which caused some heartburn with the troops going on the mission.

The transportation plan was a little more complicated, but we managed to get the troops from Houston and Austin and the BP agents from McAllen to El Paso on time to start isolation. We also had several active duty soldiers from 18th Airborne Corps at Fort Bragg who were bringing the SATCOM radios. Several of the same guys who helped us during the last operation showed up again. The Air Force opted out of this one, so we never saw Sgt. R again.

Tom had decided that I would run the forward support element this time so I spent the time getting things ready for the move to McAllen. I would handle resupply and any emergency resupply situations. Several regulars and some Guardsmen were going with me to run a small alternate CP. I briefed them and gave them tasks that kept us busy until we left for Corpus Christi the day before the troops would fly there. .

Tom ran isolation. We used the SMU headquarters again for the CP. Operation Alliance had ordered them to assist us, but they weren't happy about us using their facilities when we weren't using their agents for the operation. The isolation went well. Due to a change in procedures, the team leader was a Guardsman, so we gave the teams more time to work out the actions that the Guardsmen would take if the agent had to take charge of the team during a law enforcement situation.

On the night of the infiltration, I drove to Corpus Christi Naval Air Station and met the plane. The INS bus was there, and lo and behold, 'Speedy,' the driver that drove the bus to Del Rio, on the last operation was driving for us again. We loaded the teams and drove three and a half hours west to 'Desert One,' the staging area where the teams would meet the BP agents that would take them to the river. The local BP agents driving the Suburbans had been briefed on the plan that day by members of Harry's ERT. This was to have annoying repercussions later in the mission. I rode back to McAllen with one of the men assigned to the FSE. We all sat down in our CP and listened for the initial entry reports from the teams. I gave everyone the shift schedule, explained the SOP, and we settled in. We would work 12 hours on and 12 hours off. I sent the ones who were off duty to a motel nearby and I waited for something to go wrong, it wouldn't be long.

Emergency Resupply

The next night the teams were making their mandatory commo checks and Team 3 failed to check in. I loaded up my Blazer with an extra SATCOM, called the BP station to request an agent to meet me, and drove up the highway to Rio Grande City. When I got to the station, I met with one of the agents I knew, and he introduced me to the agent who would drive me to Team Three's location.

We drove out to the place where the team should have had someone waiting to do the radio exchange but there was no one there. The agent assured me that we were at ___Crossing. We whispered, we whispered louder, calling for them to come out, nothing. Finally, the agent got on his radio and called his supervisor.

The supervisor drove up about 15 minutes later. He wore a set of Master Blaster wings on his ball cap. He was a retired Army Captain who had joined the BP as a double dipper. You can't do that anymore but there he was, and he proceeded to show me a new technique for low visibility link-ups; he started shouting at the top of his lungs for the team to come to his location, NOW!

Then, just to make sure that anyone watching the area had no doubts that this was a secret operation, he turned on a million candlepower light that he brought with him and shined a very bright beam back and forth in a kind of one man light show that even Stevie Wonder would have appreciated while shouting at the top of his lungs.

I was stunned and stood there with the other agent while the BP supervisor, showed us how to find the lost team. He finally got tired and stood there with us wondering why they didn't come out and get the radio. I wanted to say that our far and near recognition signals were different than the ones he used but decided to see what he would do next. He asked the agent where they had dropped the team off and the agent told him, ___ Crossing. "This isn't ___Crossing," he exploded. "Follow me" he said as he jumped in his vehicle and took off into the dark. I jumped into the BP vehicle with the agent who said a number of unkind things about his supervisor as we tried to stay with him as he led us up to the highway. We followed him a few miles down the highway and when he turned off on a dirt road going toward the river, we followed him until he stopped. "This is the Crossing," he said.

We looked around, I had never seen this place before and I had been to every drop off and link-up point that we were going to use for this operation. After a few minutes, the supervisor started up his shouting and lighting show but no one responded. I was starting to get a real bad feeling that something might have happened to the team.

I asked the supervisor if we could talk to the agent who drove the team in the night before. I had his name and when I gave it to him, he said that the agent had just come on shift and called him on the radio. The agent responded and the supervisor told him to meet us at a nearby location. He did and when the supervisor told him to take us where he had dropped the team off, he locked his vehicle and rode with the supervisor while we followed them to a third location where he had dropped them off.

There sitting in the dark was my old friend SSG K, the commo man for my team when I was in G Company. He was a staff sergeant now and the team leader. I heard the supervisor and the agent having a whispered argument about where we were but I ignored them as I was glad that we had found the team and now I could get them on the air and get out of there before the supervisor burned this area too.

I exchanged the radios and asked SSG K to make commo with El Paso before we left which he did. That was all I needed to do, and I interrupted the supervisor and the agent's argument to thank them for the help and told them we were leaving to go back to RGC station. When we got there, the supervisor was pulling in too. He took me aside and explained that there were several new agents in the station who didn't know what the names of the various crossings were. I thanked him for his help and after saying goodbye to the other agent, I drove back to McAllen. When I got there, Harry asked me why it took so long and when I told him, he got angry then he laughed. "We have had so many new hires lately, the new ones are being trained by the ones who have been here for a year and don't really know the area; now we have several different names for many of the crossings on the river."

Give us Stingers!

The next night there were more problems with Team Three's radio. I drove back to Rio Grande City with a fill device for the encryption and some extra batteries. When I arrived at the Rio Grande City station, I was met by the BP agent who had taken me to recon the sites and with whom I had become friends. When I opened the door I saw his angry face and heard him shouting as he walked toward me.

"What are you people doing," he shouted. "The only help you can give us is some Stinger missiles to shoot down your so and so airplanes that are screwing up our interdiction!" I just stood there, not knowing what to say. I had no idea what he was talking about. After a few more rants about the military and their stupid activities in his world, he finally stopped and stood there staring at me.

"What happened," I asked him. "We were set up at ___Crossing watching a smuggling operation," he replied. "The smugglers sent their scouts over to check the place out and they didn't see anything, so they signaled back, and the smugglers started loading several large boats with bundles, it was a lot of dope," he said, wistfully. "Then the boats were pushed out into the water and they were about halfway across when your airplane came down the river at about 200 feet and right behind it at about 500 feet came a big four engine Customs AWAC, the kind with big dish on top. The smugglers heard then saw them coming, turned around and were paddling back to Mexico when they flew over." The U-27 lick-ass strikes again.

I began the humiliating but necessary apology process, even though I was not responsible for the actions of these guys who were part of the operation but not under anyone's control. They were supposed to be in a silent, invisible airplane that could see in the dark and transmit live video back to our headquarters. We were going to test the video feed later in the operation but now I realized we were going to have to talk to them as soon as I got back to McAllen.

I told the BP agent and his supervisor that this was all our fault and that I would make sure it didn't happen again. This seemed to defuse the situation and they agreed to help me do the resupply to Team Three

There were no problems this time because we knew where they were and we got their radio up and running again.

The BP agent had heard about the fiasco that took place the night before and was a little sensitive about what had happened, so he said, "I guess we all have some things to learn about this business." This was his way of apologizing to me and I assured him that we appreciated the BP's patience with our learning about their world.

As soon as I got back to McAllen, I called Tom and told him what happened. The plane had finished flying for the night so he told me that he would get with Bil and they would call me in the afternoon. Everything else was OK but I had to listen to Harry and his guys chide me several times about, "When are we getting those Stingers?"

Sound Check

The next afternoon, Bil called me and I told him the whole story. He had already heard about it from his buddies at Operation Alliance. I told him that I thought we should have them do a signature test at the airport where one of their pilots and one of our people stood outside the hanger and watched and listened as the plane flew back and forth at different altitudes. Bil agreed and told me he would call them to set it up that evening. A little while later the General Dynamics chief pilot showed up at the BP Air Operations and he was pissed.

I went over the incident the night before and he seemed a little embarrassed but put the blame on Customs. When I asked him why they would bounce him when they knew we were conducting operations on the river, he said he didn't know. We let that go and I told him that we would like to do a signature test that evening before they went back to the river, "It'll help you guys with your presentations. You'll be able to say with absolute certainty what the signature of this airplane actually is," I said brightly. We both knew that this plane could not do what they said it could, but we weren't going to let them fly again on the operation until we did this, so he agreed to meet me at 10 PM for the test.

Harry decided to come and watch along with the chief BP pilot, so we were all standing out on the tarmac in front of the BP hanger when the chief pilot from General Dynamics showed up. He had a radio with him and told us that the U-27 had just taken off and that they would do the first flyby at 500 feet. A few minutes later he said, "Here they come," but there was no need as we could hear and see the plane from a long way off as it approached. The plane passed over us like Richthofen's Flying Circus, as Harry put it, and the face of the General Dynamics pilot looked like one of the Baron's victims as he spiraled down in flames. The General Dynamics pilot stepped away from us and talked softly into his radio before rejoining us. "I told them to come back at 750," he said. We didn't say anything and waited for the next pass. I'm pretty sure he told the pilots to idle the engine as they passed over us, but they were as loud and as visible as they were at 500 feet. This went on until they passed over us at 5000 feet and they were still visible and noisy. The General Dynamics pilot threw up his hands at this point and left. We never saw him or their airplane again.

Operation Unity 5-The Wrap

The rest of the operation went without incident. The teams called in several suspected smuggling activities, but the BP was either unable to interdict the vehicles that were described, or they were unable to find any drugs and let the suspects go. They did apprehend a number of illegal aliens due to our reports and the teams saw a lot of suspicious activity. The teams heard a lot of gunshots, and they saw a lot of flares in the sky, but we were unable to catch any dope. When the time came to extract the teams, the BP agents who brought them in picked them up and took them to McAllen for a debriefing.

One of our teams was being led by Captain C, our new LRSD commander. He seemed ok when we first met him, but he really surprised us when he and his team were debriefed. The debriefers were some of our Intel guys and I heard SSG R ask Captain C to describe the bundles they saw being loaded into a truck. SSG R went back to those bundles about five times, asking what their measurements were, what their shape was, what color they were, etc.

Captain C was unable to shed any more light on these mysterious bundles so SSG R gave up and proceeded to ask about the tower they reported. This tower, they had reported, was manned by several men who had an infrared spotlight that they were shining on the river and at various points along the bank on the U.S. side. This was of great interest to the BP and their Intel guy jumped in and asked them a number of questions about this tower, mainly, where was it? Captain C had an azimuth from their OP to the tower which he gave, and showed the location of their OP on the map. The agent was writing all this down in his notebook and I was sure the BP would be happy with this 'intelligence find' and overlook the fact that we hadn't caught any dope. Captain C also mentioned that the team had been triangulated by coyotes which had us all scratching our heads. No attempt was made for further details of this phenomenon.

The rest of the teams were debriefed and fed a hot meal. We had a brief ceremony where we gave a plaque to the Harry's SOG team and to the BP Air OPS and Harry gave our guys some certificates from Operation Alliance. We packed up everything and after the troops got on the bus to Corpus Christi for a C-130 flight back to Austin, I loaded our guys in the Blazer and we high-tailed it home.

We did the usual wrap up but this time, E-Systems wrote their version of the Intelligence Summary using their B.I.A.S. system. The Intel guys under Captain L were not about to be outdone by these outsiders so they wrote their own and gave it to Bil. Both of them were filled with intel jargon and not much substance. A copy of each was sent to the BP in McAllen and I figured that that was that. Wrong! McAllen Sector requested that Bil send some of us back to McAllen as soon as possible to help the agents in Rio Grande City locate the tower with the infrared searchlight that was reported by our Captain C.

Bil sent me, SSG R and a major, back to McAllen the next day to help them find this tower. We flew the river looking for it and we drove out to the team's location and walked all over the area with our compass and binoculars. We couldn't find any tower. We did see a large storage tank that could have been used as a place to watch the river and we saw a building across the river that men were standing on watching the river. There was a power line pylon behind the building and when we went to the team's OP it looked like they were on the pylon.

As far as the infrared searchlights on the tower went, we never saw them. When we had thoroughly searched the entire area, we reported back to the BP Intel man that we were unable to confirm what our team had reported. The BP agents on the team were pleading ignorance of the whole matter so it died an ignominious death.

McAllen Airport

I stopped in and talked to Steve before I left town. There had been an accidental shooting incident in the valley that week. We were discussing negligent handgun discharges and I asked him what the BP procedure was and he told me, "well, if the bullet hits your vehicle, you drive over to the nearest convenience store where you buy a certain brand of bubble gum that is the exact same color as the paint on our vehicles. After you chew it for a while, you put it in the bullet hole and hope that no one notices it for a few days."

The major took me to the airport in McAllen where I proceeded to do one of the most gut rending and stupid things I have ever done when I fired my 9mm pistol into the ticket counter while I was giving them my pistol for transport. I fired it straight down through the counter into the space where you put your bags to be weighed so no one was injured but it scared the hell out of the young lady who was checking me in. The police officer who was stationed at the airport was summoned and he took a report from me.

I knew I was in deep shit and would be fired when I got back but I decided to tell the police officer that I was a police officer rather than a member of the National Guard. I did this because I didn't want the Guard to end up in the local paper associated with a man in civilian clothes carrying a gun. That could cause big problems for the program. When he found out I was a police officer he become very sympathetic to my situation. I gave him a short statement for his report. He put the gun in my gun case which now had a 9mm hole through it, gave it to the airline people and cleared me to leave on my flight.

I went out to the curbside parking where the major was waiting to find out if I could get the flight. He looked a little worried and asked me if I had heard a shot fired in the terminal. I told him that I had and that I was the one who fired it. I then told him that I would make a full report to the headquarters when I got back to Austin. I also told him that he should make a report when he got back, and he was relieved that I didn't ask him to keep it to himself. We shook hands and he left.

I went back into the terminal and waited for my flight. I started reading a newspaper and from time to time I would scan the area with my peripheral vision, and I noticed that most of the people in the terminal were looking at me. If I looked up, they would turn their eyes away until I went back to reading the newspaper then they would check me out again. There was nothing in the newspaper that I could focus on because my mind was in such turmoil. Finally, I just called "Oh Lord Jesus" quietly under my breath until I touched the Lord in my spirit and my mind was quiet. I sat there waiting for my flight and calling on the Lord until it was time to get on the plane.

After a short flight I was back in Austin. My wife picked me up at the airport and I told her about my screw-up and the most likely consequence of my actions which would be losing my job at Counterdrug and going back to the police department where I was on the top of the chief's shit list. She thought about it for a moment and then said we should give the whole matter to the Lord Jesus. We prayed and gave it all to him and went home. I called Tom and told him what happened, and he said he'd talk to me in the morning.

The next morning, I met Tom who told me that Bil wanted to talk to me. I went to Bil's office and he told me to close the door. I did and he told me to tell him everything that happened. I told him the whole story, ending with my reporting to the police officer that I was a police officer and that no one knew I was involved with the Guard.

This saved me from being fired because my actions had not been associated with the Guard but with the Hospital Police with whom I was still employed. Bil had had a similar incident sometime in his past and decided to keep me on despite my pathetic action that day. Sam, my boss at the Police Department was pissed but had to agree that I was still employed with the Hospital Police, so I hadn't lied about my status as a police officer. Bil told me to go and sin no more.

I didn't hear any more about it except for one time when Tom and I were checking in at an airport and I told him, "Put in your earplugs, I'm going to check-in." Tom didn't think it was at all funny and stopped me right there and told me not to joke about what had happened. "If you joke about it, you won't remember your mistake and you'll do it again." Words to live by.

Bil Enney

SFC Thomas Payne

Captain L

Author

C-27 LICAS

U. S. Customs AWAC

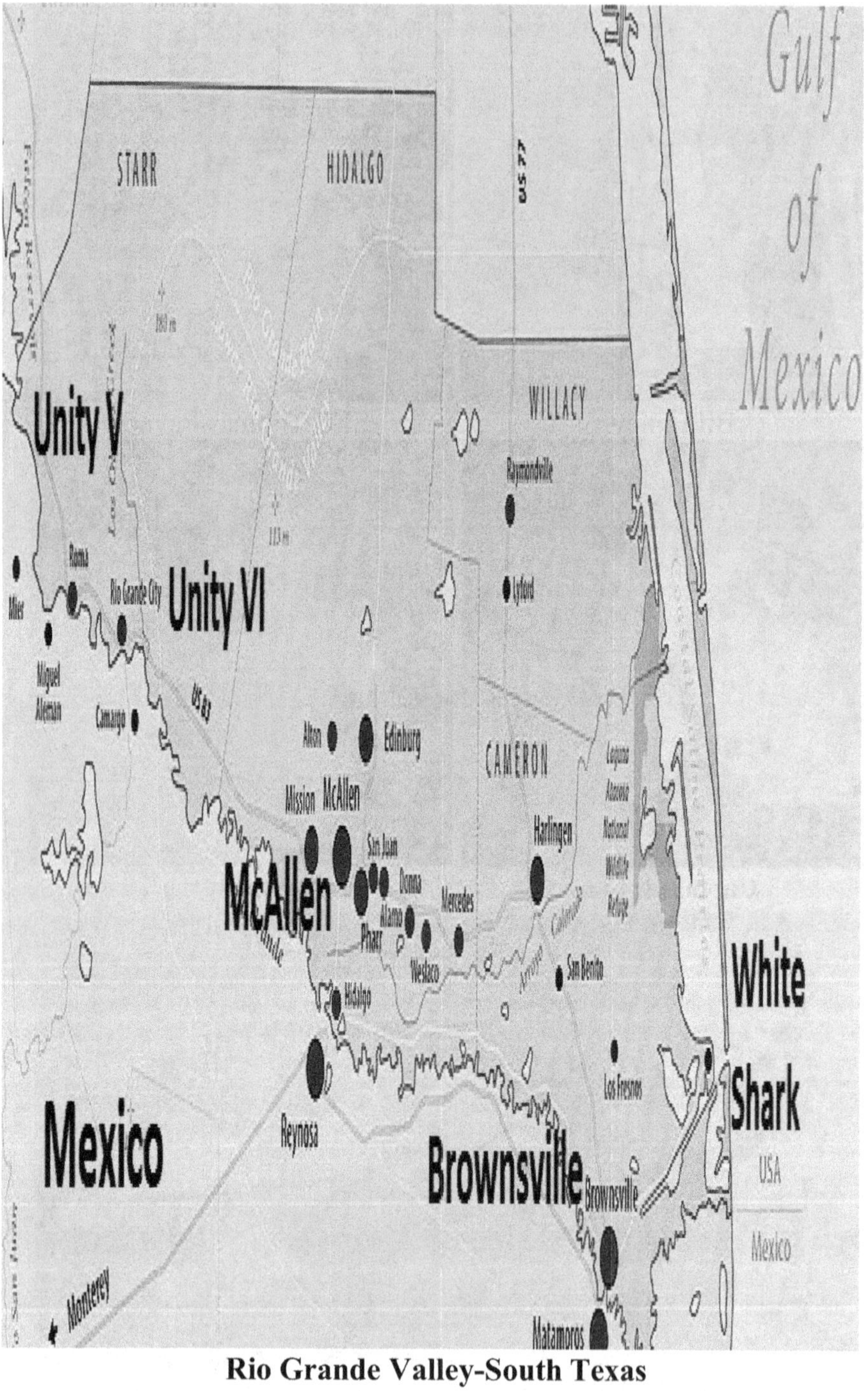

Rio Grande Valley-South Texas

Or what king, going to engage another king in war, will not first sit down and deliberate whether he is able with ten thousand to meet the one coming against him with twenty thousand? Luke 14:31

Chapter 8-December 1989

Laredo

One morning I saw Captain C in the headquarters. He was going into Bil's office. "What's he doing here" I asked Tom. "He's trying to get a job here so he can run these operations," he replied. "No way," I said. "He doesn't understand this mission!" "Don't worry," he told me, "Bil knows it." Captain C came out with a frown on his face and that was that or so we thought. Bil called Tom in and alerted him about an operation in Laredo. Tom and I had wanted to operate in all five Border Patrol sectors in Texas. So far, we had worked in the McAllen, Del Rio and Marfa sectors. That left the Laredo and El Paso. El Paso was a long way away and we hadn't heard much about the drug threat there. Laredo, however, was the closest sector to Austin and was known to be the center of a huge drug trafficking corridor.

Enter the Marine Corps

We talked to Bil about Laredo early on and he said that the Marines had gotten into bed with the sector's Emergency Response Team and there wasn't any interest in Guard support for surveillance operations. Sometime later, we were in El Paso for a meeting at Operation Alliance. After the meeting we went out to dinner at a Steak House and there we met Colonel M, a highly decorated Vietnam veteran and active duty Marine assigned to a Marine Reserve unit in San Antonio. These kinds of dinner meetings were a little tense at the beginning. There was some ball sniffing, name dropping, and history lessons that had to be taken care of before people would show a card or two.

After that was taken care of, Colonel M was in fine form and delighted in telling us that he had two, not one but two, Letters of Condemnation in his personnel file from previous Commandants and that he was still on active duty. He was a big guy and very outgoing. He had been assigned to Fort Bragg before his current assignment and he knew some of the people that Bil knew. The colonel was a hard charger and told us that he had a mandate from the Commandant to get the Marines into the drug war. That seemed reasonable except for the fact that most of the Southwest Border was far from the ocean where Marines normally operated. I decided to withhold that observation from Colonel M for the obvious reasons: he was a colonel and, he was a genuine bad ass. Later I found out that Colonel M had received a Navy Cross for his valor in the hills above Khe Sanh and had been on the Mayaguez rescue operation in Cambodia at the end of the war.

He also let us know that he was ramrodding something called the Pegasus Group that our Intel people might be interested in. He pitched the Marine Corps' Southwest Border Intelligence plan with an Intelligence 'Fusion' Cell in San Antonio. Tom and I were familiar with 'Fusion' cells after our visit with E-Systems but we kept that to ourselves. A few months later Captain L told us about the Pegasus Group in San Antonio after he was invited to attend one of their monthly meetings. There were representatives from all the Federal Law Enforcement agencies. The CIA had someone there, although, according to Captain L, he didn't say anything. Tom and I laughed, and Tom said that they had misspelled 'fusion.' It should have been spelt with a 'Con' in front. Captain L, however, was not amused. He determined that any military intelligence operations in Texas would be under his control. He had some closed-door meetings with Bil and we didn't hear any more about the Pegasus Group.

Exit the Marine Corps

Shortly after that, Bil told Tom and I to go to Laredo and meet with the sector people and set up an operation in there. When we asked him about the Marines, Bil laughed and said, "Colonel M and his boys stepped on their dicks; you'd better watch your step." We were directed to meet with a Marine Gunnery Sergeant at the 4th Marine Recon Battalion next to Fort Sam Houston in San Antonio. He gave us no information about the situation with the Border Patrol in Laredo and we would have to wait for the drums along the river to find out.

The Marines had had their problems in Laredo. They showed up with a large headquarters which they proceeded to hide under camouflage nets. They showed themselves to be way ahead of the times by deploying an unmanned aerial vehicle, a drone, to watch the river. Due to strict aviation rules, they were required to have a helicopter fly behind the drone for the visual safety of other aircraft crossing the river during their flight. Eventually the drone crashed, in Mexico, we heard. There were also some inter-service problems with the boys in blue who brought the circus to Laredo on unauthorized C-5a flights.

1989: Laredo-Dangerous Don

We went to the Laredo Sector Headquarters which was near I-35 and met with the Assistant Chief who oversaw the Sector Emergency Response Team. While he was talking to us, Don, the commander of the ERT walked in and we were introduced to one of the true 'characters' of the Border Patrol. He was wearing a 'Scuba" badge and some Marine Corps Jump Wings on his Border Patrol uniform. The assistant chief introduced us and suggested that we go to lunch.

We were still under the illusion that the cops had the same concerns that we did about operations security. Being seen in a restaurant with Border Patrolmen seemed like a bad idea. Tom mentioned his concern and Don gave us a smile to convey his knowledge of all things special in both the military and law enforcement. "Don't worry; we're going to a place where we won't be noticed."

Tom rode with Don's boss and I got into a BP suburban with Don who drove us to an 'out of the way' restaurant on the frontage road of I-35. Apparently, all the cops on the north side of Laredo ate there, no doubt due to the 50% discount that the restaurant gave them. Tom and I didn't like the idea, but we were in the initial stage of our support plan and that altogether depended on the establishment of rapport with them.

We went inside and joined all the law enforcement people in the restaurant, sitting with our backs to a wall and checking out everyone in the place. There were several BP agents, two sheriff's deputies and several men in civilian clothes with pistols in their belts occasionally covered by their open sports coats. Don and his boss said hi or nodded to most of the cops as we waited for our meal.

Don did most of talking, telling us about how he had trained his men to Force Recon standards and how they were the best there was when it came to operating along the river. His voice was pretty loud, and I was sure that the people sitting near us could hear every word he was saying. He would raise his head slightly and make eye contact with us and lower his voice when the waitress approached, showing us that he was on top of OPSEC. When she left, he would resume his monologue at the same level.

He asked us about our backgrounds and was impressed when Tom told him of his SF and Ranger history. When he asked me, I told him that I was a cop on a leave of absence. Tom told him that I was in Vietnam. "What unit were you in?" he asked. When I told him I was with the 1st Infantry Division, he told us that he was a helicopter pilot with them at Fort Riley in 1966 which was interesting to me because the entire division was in Vietnam in 1966.

After lunch, Don's boss returned to the Sector Headquarters and we went with Don to his house. He wanted to take us down to the river and he stopped at the house to get some equipment. We were invited inside where he introduced us to his wife, daughter and son. Don's son adored his dad and sat at his feet while Don continued to inform us about his past exploits and the current capabilities of his ERT. After visiting with his family, we left and got into his BP Suburban and headed for the river. He took us to a few landing places and told us that we would return to these places that night when we would accompany him and his ERT on a patrol along the river.

After Don took us back to his house, Tom and I agreed to be there at 7:00PM that evening for his team briefing before we went out. We drove to our motel and checked in. I went to Tom's room and we discussed the events of the day. I told Tom that I thought Don was crazy. He laughed and said that Don was definitely the most unique individual we had met so far in the 'War on Drugs.' We weren't supposed to be conducting an operation without completing the process that ended with a joint operation's plan and a signed Memorandum of Understanding between the Guard and the supported agency. Tom said that we could consider this an observation of training by the ERT and not worry about it. I didn't argue about technical details because I wanted to go out in the dark and look at the river with a bunch of armed BP ninjas.

We got to Don's house at ten minutes to seven. There was a pickup truck parked on his front lawn and several guys dressed up in camouflage with tactical gear were climbing out of the back. Several of them had MP-5 submachine guns, a popular weapon among special operations units. We followed them into Don's house and sat on the floor.

He introduced us to the team and told them that we were part of the special operations unit of the Texas National Guard. He told them that we were there to see how to do a patrol on the river.

Some of the team members were smirking at us and I figured that since they had been training with Marines, it was beneath their dignity to be doing anything with the National Guard. This was a consistent perception, and we were used to it. We dove into our rapport establishment routine with Tom telling them how we had heard that the Laredo ERT was the best in the Border Patrol and how we hoped to learn from them. When he got done with his spiel, I looked around the room and saw that they were pleased with our humility in their presence and the atmosphere lightened considerably.

The doorbell rang and Don, who was orchestrating this meeting, then informed us that we would have pizza which had just arrived. He went to the door and invited the pizza deliveryman into the room. The deliveryman stood there looking at ten heavily armed ninjas sitting in front of him. Tom and I were appalled but Don had more surprises for us because after the pizza guy left, his next-door neighbor, a female schoolteacher, walked in the back door and joined us for our pizza and combat patrol briefing. Don saw the surprise on Tom's face and told him that "she's ok guys, I trust her."

Don's briefing was informal. His time with the Marines at Camp Pendleton had taught him how to reduce patrol planning to the most essential tasks which were given to us between bites of pizza. "After we finish here we will load up and drive to Santa _____Crossing" bite, chew, "we'll unload and patrol along the river path," bite, chew, "when we're done we'll go back to the vehicles," bite, chew, "and come back here for a debriefing," bite, chew, "any questions?" He looked at Tom and said, "You and Greg ride with me."

I was impressed. It looked like Don's team was a well-oiled machine. The number of tasks that were left unsaid would have taken an LRS team several hours to cover and Don took care of it all in 10 minutes while he and his team ate dinner as well. We finished the pizza and went out to the front of the house. Don was at the door saying goodbye to his wife, his kids and the next-door neighbor. Don's boy was hugging his waist and looking up at him with intense pride. It was quite a scene given the conflicting information in my mind.

We got into Don's Suburban and drove through the subdivision with the rest of the ERT guys following us in their pickup trucks. Some of the team was sitting in the back on the sides of the pickup bed holding their MP-5s and shotguns. It must have looked like a parade of third world rebels to the people standing in their yards.

Santa ______ Crossing was about a mile from Don's house. We stopped on the edge of a bluff overlooking the river bottom. There was a lot of vegetation and you couldn't actually see the river from there. The team gathered around Don and when his 2IC told him that they were all there, Don told them to move out on a trail that led to the bottom of the bluff. There was a ¾ moon that night and no clouds so the movement on the dirt trail was easy.

When we got to the bottom land, the team gathered around Don again and he whispered to the 2IC and pointed down river. The 2IC started in that direction and the rest of the team followed. They spread out a little and followed in a single file. Tom stayed with Don and I waited until the last man and entered into the formation. I had my big video camcorder and I wanted to take some video of this patrol. I tried the low light function and I was able to get some video of the entire Laredo sector ERT on patrol on the Rio Grande.

We did this for an hour and then returned to the vehicles and went back to Don's house where we conducted a debriefing with beer and the rest of the pizza. Don did most of the talking and it was more of a praise session than a debriefing which made sense because we didn't see anything worth talking about. I would have liked to mention a few glaring tactical problems, but I waited instead for Tom. When Don asked him for his observations, Tom turned on the sunshine gun and praised them as the tactical standard bearers of the Border Patrol. Actually, they hadn't done too bad with their tactical movement to the river but after they got there, Don told them the smoking lamp was lit and the rest of the patrol was a gaggle.

After the love fest, Tom and I excused ourselves and went back to the hotel feeding each other's laughter with fresh observations about the night's activities. The next morning, we went back to Austin and told Bil about our trip. He laughed and told us that we needed to be very careful when we went back to Laredo.

On New Year's Eve, I did a quick calculation of my first six months on the CD Task Force. I had spent 29 days on LRSD business, 72 days

TDY, and we had conducted three operations. Things were going to get a lot busier.

Planning for a Unit

After Operation Unity 5 was finished, Tom and I talked about how things had gone. There were a few good things that came out but there were a lot more problems. Unity 5 was conducted in Starr County, the most intense drug trafficking area on the entire U.S./Mexico border. We had sent five teams into the river area and we hadn't caught a thing. The team led by the intrepid CPT C reported the observation tower in Mexico that was manned 24/7 and the people in the tower that were using infrared searchlights to watch the U.S. side. They also reported the loading of bundles into a vehicle that was never found.

Another team's members reported to us secretly that their team leader had acted very strange while they were on the river and they wouldn't go back out with him again. Other guys told us about other things that happened out there that made us feel that if we didn't correct the problems with leadership we were going to piss off the Border Patrol agents that went out with us; they would tell their supervisors and soon we wouldn't have any requests for support.

As we processed these things, Tom was writing them down. He didn't like too much talk if there was no action. "Maybe we should ask Bil if we could hire six guys to be team leaders for these missions and we could run small missions with them for requests of an immediate nature," I suggested. He thought about it for a few seconds and wrote some more. He eventually completed a memo that covered many of the things we thought should be done to improve our operations.

After he printed the memo, he took me with him to Bil's office. He gave him the memo and told Bil what we wanted to do. Bil told us he would get back to us later. That afternoon, Bil called us to his office, and told us that we could hire six people and the program would be able to pay them using unspent TDY money for the staff.

I never really gave it much thought that we were starting a specialized unit and it didn't occur to me that this was something historic in our lives. We were so caught up in doing what needed to be done on a day to day basis that this new endeavor didn't seem like anything more than the next thing on our crowded plate. We were working 15-hour days, six and seven days a week.

I was so caught up in the mission that I didn't take any leave for the first year and a half on the program. Due to the funding and 90-day orders I ended up losing the 45 days of leave and I never noticed. After we had the new unit up and running, one of the guys said we had to pace ourselves to prevent burnout and someone laughed and said that Murry was burned out before he got there.

December 1989: LRSD The Train Wreck (Part I)

We had drill two weeks after the end of the McAllen operation. I was still worried about Captain C's visit with Bil, trying to get a job. He called into his office on Saturday night after the troops had been dismissed. We sat down and listened to him turn on the charm. He told us that we were the best NCOs he had ever worked with and that he hoped to learn from us on future operations.

He went on in this vein for an inordinate amount of time and I was feeling really uncomfortable. We were in the presence of an angry man who was capable of causing us a lot of trouble, he knew it and he wanted us to know it. He was waiting for the right moment to show his power and when he finally stopped, he asked us if we would like to say something. We were quiet so he said, "Come on Sergeant Payne, I really would like to hear what you have to say."

Tom looked at him and began to speak. "I appreciate your kind words, but I want you to know that I hate all officers." Oh No! Now what have we done? Captain C looked at Tom and then began to blabber things like, "I know I'll have to gain your respect" and "I'm sure we can work together" before he finally dismissed us, and we left. I asked Tom, "why?" Tom just laughed and described what the captain could do if he couldn't take a joke. The next day he fired Tom.

Bil got Tom a job at STARC HQ so we went on with our work but not having Tom at LRSD made the drills seem flat. He was a good leader, the troops respected him and he would be hard to replace. Captain C made the readiness NCO the detachment sergeant. He was an older guy, a Navy veteran of Vietnam. He had been a salvage diver in the navy, dangerous work but not infantry, and he brought a new flavor of leadership to the detachment. Rather than continuing the meritocracy, he brought in an atmosphere of fear. The people who sucked up to him got better treatment than those who did a good job but didn't go out of their way to be nice to him. The morale in the detachment went down and it didn't recover for as long as he was there.

1990: In the Beginning-The Go-rillas

They say that the best way to eat an elephant is one bite at a time, so after Bil approved the new unit, Tom and I went back to the office and started making a list of possible candidates. We both thought that Chris and George, two former 2nd Ranger Battalion members, currently in the LRSD would be good prospects. Chris Jay had served his entire enlistment with the Rangers and was an excellent soldier. He mixed well with others and had a friendly disposition. He knew his stuff and could be relied on to accomplish any task he was given. George Huseman was a skillful NCO. He had been sent away from the Rangers for some youthful indiscretion, to the 9th Infantry Division (Test). He served in an experimental recon unit where he was HALO qualified and proud of it. He was a total extrovert and needed to be the center of attention in whatever situation he found himself in. He had a twinkle in his eye and always had something funny to say about whatever subject was being discussed.

Tom recommended several typical National Guard soldiers; one was Clayton Hardee, a young Spec 4 from LRSD who had joined the Guard two years before and had completed airborne school. He was a country boy and ready for whatever the Army could throw at him. He would go on to get his Ranger Tab and become a jumpmaster. Another was Larry Rayburn, an older corporal also from LRSD. He had come from G Company where he saved some men from drowning while on an airborne operation in Germany. He had a little difficulty blending with the others, but he never seemed to notice it. He became a jumpmaster and would amaze, if not amuse, everyone with his 65th Jump celebration during the next AT. Then there was Bill S, from G Company, who was rumored to be an oil millionaire in a middle-age crisis. He really amazed me one day, because he had been one of the biggest complainers about any and everything that came along, when he said, "I'm tired of complaining about everything, I'm going to stop," and he did. Bill continued to deal with his middle-aged crisis by going to

Ranger School when he was in his mid-thirties and getting his tab. These three guys were traditional guardsmen with no active duty experience, but they were good troops and we thought they were trainable.

I nominated two guys who I used to call our first mercenaries. Steve, Tom Stephens, and I had worked together in the Police Department. They both had a Special Forces background and weren't doing much at the moment. Steve had just completed his master's degree at the University of Texas in Soviet Studies and was trying to get into the FBI and Tom Stephens had finished his degree in Radio, Television, and Film, also at UT, and was working as a cop.

Steve had started his military service with the Texas National Guard's Pathfinder unit. When he was a young soldier, he told my buddy Alan that he wished he bore the scars of an AK 47 bullet on his body. Alan said, "Well, we're going to the range soon, would an M16 bullet do?" After leaving the Pathfinders, he got into Special Forces and he was now a Special Forces weapons sergeant and had recently completed the Special Forces Operations and Intelligence (O+I) course and was serving with an Army Reserve Special Forces Group.

He had trained extensively in martial arts and was an expert marksman with both long guns and handguns. He was well read and had an answer for anything that might come up in a conversation. I remember one night when our exasperated shift sergeant, Mike Hargett, told him, "It must be nice to be the man who knows everything!" Steve had a way of ignoring these types of comments and would continue talking as if nothing had been said which would infuriate Hargett even more.

Tom Stephens was a Special Forces Medic. He had served on active duty with the 5th Special Forces Group and had been a member of 'Blue Light;' the interim hostage rescue force that had been raised up as a stop-gap measure while Delta Force was being stood up. He was HALO qualified, an expert with sniper rifles and pistols and he spoke German and Latvian. He had also been a member of the Gabriel Demonstration Team, the full time SF Dog and Pony show that was featured in the beginning of the movie, 'The Green Berets.'

Tom had been badly injured in a parachute accident and spent a long time recuperating at Walter Reed Army Hospital. When he was fit for duty, after returning to Blue Light, he was participating in a hostage rescue training exercise which required him to grab a 150-pound dummy and slide down a rope from a four-story building. The rope broke and Tom went back to Walter Reed for a long visit. He also expressed interest and agreed to come over to the camp and talk to Tom.

Tom Stephens had a great sense of humor. He taught me a lot of police things and on slow nights, I use to watch with amusement as he would string together all the paper clips in our report writing desk. When I transferred to the day shift, I would watch Jim, one of the day-shift officers, as he cursed Tom's handiwork and spent 30 minutes undoing the chain, only to have Tom do it again the next night.

When Tom Payne agreed to talk to them, I called them and asked if we could meet so I could tell them about some work that they might be interested in. They both agreed. I met Steve for lunch at a nearby restaurant and told him about what we were doing. He expressed some interest in what I was proposing but wanted more details. I met Tom Stephens where he was working and told him the deal. They both agreed to come over to the Camp and talk to Tom.

After Steve and Tom Stephens interviewed with Tom, they both said that they would be interested in coming on board. The first thing they had to do was get in the Guard. Steve thought he might like the LRSD so he came to the next drill and talked to Captain C who offered him a job as a cook! This didn't set well with Steve, so both he and Tom Stephens ended up at the Division G2 and were offered the opportunity to train as Counterintelligence agents. Tom had medical problems and was collecting disability. They were not going to let him enlist, but Bil, who had also been an SF medic, tapped into the Special Forces medical underground at Fort Sam Houston and everything was taken care of.

Tom Payne had contacted the other guys and they had all agreed to come on. The headquarters cut their orders and they were to report on the 15th of December 1989. We went back to Bil and told him that we needed a team room. Tom and I were crammed into a small space with the rest of the Counterdrug Projects and there was no way that we were going to fit six more 'Go-rillas' as Bil liked to call us, in that space. He told us he was working on it and the next day, George C came to our cubicle and told us to follow him. He took us across the camp to an old abandoned wooden building near the old PX. It had been a horse barn in the real cavalry days.

December 1990: The Team Room

The building had been converted into some kind of maintenance shop years ago and now the only thing it was being used for was to store a Hetzer, a Czechoslovakian Assault gun that had been used by the German Army during WWII. It belonged to a group of reenactors who liked to dress up in German SS uniforms and ride around on top of it for Muster Day which took place every year at the camp. They would have mock battles and the Hetzer was a crowd pleaser.

The reenactors had painted their vehicle with authentic paint patterns. It had the symbols and vehicle identification marks of the 2nd SS Panzer Division. I talked to their leader about it one time and asked him why they had chosen to be identified with the unit that had murdered 600 French men, women and children from the village of Oradour-sur-Glane, putting them in a church building and setting it on fire.

He told me with a straight face that the reason they had done that was because that village was full of communists and because a French maquis unit nearby had shot one of their officers, they had to make an example. Every time I saw that guy with his reenactors, I got the feeling that he really wished he could have lived during WWII and been an SS man. He wore the uniform of an SS officer and when he was on the parade field with his crew it looked like they were in another realm.

Anyway, our new building had a lot of rooms. George C unlocked one and said "This is your new home." Tom and I walked in. There was a lot of room alright. It was totally trashed and it smelled like rot and mildew. There was a huge M-60 tank painted on the wall at the far end of the room. George C said that he had found some old desks for us and that he would get a flatbed truck for us when we were ready to move.

The new guys reported in the next day and the first thing we did was to clean up the trash in our new room. We hauled everything out of there, swept and mopped the floors, and opened all the windows for the place to air out. We got the flatbed and went to the furniture storage building. There was a desk for each man and plenty of chairs. We moved everything back to our new building and began setting up. The next thing the boys did was to get a can of black spray paint and draw a large circle around the tank with a diagonal line running through the tank, 'No Tanks!' Tom grabbed a hammer and started tearing out a bunch of cabinets that were built into the walls. He stayed there late that night continuing his project. The next morning, I asked him if he was all right. He said that the project was a good stress reliever.

Everyone was busy as we tried to put the new unit together before the next operation. Tom had discussed selection and training during several meetings with Colonel Charlie Beckwith who gave us the benefit of his experience with both Project Delta and Delta Force. During the first team meeting I suggested that we call ourselves the Border Rangers. George objected and said we should call ourselves the 1st Texas Special Operations Detachment or SOD. The guys liked that and from then on we were known as SOD.

Tom and I had come up with a training schedule for the new guys. Tom would put either Steve or George in charge of PT. They would come in at 10:00AM after a 2-hour PT session, shower, and breakfast. Tom Stephens would stay and work on medical and logistical issues. We gave each of the new guys a class to teach based on the skills they brought with them such as helicopter landing zone operations and advanced first aid. Tom and I gave classes on rapport building, Spanish phrases, and an area study of Mexico. We covered reporting using SATCOM communications, Border Patrol history and customs, and lessons learned from the past five Unity operations. This was the first OTC or Operator Training Course although we wouldn't call it that for another couple of years. We probably devoted 18 hours of classes to the first train-up but as we learned more and more equipment was given to us, the amount of time increased so that 10 years later OTC was taking up to 300 hours of classroom instruction and many days and nights of practical exercises.

The missions required that we perform a close facsimile of our wartime mission, but I was using my spare time to find out about photography and videography for surveillance. I had a feeling that this would eventually play a big part in our support to investigations of drug trafficking. I had attended a Texas Department of Public Safety Photography course several years before and I still had the handouts and brochures they gave us. We had no equipment except our own; I was using my own 35mm camera for aerial photos and my Super VHS camcorder to video tape for site selection. Tom Stephens was also interested in photography and began researching camera systems and costs.

I also took video at home. If I had any spare time I passed it playing my pawnshop guitar and scratching out lyrics and chords for songs that

I would use someday for my videos. I used an instrumental that I had recorded using the rhythm machine on a cheap electric piano and my own lead guitar to add to a burning map scene that I put at the beginning of each site video that we used on the last operation; kind of like the opening scene of the TV show, Bonanza, that we used to watch when I was a kid. Bil liked it.

SFC Thomas Payne, US Army Special Forces (Retired)
Commander, Special Operations Detachment

And there was war in heaven: Michael and his angels went to war with the dragon. Revelation 12:7

Chapter 9-January 1990

OPERATION UNITY 6

The next operation was the first one that we planned and coordinated without any guidance from Bil and the command group. We would still have to brief the plan, but we were on our own. In the middle of January, Tom told me to go down to McAllen and meet with Harry, the Border Patrol Emergency Response Team leader for the McAllen sector. They wanted us to come back and try another mission. I drove down to Harlingen. Harry was the station chief there in addition to his duties as the leader of the Special Operations Group (SOG) which is what they now called the Emergency Response Team for the sector.

I went to the Harlingen BP Station and Harry met me in his office. We talked about this and that until he reminded me of our meeting with Chief Reyes the last time I was here. "You ready to go back and see the Chief," he said. I looked at him in alarm. He laughed. "Don't worry; he was happy about the overall performance of your guys during the last mission. If he wasn't, you wouldn't be here."

I told him that we had hired six men and that we were training them right now to be team leaders when we brought in traditional Guardsmen, or they could combine to make two small teams that could respond much quicker to law enforcement requests. Harry was surprised but happy. I asked him if his agents could give them some classes in combat tracking, Spanish phrases and use of force scenario training. He assured me that he would arrange for that training during our isolation phase. With that we started to discuss the upcoming operation.

Remodeling the Model

There were a number of things that Harry wanted to address for this operation. He wanted to have the CP in McAllen rather than in El Paso. Though the satellite radios were able to provide clear voice communications, it was psychologically challenging for the Border Patrol supervisors to make decisions about things happening in their sector when they were 800 miles away. He wanted to bring the whole show to McAllen and put us in the Border Patrol Air Operations hanger at the McAllen Airport. Asking a supervisor from McAllen to go to El Paso to command an operation in McAllen didn't make much sense to him and surely took him out of his familiar surroundings. As a former police officer, I could understand where he was coming from and I felt that this was one of the procedures we needed to change.

He wanted to have the isolation training closer to the Valley and asked if we could fly down there in a couple of helicopters, pick up his agents and fly them to the Isolation Facility. I was taking notes as he talked. Moving the isolation activities and the Command Post from El Paso was a big change but I thought it was a good idea. He wanted to break the pattern up and run shorter iterations of three or four days at a site and then bring the guys in for a short rest and then send them out to a different spot. He was short of people and he could send out only two, which meant that we could send out only two teams, perfect!

This would be our first operation using the full-time organization that we were putting together. This was really good! Harry and I were getting along well; my cop background helped me to understand where he was coming from. Cops are operational every day and we had a lot of work to do requiring training and experience before we could even approach the flexible mindset that most cops have.

I discussed with Harry some of the obstacles we faced in making these changes. He might get some flak from Operation Alliance where his fellow Border Patrol agents were overseeing the military support to law enforcement along the Southwest border. Harry knew Lee Justice who had the oversight responsibilities in Texas, and he thought that Lee would be ok with it. I told him that I would try to convince Bil of the merits of what we had come up with. When we had hammered out a basic concept of the operation, Harry hooked me up with one of his SOG supervisors for site selection to look at quite a few sites from the Starr County line down river to the Gulf.

Enter the Intel Geeks

I got a call from Tom that night telling me that Bil was going to send down three or four Intel people who would accompany me on site-selection to do the Intel part of the target folder for each site. Great!

Now I had to drag around a bunch of people with me and because they were Intel geeks, there was no telling what problems they would cause with the BP agents. I told Tom to have them meet me at my motel the next morning at 7:00AM so that I could brief them and give them some specific do's and don'ts in their dealing with the BP.

The next morning the Intel guys showed up at the restaurant and we went over the things that I felt were important for them to understand. Fortunately, I knew several of them; one of them, SSG R, had been a team leader in LRSD and the others had been involved with Unity 5. They were led by a captain, an old Vietnam infantry vet who was close to retirement. He wasn't really into the Intel business, wasn't very motivated and he didn't do much except take up space in one of the Border Patrol vehicles. We ended up having to use three vehicles to take everyone from site to site.

We went to the McAllen Sector Headquarters where we met one of the SOG supervisors who put us in Suburbans and we took off up river to the first site. The upriver limit for this operation would be the Starr/Hidalgo County Line. Our first two sites were pretty far from the river but overlooked flat farm fields. It was wintertime so nothing was growing, and the operators would be able to see a long way across them. One of the sites was on the top of a small hill and they would be able to see everything for miles during the day, and nighttime would depend on the phase of the moon, clouds, ambient light etc.

The agent told us that because the operation was a month from now, there was no problem with us being seen at these places. I wasn't sure about that but he was the area expert and we had to trust our counterpart's judgment. Regarding safety and tactics, we together would formulate the mission plan and they would have to trust us regarding our rules and the need for contingencies.

At each site I determined the grid coordinates for the drop-off point and the hide site. I looked for a couple of places that might be good for OPs and wrote the grid coordinates for them on my site selection work sheet. Then using my personal camera and video camcorder, I took a series of panoramic photographs of the area around the site while the agent stood next to me describing the law enforcement significance of what we were seeing.

The Intel guys took notes as they listened to the agent. When I was done, they asked a few more questions and then we moved on to the next site. Later, we would have the team leaders doing this because they would be the ones who would be coming here to perform the mission.

It took us three days to survey all the sites that Harry wanted to use in the McAllen Sector. When we were finished, the Intel guys took off and went back to Austin. Harry had arranged for me and SSG R to fly with a Border Patrol pilot and take video and photographs of the sites that we had surveyed the last few days.

Flying the River

I went with SSG R to the Border Patrol Air Branch hanger at the McAllen Airport. We met the senior pilot and he introduced us to the pilot flying the mission. He asked me where I wanted to go. I pulled out a Joint Operations Graphic map on which I had each site marked with a circle. The BP wasn't big on maps; they passed location information to the new agents by word of mouth, and they didn't have the common reference system provided by maps.

I saw the pilot turning his head back and forth like the map wasn't helping him a bit so I said, "I need for us to go up to the Starr County line and shoot video of the sites upriver from the airport at an altitude of 1000 feet. When we get to the furthest upriver site, I want to go up to about 3000 feet and take photographs of the same sites as we fly down river. We'll continue past the airport to the furthest down river site where we can turn around again and I will video the sites on our way back upriver to the airport. What do you think?" I asked him. "It sounds like a plan," he said, "let's go."

We walked out of the hanger to the flight line and past the A-Star helicopter that was sitting on a small trailer. I asked him why it was on the trailer and he said so they could pull it into the hanger when they needed to work on it. I was impressed that the pilots could land perfectly on that small trailer and I was getting the impression that the Border Patrol pilots were good at what they did. I have made many an aerial photo flight with them and I never had a bad ride, I had some exciting ones, but never a bad one.

I asked him if he was hired as a pilot by the Border Patrol and he told me that all the pilots start out as regular agents on the ground and they can apply for service with the Air Branch after a few years. If they are selected, the Border Patrol sends them to flight school.

The plane was a small single engine Cessna. He walked around it as part of his pre-flight routine and then told us to climb in. I sat next to the pilot and SSG R sat in the back seat. The pilot handed us some headsets and I put one on. When he tapped his earphone, I pushed the push to talk button and made a communications check with him, SSG R did the same. The pilot started the plane and we took off. He flew us upriver toward the Starr County line. We talked a little while he flew but mainly, I was taking video and keeping track of the features on the ground and trying to match our position to the map in my lap. This was before we had GPS and aerial navigation was fairly new to me.

The flight took us about an hour and a half. As we were flying down river it became harder and harder to follow the actual river because there were so many ox-bow lakes. These had been formed years ago when the course of the river changed during floods that came through periodically. The pilot was in no hurry to return to base and when he asked us if we wanted to go to the Gulf we said yes. I kept looking at the terrain and trying to match it to the map, but I couldn't spot anything that should have been there. After while I gave up and just enjoyed the ride. I could see the Gulf ahead of us and was hoping that there might be a surfing spot that I would discover when we crossed the coast.

The pilot asked us if we could see the lighthouse that was on the Mexican side at the mouth of the Rio Grande. We looked to the right as we crossed the beach and flew over the Gulf. No lighthouse. The pilot turned south, and we all strained our eyes looking for the lighthouse. "Maybe I flew a little too far south," he said, banking the plane until we were flying north up the coast. Ten minutes later I saw the lighthouse and the hotels on South Padre Island, a few miles north of the lighthouse. We crossed the river and were back in the U.S.A.

Drums along the River

When we returned, I had another meeting with Harry. He took me in to see Steve, who had an interesting bit of information that he heard from the 'Drums along the River' that keeps the BP informed. "A team of spec ops soldiers was inserted deep into Mexico the night before last," he told us with a smile. "I guess no one told the Army helicopter pilots that there is no river between New Mexico and Mexico. They didn't figure it out until the team called in and told them that they were far from where they were supposed to be according to their GPS. It was too late to get them out before the sun came up, so they had to spend the day hiding in Mexico and the helicopters went back and got them last night." That was pretty interesting news so I asked him how he had heard about this and he just winked and smiled.

I told him about the sites we had visited and about the aerial photo mission, 'no problemas,' and he was happy. I decided not to mention our own side trip into Mexico. We made sure we were both on the same sheet of music about the proposed changes to the operational model and then I left and drove back to Austin.

Changing The Model

When I got back, I met with Tom. We took a ride down to the back forty by the pond and I told him about the changes Harry and I had talked about. I really laid out my feelings about how we had conducted operations up to this point, and why I thought the changes were important. I told him that if we didn't become more flexible one, we weren't going to catch any dope and, two, the cops were going to decide that our 'support' to their operations was too much work for no gain.

"I'll tell you what, Tom," I said, "if we don't catch some dope pretty soon, they won't want to use us down there." "Why's that?" Tom asked. "Because they already know that tons of dope are crossing the river every night," I said. "All they want to know is when and where it's crossing so they can go out and catch it. The only kind of intelligence they are interested in is the intelligence they can act on to catch dope."

Tom thought about it and then he said he agreed with me. I told him what Harry and I had come up with and he said he would go and tell Bil that we wanted to make some changes to the current mission model. We both knew that Bil could be pretty volatile when we touched his baby, but I felt we had good reasoning behind our requests and all Tom had to do was to make a clear presentation to Bil and he would see the merit in our suggestions. Tom went in and got Bil to agree with Harry and our proposals. It wasn't a pleasant meeting but those kinds of meetings with Bil seldom were.

Tom came back from the meeting and told me that Bil had agreed with everything. Harry had given me the dates for the operation. The mission HQ would be in McAllen and it would be up to the Border

Patrol to come up with a suitable location. We would send two Hueys down to pick up Harry and his boys and bring them to Camp Swift which was about 30 miles from Austin for isolation.

Tom told me to write a training schedule for isolation and fax it to Harry. He called me when he got it and recommended a few minor changes which I made, and he told me he was very happy with our flexibility. He had heard from friends in the other sectors that Bil was very inflexible. I told him that Bil was ready to entertain a logical recommendation, but he always made the person who made them defend it against his attacks on their logic.

Most people couldn't handle the pressure of Bil's 'Murder Board' methods but Tom and I were veterans and we just considered it part of doing our work. Bil didn't want yes men but if you decided to question one of his directives or change one of his policies, you better be prepared to fight the battle of your life. After you did it a few times you either learned to make sure your shit was together, or you moved on because Bil wouldn't tolerate fools or amateurs.

We briefed the guys on the general scheme of the operation, gave them the dates, and told them to make sure they had their gear ready to go on the day isolation started. I went to work building the target folders for each team. I had to complete a site specification sheet which had the grid coordinates for the proposed hide sites, OPs, drop-off points and pick-up points and it included the recommended special equipment and camouflage considerations. Last was the route they would have to take to their hide site. The team guys would be carrying rucksacks that weighed up to 100 pounds and sometimes they would have to carry another 45 pounds of water in a 5-gallon jerry can. The drop-off point and the route they took were major considerations and attention to detail was critical when preparing these target folders. Then I made maps of each site and E+E maps for each team member. I also made a map of the Infil/Exfil routes for each driver starting from a major road on the map with checkpoints and the drop-off point. I contacted the Intel guys and they gave me their Intel products for each site. When I was done with the target folders, I put them and the maps into mission bags and locked them in a large file safe we had in the team room.

There was too much to do in the normal eight-hour day, so we all stayed in a kind of wartime mode in the middle of a peacetime National Guard Base. Bil called us 'Spartans in an Athenian Camp,' and we were.

He would caution the 'Go-rillas' not to expect the other people around the Camp to understand us and our sense of urgency and precision. "Don't talk to them if you don't have to and let us deal with any problems you run into when you have to get something from them."

The day for isolation came like a fast freight train. We were busy up to the last minute, trying to take care of all the minute details that would mean the difference between success and failure. Colonel Charlie Beckwith had told us that it was the little things that caused big things to go right or wrong and we figured he ought to know so we checked and rechecked everything we did. Sometimes Tom would ask me to read something he wrote just to check it before he sent it out and I would ask him to check my stuff. This was our busiest part of the operation. Once the teams were inserted, we were able to set up shifts in the CP and get a little rest.

Operation Unity 6-Isolation

The morning to begin isolation came and we all met in the team room. We loaded all the CP boxes into the Alliance Chevy Blazer. The team guys loaded their gear into a couple of vans that we got from the motor pool and we convoyed out to Camp Swift. We were given some old tin huts to use for isolation and someone had placed a row of concertina wire around the whole area. We unloaded our equipment into the CP building and took our personal gear to one of the other buildings where the bunks were. Each team had their own hut where they could sleep and do their mission planning undisturbed.

We got a couple of Guard guys come up from McAllen's Border Patrol Intel section, to help us put on isolation. They each stayed with a team, taking care of all their external needs and acting as a backup in case one of the team was injured or sick. They were both from LRSD and had gone on one or more of the earlier missions. One of them was James F, who was about 19 years old and had wanted to join us as soon as he heard there was a unit being formed. He had brought his field gear just in case there might be an opportunity to go to the field. Steve told us that Larry was sick, so James got the nod to join Steve's team although he was young and inexperienced; he turned out to be one of the best operators we had.

Around 11:00AM I stepped outside of the CP with my video camera. I wanted to get some shots of the helicopters flying in from McAllen with the Border Patrol contingent. I had decided to video as much of the operation as I could, and Tom had given me the ok. Bil used to look with suspicion on my video camcorder but eventually he came around and would ask me if I had any new footage. I don't know why I was so fixed on shooting video. Perhaps it gave me another thing to do which kept me interested in the overall mission even when there was a lot more paperwork than action.

About five minutes later I heard the faint but familiar sound of the Hueys approaching from the South. That sound had been embedded in my brain since Vietnam and I can still identify the sound of one even if I can't see it for a while. I saw two black dots in the sky, which quickly became larger and larger until I clearly recognize them as UH-1H helicopters, the Huey. The pilots flared out about 75 meters from the concertina wire and brought the birds to the ground. Within seconds I saw Harry and his assistant jump out of the closest one. I shouted Harry's name and waved him toward the entrance to the compound. His agents were unloading their gear and Tom sent some of the team guys out to help them. The rapport building process had begun.

The team guys brought the gear to the front of the CP where Tom, Harry, and I were talking about team assignments. When the two agents got to where we were standing, Tom called Steve over and Harry introduced Juan, the agent assigned to his team. They shook hands and Steve called one of his guys over and they picked up Juan's things and he followed them to Steve's team hut. Tom then called George, the other team leader and he met Mike, the agent assigned to his team. Tom told both team leaders to have their men in the briefing room in 30 minutes.

About this time, Bil drove up and he and two other men got out of the car. Tom walked out to meet him, and I saw Bil introduce them. They talked for a minute and then Tom led them into the compound and after offering them some coffee, he took them all to the conference room where they could sit in a chair and be out of the sun. He came back to the CP and motioned for me to follow him toward the back of the compound. After making sure that there was no one within hearing distance, he told me that the two men who came with Bil were officers.

One was a lieutenant colonel and the other one was a captain. Tom said, "Colonel Benny wants to have an officer present during the operation. He's worried that if something happens and there are only NCOs present it will make him look bad in front of the Chief of Staff and the Adjutant General. Bil told me not to worry about it. He told the captain that I was in charge and that the captain should stay out of our way and be a figurehead."

Of course, Colonel Benny had told the captain just the opposite, to keep a tight rein on things, and now the captain didn't know what to do. He looked really nervous when Tom introduced me to him. His name was Captain John. Tom told him that he should start the briefing by introducing everyone and then turn the briefing over to Tom. Tom gave him the names of the big shots that he should introduce and directed him to the briefing hut to wait with the others.

McAllen 1987-Border Patrol Justice

I heard a car drive up outside and went out to see who it was. The car door opened and out stepped Lee Justice, the Border Patrol representative at Operation Alliance. I really liked Lee and walked out to his car and gave him a big abrazo, or hug, and after greeting him, I helped him carry his bags to the hut where he would be staying.

Lee and I went back a few years to when I had met him at the Texas Police Olympics in McAllen. Tom Stephens and I had driven down on a whim and entered the pistol-shooting contest. We were both broke and instead of firing wad cutters, the normal match round for revolvers, we were firing a mix of rounds that we had scrounged up before we left. My strategy was to fire some really cheap rounds with aluminum cases when we were close to the targets, at the seven and ten yard line. Then we would switch to the Plus P Plus rounds that a Secret Service agent gave us at the Austin Police Department range. Tom was a good pistol shooter and he was a bit skeptical regarding my strategy, but we didn't have any choice because we barely had enough money to get a room at the Motel 6 and buy some burgers at MacDonald's.

We went up to the firing line when our names were called, and advanced to the closest point from the target. I was sure that we could get the maximum number of points at the close range positions and when we got back to the 15 and 25 yard lines, our Plus P Plus rounds from the Secret Service would kick in and we were sure to have a good score.

The whistle blew and we fired at our silhouette targets. The whistle blew again, and everyone stopped firing. The range officer told us to move back to the ten-yard line, but before I did, I took a close look at my target. There were no holes in the x-ring and there were no holes in the ten ring. All my bullets had struck the target in the center and about six inches below where I was aiming. I couldn't believe it.

"All shooters move back to the ten-yard line," the voice on the loud-speaker said again. I looked around and saw that I was the only shooter still on the seven-yard line. I hurried back to the ten and whispered to Tom, "did you hit where you were aiming?" I asked. "No, I was hitting way low," he said. "Let's aim a lot higher this time," I told Tom as the whistle blew to start the next round.

I fired the correct number of bullets at my target and this time moved back to the 15-yard line with everybody else where we loaded our pistols with the Plus P Plus bullets and waited for the command to fire. Ka-Boom! went Tom's pistol. Ka-boom! went my pistol. I could see the guys next to us looking at me and Tom, but I ignored them. We fired the rest of the course and after getting our targets scored and turned in; we walked back to the bleachers and sat down.

A rather distinguished looking gentleman walked up to us and introduced himself as Lee Justice of the Border Patrol. "What kind of rounds were you guys firing during the second half of the match?" "Plus P Pluses that we got from the Secret Service," Tom said. "You should have seen the people in the stands when you started shooting them," he said. "Every time one of you fired a shot, everyone in the stands turned their heads toward you. It was like watching a bunch of people at a tennis match. Why are you shooting such powerful ammunition at a pistol match?" he asked. We told him about our limited budget for this match and he laughed. "Wait here" he said, "I'll be right back." He came back with a brown paper bag. He reached in the bag and started pulling out boxes of bullets. He gave us each three boxes of wad cutter match ammo. "I get so much ammo from the BP that I can't shoot it all."

I was blown away. Here we were, a couple of low rent cops, and a Federal Agent that we had never met before was giving us 300 rounds of match ammunition. We both thanked him profusely and I never forgot him. Imagine my surprise when I ran into him in our isolation facility in El Paso during that first Unity mission. Our connection at that match gave me a little credibility with Lee during that first mission and we spent a lot of time together talking in the CP while the teams were in the field. One thing he told me I never forgot. His father had been a rifleman in the 8th Infantry Division in France during WWII. He was killed in action. After the war, two of his father's buddies visited Lee and his mother and told them about their friendship with Lee's dad and how he died. Then they gave Lee the leather sling from his dad's M-1 rifle. To me, that was the essence of the 'old' Army.

After we dropped Lee's bags off in his hut, I took him up to the CP where Tom greeted him and gave him a copy of the isolation schedule. Tom told him about the in-brief in 10 minutes and took Lee over to the briefing room next door where he sat down with Bil and the others.

Operation Unity 6-Mission Briefing

I gathered up the mission packets and took them over to the briefing room. I saw Harry and Lee talking together; Bil was in the corner talking to the LTC who came with him to the Camp. The team guys filed in and took their seats in the front. Tom walked in with Captain John and the briefing started with Tom asking Steve and George if they had all their people. They each gave him thumbs up and Tom turned to Captain John and told him, "Go ahead, Sir."

He introduced himself as the National Guard's command representative for Operation Unity 6 and made his way through the introductions before turning the briefing over to Tom. Tom introduced himself and announced the beginning of isolation for Unity 6 and gave the rules for our stay at Camp Swift. No one was to leave the compound, no telephone calls, "if you need something, just let us know and we will try and get it for you." The Border Patrol agents didn't look too happy with these rules but Harry had agreed to them so they were stuck.

The isolation concept went back to WWII when the Special Forces of those days were kept in complete isolation before parachuting into Europe to conduct espionage and sabotage missions against the German occupation. The Germans and their French collaborators had a very competent counterintelligence operation and there were many double agents and traitors within the allied countries. The identities of the agents were kept secret from everyone except their handlers and their commander. They would prepare for their mission in an isolated area with no contact with the outside world. When it was time for them to be dropped into Europe, they would be taken by night in a covered vehicle to a nearby airfield where they would get into an airplane flown by a special duties squadron crew. They all had code names and they were flown to their drop zone and dropped into a very dangerous environment to conduct their missions.

The isolation concept had evolved from those early days into a doctrine driven activity used by Special Forces and other behind the line units to prepare for their missions. Bil had spent his entire career in Special Forces and he felt that this was the only way to do things. Tom was a Special Forces man and felt the same way at the time.

I was a cop in my thinking about these things and just seeing the disappointment on the faces of the Border Patrol agents I felt that we needed to become a lot more flexible. These guys had come up from small towns in the Rio Grande Valley and were probably hoping for a night or two on the town to eat in a restaurant and a chance to go to a big store and buy some things. A few years before, G Company hosted a squadron of 21 SAS for two weeks at Camp Bullis in San Antonio. These Brits came all the way from England and were forbidden to go to downtown San Antonio for 4th of July weekend. That was the last year of the SAS/G Company exchange program.

In the police world it was understood that you didn't tell anyone about police activities that were eminent. Border Patrol agents did surveillance on the river routinely and for these guys to be restricted from calling their wives and going to the store seemed ridiculous. Fortunately for us these guys were able to take the restrictions to do the mission. In later missions we would not be so fortunate.

Tom told me to hand out the mission packets, which I did, and he directed the team leaders to inventory them. They did and everything was there. He told them that everyone should go to chow and be back in the briefing room for the first class which would be on an orientation to the Border Patrol, Law and Liability, and use of force. This class would be taught by Lee Justice. and I set up my video camcorder on a tripod in the back of the room and prepared to tape it.

Operation Unity 6-Rules of Engagement/Use of Force

Everyone went to lunch and was back in the classroom at 1:00PM. Lee gave us the same briefing that he gave on Unity 3-1. He went into the Tort Laws that basically say that if you are in the scope and performance of your duties the government will defend you in a civil lawsuit. He urged everyone to stick with the plan and stay within its boundaries. He concluded with a serious presentation of the use of force policies of the Border Patrol, which we would be following. Since the first border operation we had been armed and this established a policy that Texas Guardsmen would be armed when exposed to the possibility of contact with armed criminals.

That policy did not end the issue and there were many attempts over the next 10 years to disarm the troops conducting these missions. The legal people at the National Guard Bureau (NGB) were always wringing their hands and telling people that if one person were shot by a Guardsman, whether it was a justifiable shooting or otherwise, the whole program would be shut down. A lot of those people were hired when the program expanded, and they were afraid that they would lose their jobs. They were always coming down with attempts to disarm the states that were doing these types of missions.

I advocated that we train the soldiers in pistol and rifle marksmanship to a much higher standard than the military qualification requirements. Pistol marksmanship had been evolving since the 1950s and people using the Government Model .45 were able to perform amazing feats with a weapon that I had been told was best thrown at your enemy because no one could hit anything with it.

Steve, Tom Stephens and I had all been avid pistol shooters with the police department and I had been getting interested in some action pistol type competitions. We all felt that if the soldiers were trained to a very high skill level with their weapons and if they participated in a lot of role playing training where they learned how to verbally confront people and control them by giving clear instructions with a commanding voice, they would have the self-confidence to be able to deal with almost any situation without having to use deadly force.

I'm happy to say that that has been the case for 20 years. The Marines had killed a teenager in West Texas and a Special Forces soldier had shot a suspected border bandit near Brownsville. Neither of these situations required the use of deadly force. Our guys would have a number of confrontations with armed smugglers and in every case, they handled the situation in a professional manner. The drugs were seized, the crooks went to jail, and nobody got hurt. The first of these confrontations was three days away but since we were not able to see the future, we conducted our training based on past police experiences and our knowledge of the law.

After Lee's class the guys took a break. George and his Border Patrol agent were already laughing and joking and appeared to be the best of friends. Steve's counterpart was more reserved which suited Steve's personality which was all business most of the time. The teams went outside for some smoking and joking. The next class was on night observation devices and one of the team guys was setting up the equipment for the class. I got Tom Stephens to help me with the videotaping so we could rotate taping these classes.

The night vision class went well. We had three night vision systems: night vision goggles, a weapons sight, and a huge scope from the Vietnam War. The AN/PVS 5 Night Vision Goggles were worn on your face like a mask. There were two image intensifiers which you had to adjust to line up with your eyes and refocus manually every time you looked at an object that was a little closer or a little further away. They used a small battery which lasted for eight hours and overall, the troops liked them. There was a new generation of night vision goggles which some of our guys had used while serving with a Ranger Battalion or with Special Forces. They were always asking why we couldn't get some of the new ones and Tom or I would tell them to shut up and remind them that this was the National Guard.

The AN/PVS-4 Individual Weapon Night Sight was designed to mount on an M-16 Rifle or a machine gun. It was simple to operate and used the same battery as the goggles. It had a larger light gathering objective lens and a greater magnification than the goggles. Some guys liked them, and some didn't. We seldom mounted them on our rifles and instead used them as a handheld night observation device.

The AN/TVS-4 was a beast. It weighed 34 pounds and was mounted on a surveyor's tripod which weighed another ten pounds. It took an entire rucksack to carry the sight and another guy would have to carry the tripod. It was a giant pain in the ass and the only reason we carried it was because it worked so well in the dark. On a night with 50% illumination from the moon we could see people who were 1000 meters away. This far exceeded anything else that was available to us.

The class was well done. The agents liked the goggles, but were used to walking in the dark and probably wouldn't use them. The weapons sight didn't interest them, but they really paid attention when the operator went over the TVS-4. At the end of the class, they had one observation and one question; if this sight really worked it would be great for where they were going; and who was going to carry it? The instructor assured them that the scope would be carried by one of the SOD operators. That brought a smile to their faces and with that, Tom said that the time between now and the evening meal was team time.

After the evening meal the teams had a class on satellite radios and communications procedures. I videotaped the commo class. It was given by a young soldier from the 18th Airborne Corps at Fort Bragg. They were providing three commo guys, three PSC-3 TACSAT Radios and two URC 110 SATCOM Base Station Radios; another one of Bil's deals involving the 'old boy Special Forces network.' After the class they went outside and practiced setting up and using their SATCOM Radios and working with their night vision devices.

I helped the operator who gave the night vision class put away the equipment and carried the TVS-4 over to the supply room and stored it. It was a backup for the ones already issued to the teams. Tom and I reviewed the day so far and discussed fixes for some of the small problems that had come up. All of our planning efforts were paying off and there were no train wrecks so far.

The next day we were up at five helping the cooks unload the chow that they brought in mermite cans from Camp Mabry. The team guys got up and prepared for the day's activities. The first event would be the mission briefing and several carloads of Intel people and some colonels from the headquarters showed up. SSG R, the former LRSD team leader would be giving the overall Intel portion of the briefing and the agents with the teams would give the site Intel because they had worked these places many times. When everyone was assembled in the briefing room, Tom read an OPSEC statement reminding everyone not to discuss this mission with anyone not affiliated with Task Force Unity, our code name for the combined Task Force. When he was finished, he introduced SSG R and the briefing started.

Intel briefs are not my favorite form of entertainment. Since the Intel guys don't have any information about the enemy's location or plans, in this case, the drug traffickers, they take up a lot of time describing the terrain, vegetation, weather, and the current situation in the area. Sometimes there is something of interest, smothered in a whole lot of filler material, so you have to stay awake through the whole thing, so you don't miss it. A few years later, when I was in the Intel business in Afghanistan, I saw how much work they put into products that were used to disguise the fact that they didn't have any usable information that would help the troops to close with and destroy the enemy forces in their vicinity. In law enforcement, the only Intel that anyone is interested in is the information they get from phone taps/records and confidential informants. It's hard to develop informants along the border because if you get caught giving information to the cops, you are dead, and probably your family also. There is much more money in drug trafficking than there is in informing.

Even the smugglers they catch are more likely to do the time, rather than inform on the drug traffickers they work for. Federal sentencing guidelines kept the prison time low depending on how much dope was captured at the time of the arrest. If they keep their mouths shut their families are taken care of and when they get out, they have a job. Besides that, not many people get caught in the valley.

If the Border Patrol shows up on the river during a drug smuggling activity, the mules all run into the river and go back to Mexico. If the law enforcement people on the highway, the Highway Patrol, county sheriff, or city cop tries to do a traffic stop on a vehicle full of dope, the driver will take off down the highway and head for the closest place to the river where they will drive through a fence, cross a field, bail out of the vehicle and run for the river. That's why there isn't much Intel about smugglers on the border. About the only things that work at all are: pattern analysis using the BP sensor system, TIPS, a law enforcement hotline for people to call in suspicious activity, and the intuition of a switched-on law enforcement person.

After the Intel portion, Tom Payne gave the operations part of the briefing. He covered the mission statement and how the mission would be executed. The devil is in the details, they say. In special operations, planning is everything. It is difficult enough to plan for an operation with the people you have been working with in a military environment, but when you add in people that you will know only for two or three days and a civilian law enforcement agency that you are in support of, the detailed planning process is critical. Every aspect of the operation had been 'war gamed' and everything that could possibly go wrong had to have contingency plans. What would we do if the helicopters weren't available at the end of isolation? We needed a ground transportation plan. What happened if a vehicle broke down? We would need a backup vehicle and a vehicle recovery plan.

Every aspect of the operation from start to finish was planned and this plan was passed on to the teams. They had to do their own planning within the framework of the operations plan. After they got out of their infiltration vehicle, everything that they did required their plans and contingency plans. Rehearsals were conducted, and before the team headed south, they would brief these plans to Tom and I in what was known as a 'briefback.' That was when they would show us what their plans and contingency plans were. We would ask questions about anything that hadn't been made clear. The goal of a good briefing was no questions at the end.

Tom Stephens gave the logistics portion of the mission briefing. We found that the logistics aspects of our operations were like a two edged sword. On one hand we had to make sure that that the teams had what they needed; food, water, weapons, ammunition, night vision and communications equipment, a variety of batteries, medical equipment, the list went on and on. On the other hand, we had to maintain accountability of all this equipment because it all belonged to National Guard units around the state. Keeping track of hundreds of items was a major hassle for us and it only got worse as we acquired more equipment and conducted more missions.

I gave the commo briefing and I was assisted by one of the soldiers from Fort Bragg who gave the SATCOM and encryption information. Commo was the center of the mission. If they saw something and couldn't tell anyone, they might as well not be there. Commo was also their lifeline. If anyone was seriously injured, the only way they could get help was with their radio. Beside the TACSAT radio system, the team carried two small radios so that they could put two people in an OP and two people could remain in the hide-site where they could take turns resting, eating and performing maintenance on the weapons and equipment. The commo plan included code words, frequencies, encryption changes, communication times and blind transmission broadcasts.

One of the most important aspects of the communications plan was the loss of commo plan. It was initiated after the team failed to come up on the air for their regularly scheduled situation report. We did these at regular intervals during the day and at night. If the team missed a two of these contacts, the emergency resupply plan went into effect.

One of us from headquarters would go with a BP agent to a designated spot near the team's area of operation. We would park and wait for the team to contact us. They knew it was us because of the type of vehicle we were in and a special signaling device as a close recognition signal. When they contacted us, we would get out of the vehicle and give them a complete radio set. They would set it up and make a commo check. When they were sure that the new radio worked, they would give us their old radio and disappear into the dark.

After the mission briefing, the team went back to their huts to begin their planning. The next time we came together was for a first aid class after lunch. Then they would conduct team tactical training during daylight hours. Here they practiced crossing fences tactically, with all their equipment. They practiced some simple hand and arm signals and the BP agents gave them a class on combat tracking and counter-tracking. The latter would be very important because in the Rio Grande Valley; the smugglers always looked for tracks around the area of the river that they were planning to bring a load of drugs from Mexico.

Use Of Force Scenarios

The next event after tactical training and rehearsals was at the range. The operators zeroed their weapons. We had to do this every time we conducted an operation because we borrowed the rifles from another unit. After the guys zeroed their weapons, they returned to the isolation compound for chow. The team guys and the agents were now sitting together when they ate and that was a good sign. Cohesion means a lot in small unit activity. After chow we assembled in the front of the compound. We were going to have a class on use of force with role playing scenarios. This was one of things I told Tom we needed to implement into our training, both with the team guys at home and with the cops and the teams in isolation.

Part of the early mission model that had been implemented by G Company for the first mission, Operation Gatorade, was standard LRRP/LRS Immediate Action Drills or IADs. These were automatic responses to enemy contact, and they were designed to put maximum fire on the enemy, which would allow the team to break contact with a superior enemy force. One IAD involved a team walking in a single file in the dark. They would come under fire from their left flank. The team would then pivot to the left and everyone would fire a thirty-round magazine toward the enemy and then bound away from the source of enemy fire, some shooting and some bounding or helping the wounded until they were out of contact.

During Unity 5, we were conducting our isolation training on the range. The teams were doing live fire IADs. One of the Border Patrol agents came up to me and said that this was stupid. "I am a law enforcement agent, if I get in a contact on the river with these guys; we're not going to do this. We're not going to Vietnam; we're going to the Rio Grande Valley. I'm not going to lose my career and go to prison because some Guard guys shot up a family of illegal aliens who were hiding in the bush. I am in charge in a law enforcement situation and we will do it my way." I told him at that I agreed with him and that I would try and ensure that the next time we did this we would train differently.

I took his speaking to heart and thought about it a lot. When I went to the police academy, we were trained in the use of force using role playing scenarios. One of the instructors played the bad guy and another instructor set up the scenario. The student would have to react to the bad guy while staying within the limits of the penal code and bring the situation to a legal conclusion. Facts and circumstance play a big role in the decision-making process of any peace officer confronted with a possible use of force situation. What we would have to do was to teach our guys to think like cops when they were operating in the counterdrug environment.

I sat down with Steve and Tom Stephens and told them what I wanted to do. Since Steve was the most experienced cop of the three of us, I asked him to run the class while Tom and I played the part of the bad guys. We went over a few scenarios based on incidents we had heard about from the Border Patrol and our own experiences as peace officers. These were basically shoot/no shoot situations with the BP agent making most of the decisions and controlling the team.

The team guys had to make decisions also and they had to use the Spanish phrases they had been learning. Before we started, we talked with Harry's deputy, a Hispanic BP agent named Carlos. We told him what we wanted to do and asked him to participate with Steve in running the training. He agreed and he and Steve got together and discussed which scenarios they might use while Tom and I got some props like pistols, knives, and bundles to simulate drugs. Steve started the class by explaining what we were going to do; Carlos described the most likely situations we might encounter on the river.

"Most of the time there is no need to fire your weapons," he said. "Many Border Patrol agents never fire their weapons, except at the range, for their entire career. Most of the time you will take charge of the situation with verbal commands and occasionally you will have to physically subdue someone who wants to fight. We go out on the river every night and we know that some of these smugglers carry guns but there are very few shots fired by agents. The smugglers might fire a shot over their shoulder when their running for the river but that's just to slow us down. We could fire back at that point but what would be the purpose? It's very hard to hit a moving target in the dark and if you miss, where is your bullet going to go? Once that bullet leaves your weapon you can't call it back. There are a lot of people who live near the river, on both sides. People fish down there, day and night. Don't fire your weapon unless it is absolutely necessary."

Carlos and Steve told me they wanted to do a meeting situation where the team was moving down the trail and the smugglers were coming up the trail from the other direction. We had one team conduct the scenario while the other team watched. They carried their gear and moved in a single file. Tom and I walked toward them after stepping out from behind a tree. The Border Patrol agent would challenge us in Spanish, and we would comply with his instructions, drop our bundles and flee, or occasionally fight. We had a lot more no shoot scenarios than ones where the team would have to shoot.

Each one was different and after it was finished, Carlos would do a quick AAR and then the other team was in the barrel. We had the team set up like they were in an OP and the smugglers would walk right into the OP. We did this for an hour; each team went through five scenarios. Toward the end, the teams were working together, some pulling security to the rear and flanks while another was confronting the situation and the fourth one was backing him up. The guys told me later that these exercises really helped them to understand the use of force policies and how to work within them.

When we finished the use of force training, we went to the range. By now it was dark. We did the same thing we had done earlier except this time we used live ammunition and the bad guys were the targets on the range. Tom Stephens would walk behind them with his own .45 caliber pistol. He would tell them what the bad guys were doing, and the Border Patrol agent would talk to the target. Every third scenario Tom S. would fire a round or two into the ground. This would indicate the bad guys had opened fire and then the two guys facing to the front would engage the target and the others would pull security. Each team went through several of these scenarios and Carlos would do the AAR after each one. We finished at about 9:30 and returned to the compound. The support people would pick up the brass in the morning.

The next morning the only thing we had to do was have the teams give their briefbacks and clear the compound. I would drive to McAllen at noon with Tom Stephens with all the CP equipment. The guys from Ft Bragg would follow us with the communications gear. Tom and the teams would fly to McAllen in the late afternoon.

The briefbacks started at 9:00AM. It was interesting to watch each team. George's background was the 2nd Ranger Battalion and an experimental Recon unit of the 9th ID. He gave most of the briefback for his team. Steve's background was Special Forces, and he had each member of his team give a part of the briefback. Bil was there along with Harry and Lee. George's team went first and then Steve's. They both did a credible job. Bil asked some questions and so did Lee Justice. When they were satisfied that the teams were prepared to go, they blessed the operation and left.

Operation Unity 6-Infiltration

I loaded my personal gear into the Border Patrol Blazer and with the guys from Ft Bragg following me, took off for the Valley. It took us about five hours to get there and we went straight to the Border Patrol hanger at the airport. Harry had gone all out and there was a huge RV that we were to use for the CP. The commo guys started setting up the SATCOM equipment and Tom Stephens and I set up the maps and other CP equipment. The teams were due any minute, so I stepped out of the hanger and looked north. There was a cold front coming in and the sky to the North was dark. I could see two dots in the sky and eventually I could make out the helicopters. They made their approach and landed right in front of the hanger.

Tom jumped out. He was all smiles. This was the payoff for all our work for the past two months. In two hours, we would be sending two teams into the heart of drug smuggling territory. We had recruited these guys, trained them, planned and coordinated the mission with the Border Patrol and now the teams were going to execute. Everyone on the teams was pumped and the agents were a little more animated.

Tom moved the teams into the hanger where they made their final preparations. They loaded their rifle and pistol magazines, put on their war paint, tested their night vision goggles and did a final commo check with their SATCOM radio. Tom talked to them as they were getting ready, showing his confidence in them. Everyone was hitting on all cylinders. I took my video camera and went outside to check the weather. It was getting cold and starting to sprinkle. I hoped it wouldn't be too cold for the smugglers. Steve was outside adjusting his night vision goggles and I videotaped him before I went back inside and videoed the rest of the teams getting ready. It seemed to me that this was an historic occasion and I wanted to get it on tape.

Operation Unity 6-Showtime

At 8:00PM, two Border Patrol Suburbans showed up. The teams met their drivers and loaded their rucksacks into their vehicle. Tom took the team leaders aside for a last word and then they climbed in. Steve's team went first because they had farther to go and fifteen minutes later, George's team left. Operation Unity 6 was on.

The teams were dropped off and made their initial entry reports. Tom set up a CP schedule and went to the motel. Harry and I and one of the commo guys settled in for another long night of listening to the rushing noise of the SATCOM and the occasional gurgle of Mr. Coffee.

At 10:30PM the commo guy on duty made the first Blind Transmission Broadcast. He gave a time hack, temperature and the weather forecast. The weather was looking bad. It was usually warm in the Valley, but a strong cold front had pushed in and the temperatures were going down into the 40s. With the rain it would be a cold night.

Around 11:30 we heard a break in the rushing noise from the SATCOM. Usually only the radio operator can hear incoming signals through the handset, but Tom had rigged up a contact microphone to the receiver end of the handset and hooked it up to a small speaker which he hung on the wall. The silence lasted for a few seconds and then the rushing noise resumed. A moment later the radio broke again, and we heard a voice calling, "One-Zero this Zero-Two, over." We all sat up, wide awake now and listening to the radio operator reply, "Zero-Two this One-Zero, over." "One-Zero, this is Zero-Two, we have three men carrying large bundles about 200 meters to our West. They are in the wood line and walking toward our position." I could recognize Clayton's voice; he was whispering as he described the situation. After a few seconds the radio broke squelch again but there was a lot of static and although someone was talking, it sounded like they were whispering, and the static was drowning them out. When the radio went back to the usual noise, the radio operator called back and told them to say again. There was silence and then the radio broke squelch with the same static covering the sound of whispering.

Harry got on the Border Patrol radio and called the interdiction units that were nearby and told them to standby. I told the radio operator to tell Zero-Two to call us after the situation had changed. Just as he was about to transmit, we heard "X-Ray, X-Ray! The smugglers walked into our position and we we're forced to interdict. We have three men in custody and one weapon. Another man ran and we're trying to find him." Harry was on the radio with the Border Patrol helicopter pilots. He told them where the team was and what he knew about what happened. We could hear the helicopter cranking up and then it lifted off and the sound faded into the night.

I picked up the phone and called Tom. I told him what happened, and he said he would be right in. I called Bil and he said he was on the way. Harry was talking to the interdiction units and the helicopter. Little by little information was trickling in. Eventually we found out that everyone was safe and there were three men in custody along with several large bags of what appeared to be marijuana. There was also a pistol that had belonged to one of the smugglers.

Shortly before George's team was compromised, Steve's team had a group of smugglers walk very close past their OP. They were not discovered, and they called in the BP interdiction team that was hidden nearby. The interdiction team was unable to find the smugglers and they got away. As a former cop, Steve was pissed, because if he hadn't been ordered not to interdict, he would have arrested them in a heartbeat; and like all hard chargers, he wanted to get the first load, but George got the glory.

Tom Stephens and I set up the debriefing room. We put a map on the wall and put my video camcorder on a tripod. About thirty minutes later George's team came into the hanger and unloaded. Steve's team was also pulled, and they arrived shortly after. Bil and the Intel guys debriefed George's team first. They clarified what had happened. The team started getting hits from the sensors that E-Systems had given us. Using a TVS-4 Night Observation Device, the team was able to see them from a long way off. They were able to aim the TVS-4 by using the compass azimuth from their position to the sensor. Four men, three of whom were carrying large bundles on their backs, appeared about 500 yards from where they were hidden in the brush facing the river.

The group of men walked slowly toward them and as they approached, Clayton took photographs of the smugglers using my camera that I had given to him on a whim an hour before they rolled out of the hanger to start their mission. He took the first ever military surveillance photos of drug traffickers and started a quest in our detachment for ID quality nighttime photographs and video. That goal was to direct all our efforts in the future. The group continued to walk right into their OP. The BP agent stood up and identified himself in Spanish and told them to freeze. The leader of the smugglers wasn't carrying a bundle. He began reaching into his open jacket. "I was pretty sure he was reaching for a weapon, so I pushed him onto the ground and handcuffed him," said Mike, the BP agent. "Then I reached into his jacket and found a .357 Magnum pistol." Mike displayed the skill and courage that BP agents consider as part of the job. A soldier would have received a medal for what the BP considered routine work. Bil asked Mike if the soldiers on his team had done what they had been trained to do in isolation and he said, "Yes, they did, I don't know of anyone that could have done any better."

Needless to say, we were happy. In less than 24 hours our two teams had been inserted into one of the hottest spots on the border, witnessed two drug trafficking events, seized 269 pounds of marijuana, a .357 Magnum Revolver and made three arrests. We reinserted the teams the next night and continued for the next two weeks but the word was out, and no more smuggling was observed. Harry was happy with the way everything was going but decided to bring an end to the operation a few days before we were scheduled to stop and this initiated a series of unfortunate events that would result in our not coming back to the Rio Grande Valley for eight years thanks to Bil's lack of flexibility.

Bil had arranged for a General to come to McAllen to glad hand the Border Patrol and give out some plaques. When he found out that Harry was shutting down the operation early, he complained to his buddy at Operation Alliance who called down to McAllen and complained to Chief Reyes who asked Harry what was going on.

Harry was infuriated when he learned that his decision to shut things down had gotten out of sector and into the ears of Operation Alliance and he decided that SOD would never work again for the BP in McAllen and we didn't. I didn't know at the time that any of this had happened and the only way we found out was through one of our friends in the BP who told what had happened, a few months later.

'Steve' James F Tom Stephens

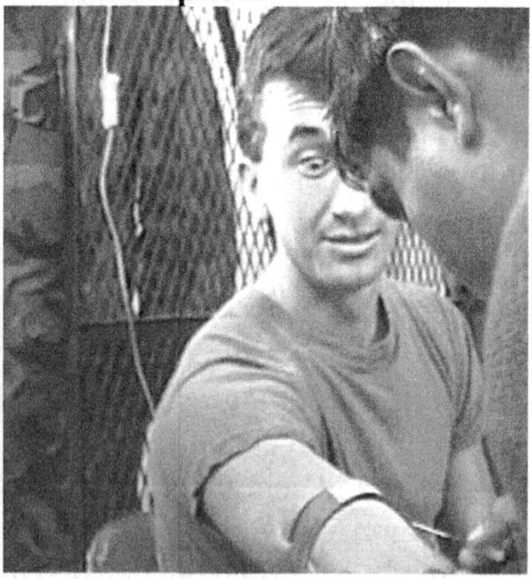

Clayton Hardee George Huseman

Unity VI-Mission Briefing Lee Justice-USBP/SFC Thomas Payne-SOD

SOD Team Leaders George Huseman and ‘Steve.’

SGT Chris Jay

SPC Clayton Hardee

SGT Bill Sparks/BP Agent Yanez **SPC James F**

SFC Payne gives final briefing for George Huseman's team

Team with TVS-4 and SATCOM Radio

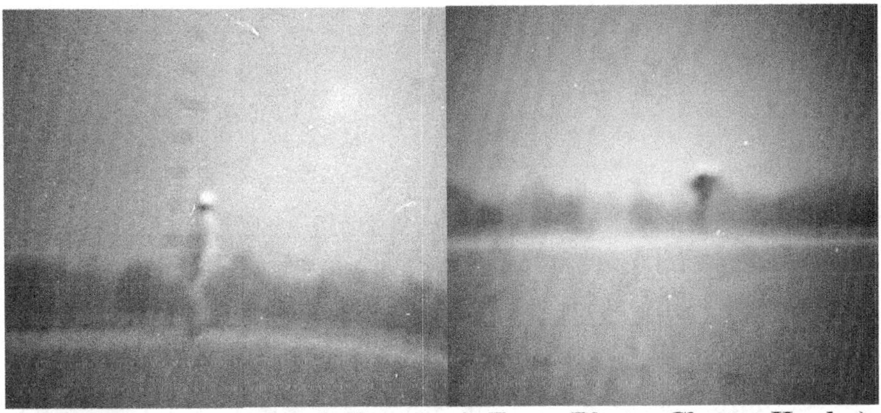

Drug Smugglers approach Huseman's Team (Photos-Clayton Hardee)

Debriefing BPA Mike Van Hook with Hardee, Huseman, and Jay

Debriefing BPA Yanez with 'Steve', James F, and Bill Sparks

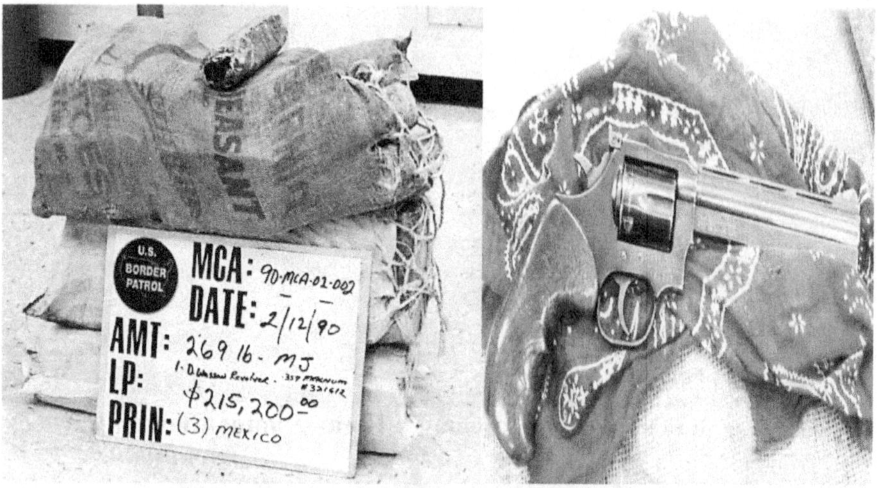

Marijuana, .357 Dan Wesson, 3 Arrests

General Area of Operation Unity VI

O Jehovah, deliver my soul from lying lips, from a tongue of deceit.
Psalm 120:2

Chapter 10-March 1990

Unity 7-8-9

Shortly after we returned from Unity 6, George C showed up at the team room and told us he needed some help to unload the truck. We all went outside and unloaded some small wooden crates and one large crate. "I'll be back later with the paperwork," he told Tom. We took the boxes into the team room and started to open them. It was like Christmas morning. The guys were chattering like magpies as they tried to get the boxes opened. In the large box was something that looked like part of a crew-served ray-gun used by Commander Corey and the Space Patrol. The smaller boxes were opened and we beheld the future.

There were six LST 5b Motorola Satellite Radios with KY-57 Encryption Devices and two URC 110 Base Stations. There were manuals that came with the radios and several of the guys who were more interested in commo started reading. The rest of the guys were beaming. "I'll bet nobody has these except Delta," one of them said. "That's BS," said another, "we had them in battalion." "No way, man," said another, "nobody has these." They went on and on, arguing about which units were high speed and what units weren't while trying to figure out how to set up these radios.

Eventually the large antenna was assembled. The guys wanted to put it up on the roof, but Tom told them no. He didn't want anyone to notice us which was the opposite of what some of the other guys wanted. Where's the fun in being cool if no one knows you're cool?

We didn't get any batteries with the radios so we couldn't try them out. Tom assigned Tom Stephens to store them in our supply room and told one of the men who was reading a manual to make a list of what we had and what we needed to get the radios operational.

CARPE PER DIEM!

Another issue raised its head at this time. Up to now we had always stayed at the SMU Headquarters near Fort Bliss in El Paso when we conducted these missions and were fed in the Army mess hall nearby. This last time we were staying in a Border Patrol hanger with no facilities for showering, sleeping, or eating. Bil decided that the HQ people like Tom and me would sleep two to a room in a nearby hotel where all the National Guard Cargo Inspectors were staying. Since we were working 12-hour shifts, we wouldn't have to listen to each other snore.

The problem was what to do for the team guys when they came out of the field. The higher headquarters people wanted them to sleep in the hanger on cots and eat MRE's between missions. Tom was telling Bil that this wouldn't work; the hanger had a small restroom and the toilet would stop up when it was used too frequently, the old sewer system being the problem.

When the guys heard about the situation they complained that the program wouldn't spend any money for them to rest after living like animals on the river in the winter time while the cargo inspectors got to live in single person rooms. The issue was per diem. The Task Force bean counters didn't want to give our guys any per diem. Per Diem in the Valley was $26.00 a day and the hotel rooms were $40.00 a day.

The battle went back and forth. George was spending most of his free time reading the regulations on travel and soon became our in-house expert. He would give Tom some ammunition and Tom would reenter the fray for another round. One of the guys came in one morning and told everyone at the morning meeting, "We've got to seize the Per Diem, Carpe per Diem!" a take-off from the name of a popular movie of the time. 'Carpe per Diem!' became a popular phrase around the team room for a long time and we decided to make it our unit motto.

Bil and I were talking about Per Diem one day and he said, "When I was in the Special Forces, I didn't even go to the latrine unless I was drawing max per diem." Eventually Bil took our side of the argument and because he was running things, he told Uncle Bennie that the guys needed to be able to stay in hotel rooms and draw per diem on the days that they weren't in the field. The Colonel said ok and Tom told the team guys. This made all of them happy but George.

He agreed to go along with it while he continued to research the regulations. George had played a big role in the teams getting per diem every time they went out of town, even while they were in the field; but that wasn't good enough for him and he found in the regulation that the guys who had to go to Houston for drill one weekend a month were authorized per diem for drill.

One day a clerk brought in a huge stack of travel forms for the guys to fill out and eventually some of them got checks for two or three thousand dollars. George had found even more entitlements but he made a deal with the Colonel that he wouldn't bring them up if the Colonel forgot about George's Government Credit Card bill that included a $500.00 dollar horse saddle which he had bought while on a mission, in direct violation of the rules for credit card use which only authorized charges for duty related food, lodging, and travel.

When George told the guys about getting the saddle, he looked like the twin brother of Sean Penn, the actor. He got this wide-eyed look with a big smile and said, "It was a great saddle, I couldn't pass it up. Besides, I didn't know it was against the rules to use your Government Credit Card to buy a saddle." George would go on in this vein until we would be laughing our heads off. "Besides, I always pay my bill on time and another thing..." George would go on and on until Tom would tell him to shut up and get back to business. George knew everything there was to know about travel rules and we all knew it. He also knew how to wear you out.

There have been a number of scandals regarding travel vouchers and per diem payments for soldiers in special operations units. I guess the optempo and the hardships that the men of these types of units have to endure in the performance of their duties causes some of the men to lose perspective occasionally. Maybe they should pay these soldiers a little more than the clerks who get the same pay and go home to their families and clean sheets every night. The Per Diem issue went in our favor, but the Government Credit Card issue was another monkey which would return periodically and bite us on the butt.

We started showing signs of bureaucratic success which was signified by permission to hire more people. SGT Jimmy K joined us from G Company. He was a quiet guy, married, and he would eventually become our administrative sergeant. He had served on the second Unity operation and was on the team that managed to avoid detection by the San Vicente Cavalry who were alerted by the beep of one of the piece-of-junk radios the team was issued. We were also assigned a medic. Nobody knew him as he came from a medical unit.

Senator Phil Graham, a senator from Texas was quoted as saying that the Texas National Guard was dressing up in 'Cactus Suits' and hiding along the border looking for drug smugglers. Jimmy K's wife made some t-shirts with the figure of a cactus with two eye holes and the objective lenses of some night vision goggles sticking out. Across the top was written, Carpe per Diem. I really loved that shirt. I never wore it because I didn't want to wear it out. My wife noticed that I never wore it and donated it to a church fund-raising garage sale.

Operation Unity 7-Marfa

In April of 1990 we got a request for an operation in the Marfa Sector. The Big Bend is a beautiful bit of the old West. The location of many a John Wayne movie although most of them were actually shot in Utah, the Big Bend was everything I hoped for as far as an adventure setting. Tom Stephens and I went down and rode around with an agent and checked out some sites. One of them was so far it took us all day to drive there and back. After we finished the ground survey we returned to Austin, arranging to come back and take aerial photos.

Tom sent us back to Marfa to do more site selection and take the aerial photos. We stayed in Alpine at a nice motel on the outskirts of town. There was a mail order distributer of interesting military history books there named Kent Spaulding and I wanted to see if we could find his place. Sure enough, on a small side street in the center of town we found a store front with windows painted so you couldn't see inside. There was a small travel decal for a hotel in Saigon on the door, so I figured that this was it. It was late afternoon when we got there but we could see a light on inside, so we knocked on the door. After a minute an older man opened the door slightly and asked us what we wanted.

I told him we were interested in buying some books about military history. He hesitated for a moment and then told us that he only sold books by mail order. "Please sir," I said. "We've come a long way and we don't know when we will have opportunity to order anything by mail." He must have been afraid of some people because he looked us over carefully, then glanced up and down the street before finally relenting and letting us into his treasure cave.

He took us to the area where the books were and told us we had about 30 minutes before he closed. I started scanning the shelves and grabbing books that I had heard of and some that looked interesting. I found several about the Rhodesian War and the special ops units that had fought there. Tom found a book about SF and there was a picture of him holding a Russian machine gun and wearing a Green Beret.

Mr. Spaulding was hovering nearby, keeping an eye on us as either a shoplifting prevention measure or a self-preservation move in case we were some of hit men that were featured in some of the books he was selling. "Come here and look at this," Tom said. Spaulding and I moved over to where Tom was standing. He held the book open and showed us his picture. Mr. Spaulding looked at the picture, and then he looked at Tom. He looked at the picture again and then he looked back at Tom. Finally, he couldn't stand it anymore. He took a step back and looked at both of us, "Who are you guys?" he asked in a plaintive voice. Well, who were we? Good question. Neither of us said anything for a moment and the atmosphere got pretty tense in the small room. Finally, I told him, "we're just passing through; can I buy these books?" He looked us over again and finally said "OK" and took the books from my hand. I paid for those treasures with the last of my cash.

I figured I would use my government Diners Club card for the rest of my expenses on the trip, bad figuring on my part. We went back to the motel and ate in their restaurant, so we were able to put the meals on our cards. It turned out there were very few places in the Big Bend that took Diners Club, a credit card whose time had passed.

The Plan

The next morning, we left and drove to Marfa to meet with the BP to do site selection. We met the Sector chief and his deputy. The Deputy Chief Patrol Agent was a man named Tom W. He was a seasoned BP agent. He wanted to do some training with the SMU and he wanted his own sector's ERT to participate. He asked us for aviation support and medical training. I assured him that we would provide him with what he needed. He also wanted to do some land navigation and airmobile training, including rappelling from a helicopter. "After we do the training, which I'm going to invite the media to attend, we'll shut everything down and leave town. Then we'll come back and run a mission on the river." He was thinking tactically which I liked, and he knew that the local drug traffickers had eyes everywhere. Four years before, Pablo

Acosta had been gunned down in a small village just across the river from the Big Bend National Park. Pablo was one of the first of the big time Mexican Drug Lords. His kingdom had been taken over by the Lord of the Skies, Amado Carrillo Fuentes. Tom W's attempt to defeat the intelligence operations of the traffickers might have worked had there not been some moles.

Execution

The Go-rillas came down a few weeks later and flew the river with one of our helicopters. The rest of us drove down with our gear and met together at the Sector HQ. The SMU agents and the Marfa ERT agents showed up and we had a small reunion with some of the guys we had worked with in the past. One of the ERT guys was a jovial fellow named Dave P. He kept us all laughing with his hilarious observations about whatever subject was being batted around. He had been in the Air Force and bonded quickly with our guys. Lee Justice also showed up and I was glad to see him.

There was a black man that the other agents respectfully called 'sir.' There weren't many black men in the Border Patrol, so I asked Lee about him. "That's Louie Barker, an Assistant Chief Patrol Agent from DC. He's originally from Panama." Some of the other SMU agents that showed up were also senior members of the BP and I was impressed that these guys wanted to go and lay-up by the river for some days in hopes of catching some traffickers in the act. They joined the BP for action, and they were still looking for it when other men of their rank were quite happy to remain in the office.

The airmobile training was a big success. The guys all rappelled several times from the helicopter. Then someone did the Australian Rappel, face first instead and that was a challenge that no one could pass on, so they all did it. Most of the guys when they braked would fold over the rope around their waist and they looked like a large sack being lowered on a rope. One of the SMU agents was like an aerial ballet artist when he came down and everyone was envious of his poise. Dave P ended upside down under the helicopter and had to endure some chiding from the others which he took with his usual good humor.

The news people showed up during this training and shot video of the rappelling. Tom W was interviewed on camera and when he saw me taking video of the interview he gave me a brief smile as he spun the cover story that we hoped would result in the word going out that the coast, or in this case, the river, was clear. Our guys were wearing large goggles to keep the dust out of their eyes and the goggles protected their identity from the TV cameras.

After the rappel training we moved back to Sector HQ where we conducted medical training. We all had to give each other IVs. Tom gave me one and I gave one to him. Not my favorite activity especially when Tom told us that he wanted us to be able to give an IV in the dark while wearing night vision goggles. When the training was over for the day, we went to eat at the Thunderbird Café which is on the highway that goes to Alpine in the center of Marfa. This was a gathering place for local ranchers from as far as two hundred miles away. Anything that was going on in West Texas would be discussed here and Tom W wanted a big presence here for our cover story.

Tom and I were joined for dinner by a Guardsman named Lee who had recently been assigned here to be an Intel support to the BP. Lee was about my age and he looked like he had been rode hard and put away wet many times. He was an old time SF guy from the Vietnam era. He knew Bil who got him into the Guard and sent him down to Marfa to do Intel. He lived at the Thunderbird Motel across the highway and kept a large ice chest full of beer in his room. One of the BP agents who worked with him told me that Lee spent most of his off-duty time at the local VFW where he kept the beer from going stale until they closed. Lee quietly told us that there was a lot of shit going on in Marfa, but he didn't elaborate. He disappeared from Marfa not long after this and I never found out what happened. Captain L was his boss and all he would say was that Lee got too close to something there and had to leave town before he was killed.

The next morning, we all met at sector HQ and had a mission briefing. The Sector Intel guy gave us a good overview of what might be encountered at each of the sites we would observe. The plan was to have each team split into two elements. The observation element would consist of one SOD operator and one SOG agent. The interdiction element would be two agents. The team would lay-up together during the day and split up after dark with the observation element watching the crossing and the interdiction element waiting where a good place had been designated for them to stop a vehicle to search it. Tom W gave most of the briefing and at the end he electrified the room when he told us that if anyone was captured, as soon as we knew of it, the rest of us would go into Mexico and rescue him. There was some 'hooahs' and high fives among the men and the morale went sky high. That night the teams were inserted to various locations on the river in the Big Bend. I had gotten sick and was running a high fever that night so Tom sent me to the motel. There were no problems during the infil and for the next four days the teams sat on the river but didn't see any smuggling.

The debriefings revealed that a team on the west side had tried to stop a vehicle coming off the river near their site. When the vehicle didn't stop, they flash-banged it. There were several illegal aliens but no dope, so they let them go. One of the other teams spotted what looked like a major smuggling operation when three pick-up trucks with campers crossed the river at the crossing they were watching. They called the CP on their SATCOM and the BP agent in charge called for an interdiction team that was hiding a few miles away with their vehicles. As soon as he transmitted the order to interdict on the BP repeater, the pick-ups stopped, turned around, and sped back across the river.

After the operation was over, a few months later I heard the guys talking about how one of them had crossed the river with a BP agent and walked about a half mile into Mexico to look at a Mexican Hacienda that had its own airstrip. There was no one there at the time so they looked in some windows and checked the place out before recrossing the river and reminding everyone that 'what happens on the river stays on the river.' Another team reported a blacked-out C-130, apparently there were a lot of things 'going on 'on the border.'

The new SATCOM radios worked well but we wouldn't have them for long. The BP agents were quite interested in where we got them and a few weeks after the operation, Bil told us that they would be kept at SMU HQ so that they could be used by whoever needed them. This was disappointing because we were quite proud of being one of the only units in the military to have these high-speed satellite radios. Ten years later I visited with a BP friend of mine and asked him about those radios. He told me that they were in a pile in a far corner of the supply room, no longer used because more modern satellites and radios had made them obsolete.

April 1990: Minnesota

In April of 1990 the Minnesota National Guard requested our advice regarding their start-up of a Counterdrug Program. George had been in contact with an old Ranger buddy of his who was with the Minnesota National Guard. They were starting a selection and training program and asked if George could come up there and give them the benefit of his experience. The head shed decided otherwise. Uncle Bennie, Captain Smith, and Bil went and they brought me along to represent the surveillance operation.

When we got there, we were introduced to Major John Plaster who was in charge of the program. He was a Special Forces veteran of Vietnam where he had run recon with MACV-SOG. He had developed a sniper school for the Minnesota National Guard which had been quite successful and now he was given command of the new Counterdrug Program. He was using a sniper's stalking course to help in selecting the soldiers that would be used for various tasks on the Minnesota border with Canada, not exactly a hot spot for drug trafficking, but the money was there and many states jumped on the idea.

I found that the sniper school cadre had some skills that would be very useful for our people. They had a block of instruction on tracking and another one on the vectoring of interdiction forces. Unfortunately, however, they were clueless in their overall understanding of operations against drug traffickers in a law enforcement environment.

I offered a few suggestions to a group of instructors and was jacked up by the master sergeant NCOIC for the placement of my rank badge on my patrol cap. It was at least a quarter inch off and he felt that squaring me away in front of the rest of the instructors was the best way to facilitate a flow of information based on our lessons learned that were formed by experience.

I still tried to give a few observations while I absorbed what the course was teaching but I soon realized that ego and regional prejudice would interfere with any learning on their part. This is a common problem in the army and probably any organization; human nature seeks self-glorification at the expense of everything else.

Major Plaster had an alleged former drug smuggler come and give a lecture on his experiences smuggling drugs across the Mexican border into the Big Bend National Park. I was interested to hear what he would say and videotaped his presentation. The guy was wearing a ski mask and seemed pretty proud of his past exploits. He gave enough detail about a certain crossing that I felt that he had been there before. He gave some details about his techniques, tactics, and procedures that I felt would be useful information.

The next time we worked Big Bend, I showed the tape to the BP agents. Lee Justice said it had a lot of information that he felt was valid. One of our Intel people who had recently been hired told me that it was a waste of time and Lee heard him. He came up to me later and we commiserated about the abundance of dummies in the Intel business.

The Minnesota Guard's program was up and running but they would hit a speed bump further down the road when the officer in charge of their surveillance element decided to entertain his troops with a self-made porno video. Someone complained and Major Plaster and the rest of his command were looking for work elsewhere.

May 1990: The Ranger Reunion

The Ranger culture in the Long Range Surveillance units in Texas was very strong. Either you had a tab, or you didn't. If you had a tab, there was a pecking order. At the top were the men who had a 'Ranger Tab' and had served with a Ranger Battalion. These were the 'BATT boys' who had been at 'Battalion.' They 'knew the deal.'

Tom was at the top of the heap in the pecking order. He had the 'Tower of Power' on his left shoulder, the Special Forces tab, the Ranger tab, and the Airborne tab which was part of the unit insignia. He wore Master Parachutist Wings and an Expert Infantryman's Badge on his chest. The only thing he was lacking was a gold star on his jump wings and a wreath around his EIB.

A few years before I met him, Tom started having Ranger reunion parties at his apartment in Austin. He was going to UT and was a member of G Company at the time. There were several of his 2nd Ranger Battalion buddies in G Company and a number of other BATT boys along with some tab wearers. All tab wearers were invited and most of them came. They would all get drunk and smoke big cigars and do some ball sniffing. Tom had a number of these reunions but the one in 1990 was to have major repercussions on our operations.

When Tom told me he was going to be having his annual Ranger reunion that weekend, I asked him if I could come by and see what they did. He laughed and told me no way. I waited for Monday morning to come so I could hear all about it. Monday morning came and when we had time, he told me all about the festivities. The usual events had taken place, they got drunk, they smoked cigars, they sniffed balls, etc. "TN showed up and the guys gave him hell and he got pissed off at us," he said with a laugh. "What happened?" I wanted to know. 'TN' was one of the guys in LRSD who had been with the 82d Airborne Division at Grenada and his claim to fame was his picture on the cover of Soldier of Fortune magazine. The picture showed him sitting in the back of a vehicle holding his rifle. 'TN' had once told me about his secret life in Central America doing 'special ops' somewhere, sometime.

He had just come back from Ranger school, proudly wearing his tab, and had gone to Tom's Ranger reunion expecting to be welcomed into the ranks of the elite but instead he found himself being badgered by everyone there. He hadn't been in 'Battalion' he was just a National Guardsman who went to Ranger school. Yeah, he had passed, but he still didn't know the deal. He argued that his exploits on Grenada and other places that he really couldn't talk about made him qualified to sit at the table. They reminded him that he didn't have a CIB from Grenada and that his records said that he was a Light Vehicle Mechanic.

"This really pissed him off," Tom said. "He started cussing at us and said that he was going to become an officer and when he did, he was going to get even with all of us. We all started laughing at him and he stormed out which made us all laugh even harder." I didn't think that much about the incident, but I did wonder how he was going to get even. I asked Tom if we should worry about it and he just laughed it off, "Screw him if he can't take a joke."

Colonel Charlie Beckwith

Before I had joined Counterdrug, I had asked Charlie Beckwith, the creator of Delta Force and a Long Range Patrol expert to come and talk to LRSD and one weekend he did. He had retired from the Army after the debacle at Desert One and was now living in Austin. He had started a security company and hired off-duty APD officers. When I was a cop,

I used to eat Sunday morning breakfast with him and Alan at a small Mexican restaurant on the East side.

When Charlie came to our Saturday morning drill, he spoke with us about some important principals that needed to be followed when conducting Long Range Patrol operations. Selecting the right kind of people was one, "I'd rather have two good men, than a whole boat of shit heads," and "some leaders wish they just weren't there" were at the top of his list. He was very adamant about the role of NCOs in these types of units; they were the leaders, he said and they bettor do their jobs or the team would fail.

I continued to see him now and then and as we had a mutual interest in military history, we exchanged books on the subject and discussed their merits. When he found out I was working at Counterdrug he showed some interest. I introduced Tom to him, and Tom told him that he was interested in joining the CIA. Charlie agreed to talk to some people he knew there on Tom's behalf.

The election for Governor of Texas was coming up and Charlie was supporting Clayton Williams, a Texas oilman and entrepreneur. He was also angling to become the 'Drug Czar' of Texas and get Bil's job. Bil and Charlie were both members of the local Special Forces Association and had not seen eye to eye about anything.

Bil had been elected president of the chapter and that pissed off Charlie, who being a retired colonel, felt that he was the rightful leader of the association. When he found out that Bil was running the Counterdrug program he was pissed again because Bil was a retired master sergeant and he was a colonel. Charlie asked Tom and I to give him a briefing about Counterdrug Operations. We knew that if we asked Bil for permission we'd be fired in a heartbeat.

With the election coming up and the good possibility that Charlie's man would be the next governor, we thought it would be prudent to meet with Charlie and give him a general overview of the drug war without any operational details. Charlie had helped us by coming and talking to the LRSD and he was trying to get Tom an interview with the CIA so we hoped that our disloyalty to Bil would remain a secret.

Charlie had set up an office in the downtown area of Austin in a new high rise near the river. We met him there one day during our lunch break. He had us sit down and told us what he was going to do after Clayton Williams was elected. Of course, he was going to fire that

SOB, Bil, then he was going to activate the Texas State Guard and put them on the border fighting the drug traffickers. He started shouting, "You guys have spent a million dollars and you haven't killed anyone!"

I was stunned by the thought of the Texas State Guard doing anything other than directing traffic at the camp once a year when we had 'Muster Day.' Tom and I just sat there while Charlie briefed us about how he was going to run the drug war. He went on for a while then he asked us what we thought. I had known him for several years and we had had some lively discussions about military history, so I dove in. "Sir, have you ever seen the Texas State Guard?' I asked him. "They're a bunch of overweight butterballs who retired from something. They march around once a month and say stupid things to each other."

Now it was his turn to be shocked. Of course, he didn't like the butterball comment because he had put on a little weight since he retired. I went on to tell him that while the thought of killing the drug traffickers was an enticing idea, in this war the military was in support and in our case, the Border Patrol was calling the shots, which meant no shots unless they could be justified by their use of force policies.

"They're a law enforcement organization and they don't have rules of engagement; they have a use of force policy that is the same as the Texas Penal Code that all cops have to follow." This wasn't news to him because he had worked with the FBI and other agencies when he was running Delta so I wondered if he really had started believing the 'Kill em all and let God sort em out' sign that he had kept on his desk for a while when he first came to Austin.

Tom told him that we didn't make the policies, we just carried them out and that if Charlie was put in charge of the drug war in Texas we would be glad to escalate our efforts to whatever level he thought was appropriate. This seemed to satisfy him that we would be happy working for him, and he thanked us for coming by. A few weeks before the election, Clayton Williams decided to have a big cookout for the press at his West Texas ranch. They were all sitting around the campfire when it started to drizzle. Clayton laughed and told them that it was like rape, "If it's inevitable, just relax and enjoy it." This remark made the front page of a lot of newspapers in Texas the next day. Even though he was way ahead of Ann Richards in the polls at that point, he lost the election, so Charlie never became the drug Czar of Texas.

Del Rio

The next thing to come up was a mission request. The Del Rio Sector wanted us to come back in May and work the river in the Eagle Pass area, downriver from Del Rio. Tom and I had done site selection in November of last year and now the BP was ready to do it. He told me that I was going to run the next operation so that he would be able to use me to run a mission if we had two running at the same time. I began working on the target folders and writing the OPLAN.

Intelligence and Intrigue

Meanwhile, Captain L was busy building his own empire. He had most of the MOS qualified Intel officers and NCOs in the state working for him doing 'secret' work. His people were embedded with various law enforcement agencies around the state. They were working for the cops and sending case specific information to his small HQ element in Austin where he contemplated building his own statewide data base.

He and his secret squirrels were so elated with their growth that they began to see themselves as having more power than CD HQ and so they plotted to get rid of Bil. They had someone with access to the National Crime Information Center run Bil's criminal history where they found a DWI conviction from years before. They planned to use this in an assault on Bil's suitability as the CD Coordinator.

They got caught and before their heads stopped spinning, they were all reassigned to work in various law enforcement agencies near Austin. Captain L spent the next 15 years working as an Intel analyst for the Texas Department of Public Safety. Sitting next to him and doing the same work was a sergeant who retired as a sergeant major. The Guard doesn't believe in capital punishment for its officers, so he was promoted several times and retired as a LTC.

Bil had to deal with this kind of intrigue on a routine basis. Someone in the HQ started a rumor that he was sleeping with the personnel officer, a captain that was involved in an attempt to get rid of a female major who was our Task Force XO. Bil brought it up in the staff meeting one morning by asking, "Captain ___, am I sleeping with you?" They say the room was silent for a long while until Bil broke the silence by telling them that he had a lawyer and if the ones spreading rumors about him didn't stop, he would sue them.

The captain lost her fight with the Major and disappeared from the HQ, but she didn't give up and went to the press. A Dallas TV station sent a female reporter down to interview Bil and several colonels from the State Headquarters. Bil asked me to set up several video cameras to tape the interview and when the reporter came into the room and saw them, she knew she was in an uphill fight.

"Oh, a three-camera shoot, I didn't know that this was that important." She looked at me, so I told her that we were getting practice videotaping interviews for our own training videos. She smirked at that one and went into her interview. She was pleasant for the first few questions and when she thought she had relaxed everyone, she turned to Bil, and thrust her right index finger in his face in 'Je'cuse' style and said, "Mr. Enney, is it true that you lost classified papers when you were arrested for drunken behavior?" Bil took several moments to consider the question before answering, "No." This went on until she gave up in disgust and we never heard any more about the news story.

Unity 8-9

When I got back from Minnesota the detachment was in high gear getting ready for Operation Unity 8-9. These would be back-to-back Cecil B. DeMille productions. We would put five teams on the river with one BP agent per team. The SMU had just finished a selection and training class and they wanted to put their new guys on this mission. We were extremely busy getting everything ready. The night before we left for isolation I worked until about 10:00 PM.

The Storm Before The Storm

It had been windy with some rain all afternoon but when I left the team room it was really blowing. I went back inside and covered my Macintosh computer with a piece of plastic and went back and got in my car. As I drove through the deserted camp, I could see the limbs of the large oak trees waving around like they were twigs.

I got home that night and went to bed while a storm passed over Austin with much clamor from the lightning and thunder that came with it. The next morning, I went into work and when I got to the team room the guys were standing around outside. "What's up?" I asked Steve. "There was a tornado touchdown here last night and it tore the roof off the team room," he said. I ran inside to see if my computer could be salvaged with all my work on it. There was debris here and there with pieces of wood laying on desks and large puddles of water on the floor. When I got to my cubicle there were papers strewn around and debris everywhere but sitting on my desk still covered with the plastic sheet was my Macintosh and my floppy disks.

A little later George C showed up and was talking to Tom. Tom turned to me and said, "You're in charge, what do you want us to do?" George C said, "I've got a place where you can store the things you're not taking." I called Steve and George over and told them that Steve's group would load our gear for the operation and that George's group should start moving our gear to the storage space that George C had found. Tom went over the schedule for the rest of the day with me and then he pitched in and helped move our things into storage. I was flying high when Steve came up to me and asked, "How can you be so calm? Our team room was just destroyed by a tornado, and you're running an operation for the first time!" I just smiled; "it ain't nothin but a thang; we operatin now."

Isolation

When the rest of the part time Guard soldiers arrived, we loaded them on a bus and moved to the isolation facility at Camp Swift. The BP agents had flown in from El Paso on our C-26 and they were picked up at the airport and driven to the camp. We followed the same isolation schedule with briefings and classes for the next three days.

During the live-fire scenarios at the range, Tom and an operator got in an argument about slings on rifles. He wanted to carry his rifle by using a sling which would allow him to hold his rifle with one hand while he carried a 5-gallon water can with his other hand. Tom told him to unsling his rifle which seemed unreasonable to me, but I didn't say anything, and he complied with Tom's order, but I could see that he was mad. The pace was fast and by the time of the briefbacks I had forgotten all about it, but they hadn't.

The move to Del Rio was by helicopter. Tom drove there with the forward support element and I flew with the main body. I had an intercom headset and was able to hear the radio traffic between the helicopters. We flew across the south Texas landscape and I followed our route on a JOG map. I noticed a notation on the map indicating a wrecked airplane. I was looking for it when I heard the pilot of one of the other helicopters report that he had sighted a wrecked airplane and was going down to search for survivors.

"Negative," I said after pushing the transmit button. "That's an old wreck" We were on a tight time schedule and I didn't want us to lose any time looking at old plane wrecks. This started a series of outraged transmissions from the pilot who had sighted the wreck. I was the mission commander, but I was a sergeant and the pilots were warrant officers and officers and they weren't used to taking orders from a sergeant. The flight commander was flying the helicopter I was in so I told him that the plane we were passing over was an old wreck noted on my JOG map and that we needed to press on to keep our time schedule. He agreed and nixed the search and rescue attempt.

We landed on time at the Del Rio International Airport. The pilots were guided onto the tarmac in front of the Border Patrol Air Operations. The troops filed into the hanger where the vehicles were waiting to drive them to their release points in the Eagle Pass area. They loaded up and left at 15-minute intervals. Tom told me later that soon after they were driving down the highway, the driver, an Eagle Pass BP agent said, "I hope you guys didn't think that this operation was going to be a secret. The station chief announced it to the entire station and their families at the station picnic last month."

I settled in at Air Operations. Tom had set up the CP and LT TN and his side kick, 'Kato,' who both worked as Intel geeks with the Del Rio Sector, joined us for the duration of the mission. Bil had made LT TN the OIC or officer in charge of the operation but he told him that we were running the mission and to leave us alone.

He was still offended by what had happened at Tom's Ranger reunion and they weren't saying much to each other. TN had just been commissioned as a second lieutenant and was feeling the power so he wasn't happy with this arrangement but there was nothing he could do about it, or so we thought.

Kato got on my bad side right away by giving me a Latin lesson regarding our unit motto, 'Carpe Per Diem.' He insisted that it was bad Latin. I was an ex-altar boy from the Catholic Church and had been required to memorize many Latin phrases, so I felt sure enough about the rightness of our motto to tell him to shut up.

Many Intel people feel that because they work in 'Intelligence' they must be intelligent and among them some are constantly working to get others to recognize their 'intelligence' and honor them. I was having none of it and so 'Kato' and I stayed away from talking to each other except when the mission required it. TN and Kato were inserted into the duty roster and Kato took a turn monitoring the radio during the day when nothing was going on.

We had one commo check when Steve's team failed to check in and for some reason, instead of waiting 24 hours, I initiated emergency re-supply that night. Tom drove out to where Steve's team was and met Steve at the right place. Steve was totally surprised by Tom being there and explained that they had just missed their commo check and that there was nothing wrong with their radio. Tom returned to Air Operations and we continued monitoring the radio and preparing for the next event which would be resupply the next night.

I left the CP each morning about 9:00AM to get some sleep at the hotel. I had instructed Kato to leave the squelch on the radio. This meant that there was a constant 'rushing' sound coming out of the speaker. As long as you heard it you knew the radio was working and you could hear weaker signals. He had turned the squelch off on his first shift and I had turned it back on and told him to leave it on.

When I returned to the CP that afternoon, I noticed that the radio was silent. I walked over to the radio where Kato was sitting, looking off into space. The squelch was off. I turned the squelch knob, but I still couldn't hear any noise. Then I saw that the power light was off. I tried to turn the radio on, but it was on, so I turned it off and then back on again, there was nothing. The lights were on in the hanger, but the radio wasn't getting power. I followed the power cable to the extension cord and then I followed the extension cord across the floor of the hanger to the receptacle; it wasn't plugged in. Someone had tripped over the cable sometime that morning and the lifeline to our teams had been cut for nobody knew how long.

I plugged the power cable back in and got the radio back up. I was infuriated with Kato and Lt TN who protected him. I told them both that the radio needed to have the squelch on at all times and if they didn't like it, they should leave. I decided to sleep on cot in the hanger from then on because I couldn't trust these guys to do the right thing. I called all the teams during the next BTB but got no replies. Either they were all ok or they were in trouble, but we wouldn't know anything until the next mandatory listening time when they would make their next commo check.

One of the Guard soldiers, a brand new private, was assigned to work in the CP. He took his advanced travel pay that was for his food and spent it all on skin magazines. He told us that he needed more money and when we found out where his money went Tom gave him some MRE's and told him to eat up. There were a few petty arguments between some of the other Guard soldiers that required occasional interventions and all of these things kept me busy until it was time to pull the teams and debrief them.

The Storm Clouds Gather

At the end of the first iteration, Tom told me that he would be going to the field with a team. That gave me pause and I asked him if this was a good idea. "I have to go to the field with these guys or they won't respect me, besides, I want to see how they operate out there." "What about Bil," I asked, "are you going to tell him?" Tom said, "Bill has approved me to go on an operation." I could see that Tom had made up his mind, so we pressed on with the task at hand and prepared for the debriefs and a day of rest for the teams before we sent them back out for the second phase of the operation.

When we pulled the teams out and debriefed them we found that there had been some personality clashes so we decided to keep a few of the trouble makers in the CP for the second phase and replace them with some of the ones who had been working in the CP.

The SMU agents went to Tom and demanded that he give them the keys to the Alliance Blazer and after a brief argument Tom gave them the keys and they took off. We kept everyone else in the hanger and shuttled a few at a time to the hotel rooms we were using and let everyone take a shower. We had a bunch of cots and the guys slept on them the next night. Everyone was told to be at the mission briefing the next day at 2PM. The agents showed up late, but we got everything taken care of before the BP vehicles showed up to drive the teams in for phase two. Tom was tac'ed out with camo paint and his field gear. He had a smile on his face because he was going to the field, but the team leader didn't have the same smile.

Since Tom was going to be gone, I decided to make a change to the overall plan. Clayton and James F were watching a crossing 200 yards away from the Eagle Pass BP station. Their BP agent was from Eagle Pass and wanted to make their situation reports (SITREPS) to the dispatcher at the station using his BP walkie rather than using the SATCOM. They were in a deep gully where they couldn't get line of sight to the satellite without moving the radio a long way up the gully until they had a clear sky. I knew that the dispatchers at the BP station were always on the air, so I gave them the OK.

The teams went in and after they all made their initial entry reports the CP settled into the regular routine. I went to the hotel the next morning when LT TN came in. I had replaced Kato on the shift roster with a reliable LRS troop. I took a shower and slept for four hours before returning to the CP. After checking on things I went downstairs and opened the door. There stood Bil and he didn't look like he was glad to see me. "What the hell is going on," he said, "where's Tom?" I was startled by his unexpected appearance. "He's in the field with a team," I said. I recovered quickly and invited him upstairs to the CP so I could brief him on the status of the operation. I showed him the map boards which showed the location of each team. I showed him the message log that showed that we had commo with all the teams. He looked everything over and then he told me to follow him outside.

LT TN got up from where he had been observing everything to follow us but Bil told him to stay where he was. We got outside and he said that LT TN had called him and told him that no one was in charge in the CP, that everything was screwed up, and that he was the only one holding things together. I asked Bil if that was what it looked like to him and he said no. He was really angry that Tom had gone to the field with the team without going through the isolation training procedures; I didn't think that Bil considered this a showstopper because he didn't order Tom to come out of the field that night.

“Is there anything else I should know about,” asked Bil, so I told him about the change in the commo plan with Clayton’s team. He got pretty angry with this and even though I argued that the BP dispatchers where more reliable than we were and that they had perfect communications with each other, he told me to get that team back up on the SATCOM immediately. I gave him a ‘roger that’ and returned to the CP and called the BP station at Eagle Pass and asked them to tell the team to call us on the SATCOM.

The team called and I told them that they would have to return to the original commo plan. I was talking to Clayton who rogered my message and then told me that the turtles weren’t going to like it. I had no idea what he was talking about, so I ended the conversation. LT TN was sitting there acting like nothing had happened. I was angry about his treachery but played my cards close to my chest.

The rest of that phase of the operation passed without incident and we pulled the teams in at the appointed time. Bil had gone back to Austin so I directed the debriefings. Tom was happy about his trip to the field until I told him what happened. I asked him to please not get into it with LT TN because we still had another day to go with a Pachanga that evening. After the meal we would send the teams out for nighttime only OPs. Tom bit his tongue and stayed away from TN.

The SMU agents came to Tom and demanded the keys to the Blazer again. Tom had an even livelier argument than before and they practically pulled their pistols on him before he gave in and gave them the keys. He said that they hadn’t filled in the drivers log that kept track of gasoline purchases and mileage and they told him that they were the BP; this was one of their vehicles, and not to worry about it. They jumped in the Blazer and spun the tires in the dirt getting out of there.

The Train Wreck In The Storm

That evening the rest of the group gathered at a picnic area outside of Eagle Pass for a BBQ put on by the Eagle Pass BP agents. Our SMU agents with the Blazer showed up late with six packs of beer and at least one young lady of questionable immigration status. I think they had been across the border at ‘Boy's Town,’ but I could be wrong because I was in Del Rio. Tom, who was there evaluated the situation, canceled the night's planned activities and called Bil. Bil must have called his buddy at Operation Alliance that night. This stirred up a hornet's nest in El Paso. The fact that their freshly minted agents had driven a BP vehicle into Mexico and brought a number of illegal aliens who were prostitutes across the river to a 'secret' counterdrug operation function didn't look good for the SMU's reputation. If these allegations were to get out of Eagle Pass, they could ruin the careers of the five agents and cast a shadow over the SMU itself at a time when they were trying to expand. Once again, Bil had failed to grasp the difference between a military organization and Law Enforcement; two different organizations with two different cultures.

The next afternoon as we were preparing to go home the next day, Bob appeared out of nowhere. He came up to me and he wasn't fooling around. "Did you have anything to do with this?" he asked. I didn't really know what was going on because I was involved with getting everything ready to move back to Austin. "No," I told him, "What's wrong?" He just glared at me and said that I better not have had anything to do with this and walked away. There was a lot of activity in the hanger and recovery of equipment that required my attention, so I soon forgot about Bob and his amigos. He gathered all of the agents together and had a big pow-wow while the rest of us stayed away. The BP agents left that evening without much fanfare.

The next day we got everything loaded on the vehicles and then we moved out to the flight line and loaded everyone on the helicopters for the flight back to Austin. I was glad this operation was over. It was the first one I had run, and a lot of problems arose during the operation that I hadn't anticipated. I had not demanded that Tom stay in the rear and I had changed the commo plan without permission.

Other than that, the operation had gone as planned so in that sense I had performed my duties successfully and I felt that given the opportunity I could run these types of operations in the future. As we flew north, I noticed that Tom was preoccupied and not joking with the rest of the guys. We flew into some bad weather and had to set down at a small airport for a few hours until it cleared up and we were able to finish the flight and get back to Austin.

Bil had some meetings with the team guys and I noticed one of them visiting Bil's office which he had never done in the past. Then Bil called me in and told me that Uncle Bennie had just fired Tom. I asked him why and he said that Operation Alliance had written a letter that was full of allegations against Tom and members of the teams and that Operation Alliance said that they would not participate in any operations with the Texas National Guard that Tom was involved in.

This was a crushing blow to me. I had become quite close to Tom and considered him to be one of the best soldiers I had ever met. Even though he was younger than me and had no combat experience I was proud to be able to have the opportunity to work with him. We weren't buddies, I don't think he ever came into my house, and I didn't hang around with him after work, but I greatly valued my professional relationship with him and now he was going away.

I asked to speak with Bil out on the porch. When we were alone, I asked him to reconsider this because I knew that Uncle Bennie wouldn't fire Tom unless Bil agreed. Bil said that he wasn't going to put the whole program in jeopardy for Tom's sake. This was a political decision and right and wrong didn't enter into it.

Change Of Command

"Uncle Benny wants to put Captain Akers in charge," he said. "Are you going to put me in Tom's place?" I asked. "No, those guys won't work for you, they all out-rank you and you're a leg." He was right and although I wished I could get a chance to run the detachment I could see that it wasn't going to be me so, "who are you going to make NCOIC?" I asked. "I'm thinking of putting ___ in the driver's seat," he said.

That alarmed me. Now I knew why he had been visiting Bil's office recently. Tom had told me about his trip to the field with him. He had made a list of 19 major violations of the SOP and operations order that ____ had committed in front of Tom in the field. These included things like holding wrestling matches in the hide sight. Apparently, he was doing these things on the first phase of the operation and was too proud to stop when Tom went with them because he didn't want the guys to think he was afraid of Tom's authority.

I proceeded to tell Bil that I was disappointed that he wouldn't let me have a shot at running the detachment but that if he wouldn't give it to me then he needed to give the operation to Steve. I ran down all the problems that Tom noted when he was out with that guy and then I told him that Steve was a much more stable soldier and he was the best person for the job. Bil told me he would take what I said seriously, and we went back to work. The next day Tom out processed. Uncle Bennie had gotten him a job at the academy so I was relieved that he would still have an income. Steve was the new detachment sergeant and Captain Akers was announced as the new commander.

Captain Akers had a tough job ahead of him. He was taking over an ad hoc special operations unit that had just had a melt-down. The men were bewildered by what had happened to Tom and why. I was pissed off about the whole thing but my commitment to the program was strong enough to follow Steve's lead in helping the new CO.

The captain had been an airport cop at the same time I was a hospital cop and we had a lot in common. He was the good friend of Alan who brought me to the church and who had helped me get a job as a police officer. Captain Akers was an old Pathfinder and a member of the airborne mafia at Camp Mabry. The members of Pathfinder Detachment considered themselves as the elite of the airborne because their wartime mission was to parachute into enemy territory first and set up drop zones for larger airborne units to use for an airborne assault. Captain Akers grew up in this unit and wore master jump wings and the flaming torch of a qualified Pathfinder.

He was a country boy and wherever we went, when it was time to eat, he kept an eye out for some BBQ. He did well with the local law enforcement agents who were inclined to eat lunch at the local BBQ as well. When we started working away from the border, he was able to interface with the Texas Narcotics Control Program (TNCP) narcs like he knew them forever and that helped us get our start in that world. His police experience was also helpful because he saw the need for SOD to know more about police things and his major contribution to the development of the unit was his connection with the man who ran the police training school at Texas A+M University in College Station.

He interfaced with the Task Force Commander and left most of the direct contact with the operators to Steve. They worked together on the direction of our organization and training. Captain Akers did well with the troops and gradually things began to return to normal.

Unity 8-9—What Happened on the River

Not long after Tom was fired, I saw the letter from Bowen at Alliance, and I made a copy of it. It had statements by the SMU agents saying that our guys told them that the Guard was going to conduct its own operations and do interdiction and a number of other allegations which just didn't seem to be credible unless someone secretly loaded up their MREs with crack cocaine. The letter did say that Operation Alliance would not work with Tom Payne.

Twenty-two years later, when I showed Tom a draft of this manuscript, he sent me a copy of the letter he wrote to Uncle Bennie that rebutted every charge made against him by Operation Alliance and Bil. Many of these details I had forgotten and have added for the sake of history. History can be useful for those who face the future.

The first problem we faced on Operation Unity 8-9 was that there was no on-scene Border Patrol supervisor in the CP. That was a violation of the procedures in place from the very beginning; there was always a command representative from the law enforcement agency and the National Guard. Since the law enforcement rep was always someone from Operation Alliance, they were the ones who attempted to solve this problem by asking the PAIC of Eagle Pass to provide someone. He couldn't send anyone to sit in Del Rio 24/7, so his station's shift supervisor was given the additional duty of being the Border Patrol's command rep. They were not in the SMU chain and knew little about the details of the command relationship between the Border Patrol and the National Guard. This arrangement left at least one of the local BP agents suspicious that they were being setup as scapegoats if anything went wrong. Bil was aware of this change in plan and approved it.

Tom had been accused of going to the field without going through isolation. During the initial coordination meeting for this operation, Bil told the Border Patrol that Tom and I were the only ones he trusted to run the Unity operations. Later, he told Tom that he wanted him to go back to UT and finish his degree and for me to run the mission so that I could get the experience I would need to run things temporarily while he finished college. This was the reason he went to the field with a team. He had developed an abbreviated isolation procedure after Unity 6 which was approved by Operation Alliance and Bil and which he followed so he was righteously indignant about this charge and rebutted it in his letter to Uncle Bennie.

Tom was accused of lacking in leadership because the maintenance on one of the Chevy Blazers that had been loaned to us by Operation Alliance. Our vehicle situation was ridiculous. Three years later we would have six 4-wheel drive Suburbans with cold plates but now we were using Alliance Blazers with maintenance procedures that were incompatible with our funding procedures.

Tom Stephens paid for several oil changes out of his own pocket before Tom told him to stop. One of the two Blazers was turned in for repairs and the other one was taken from us and given to the helicopter medevac crew. They failed to perform the maintenance and Tom was blamed for this as well.

The incident regarding the SMU agents showing up to a BBQ with a girl or some girls was actually a meal before a nighttime only insertion. Tom saw that they had been drinking and decided that it was a safety thing and called cease fire. When he informed Bil, Bil immediately contacted the big shots at Alliance which resulted in my less than friendly encounter with Bob. Bob had been with the BP for a long time and knew what to do. Under his guidance the agents employed a classic strategy: 'Admit Nothing, Deny Everything, Begin Making Counter-Accusations Immediately,' and it worked.

This was inter-agency politics at its worst. One of the 'dirty' little secrets of police work is the inability of the various agencies to get along. I don't think Bil and Uncle Bennie really understood this. The armed forces all have the same chain of command under the Department of Defense. There is no Department of Law Enforcement, so the various agencies tend to go their own way and cooperation between them is minimal except when based on personal relationships.

This was the second time that Bil's complaint to the cops blew up in his face but rather than put some effort into establishing the truth, he and Uncle Bennie resorted to a typical National Guard maneuver, making Tom the scapegoat but keeping him on the payroll so he wouldn't force the issue. Tom demanded a court-martial and Uncle Bennie refused him, telling him that he knew Tom would be exonerated. He arranged for Tom to remain on active duty at the academy where he trained soldiers preparing to deploy to Operation Desert Storm.

The captain and Steve had their hands full trying to settle us down and get us focused on operations rather than on recrimination. The younger guys who worshiped Tom were totally dismayed and wanted to know what happened. I was very bitter and I didn't hesitate to let Captain Akers know that I thought he was wrong when he said that Tom deserved to be fired because of his incompetence.

The team leader who didn't get the Detachment Sergeant's job was shooting off his mouth to the guys in the team room about how he would handle things. He came to me after Steve counseled him about the list of 19 violations of the SOP that Tom had observed and I had passed to Steve, insisting that he call him on it. "What is it that you don't like about me, Sergeant Murry?" he asked. I was caught off guard because Steve didn't tell me that he was going to tell him that I was the source of the derogatory information that he had assumed was gone when Tom left. I said the first thing that came to mind, "I don't trust you when you go to the field." He looked daggers at me for a moment and then he left.

August 1990-Saigon 90

On the 2nd of August 1990, a day that would ever after be known among us as 'Saigon 90', we received a phone call from CD HQ and Captain Akers went down to talk with Uncle Bennie. He came back and called for a meeting in the team room. The nut of it was; we were all fired. "They ran out of money so everyone has to give me their government credit card and ID card. When you go home tonight, start looking for a job; if anything comes up, I'll call you."

This was a major WTF moment as they say. Everyone was speechless. "You're kidding, right, sir" said George with a big grin like he was in on the joke. "I'm as serious as a heart attack," said Captain Akers with a sad look on his face. This situation affected him too. "I want you all to give Jimmy your credit cards and ID cards now," he said, pulling out his wallet and searching for his cards. He laid them on the table and looked at us, so we did the same. He had them collected and he directed Jimmy K to cut them into small pieces with some scissors. After this he said goodbye and left to take care of some business.

Everyone was bummed out. Some of the guys were broke and they had families to support. George started talking about how we needed to get some poster board at the print shop to make signs that we could hold up when we stood on the side of a highway at an intersection. “Mine’s going to say, ‘Will perform surveillance for food, God bless,’” said George. This brought a bit of laughter into the room and before you knew it the team room was the rollicking place it usually was. I was loading my files into paper bags to take to the shredder in our HQ and the other guys were gathering their gear and taking it out to their cars when Captain Akers came in and called another meeting. “Guys,” he said with a sad look on his face, “Division,” the smile starting to break out, “has loaned us a million dollars so we’re back in business.” Suddenly, we all turned and looked at Jimmy, who shrugged and pointed to the small pile of what remained of our credit cards and IDs.

First SOD T-Shirt

The Go-rillas Fly The River

The Thunderbird Restaurant

SMU Agents, Huseman, Sparks, Jay, Kipkak

SMU/SOD Rappel Fest-Marfa, TX

Therefore having put off the lie, speak truth each one with his neighbor, for we are members one of another.
Ephesians 4:25

Chapter 11-October 1990

SOD Continues To Evolve

With Captain Akers, Steve, Tom Stephens and I, all former police officers, in the leadership, this was when we began training the guys to think like cops—not to make arrests but to look at situations through a cop's eyes. I started working on a training program that would require them to understand the Fourth Amendment and their role in providing reliable information that would give a law enforcement agent probable cause to suspect that a crime was in progress or had been committed. I found a good video presentation of illegal search and seizure history and we used some of the material I had saved from the police academy for these classes.

In April, just before we went to Del Rio, Uncle Bennie authorized us to carry side arms and we were issued M1911A1 .45 caliber pistols. Steve put all the guys through a pistol marksmanship program. He was an expert pistolero and had refined his skills with a M1911A1 so that when he handled it you knew that anything he told you was worth remembering. He said we would carry the .45s with an empty chamber and a full magazine in the pistol. The weapons had been with the Guard since before WWII and some of them were pretty loose. "We will use the Israeli draw," he said as he drew his pistol from his civilian holster, pulling back the slide and releasing it as he assumed a classic Weaver stance. He moved so fast that the guys were stunned and one of them asked him to do it again slowly so that we could see it.

Steve holstered his weapon while keeping his eyes where he had been aiming and drew again, slowly going through the steps of taking an unloaded pistol and presenting it ready to fire. When he had demonstrated the Israeli draw, he showed us the proper way to grip the pistol. Tom Stephens and I were his assistants. Tom was also an excellent hand gunner and I had managed to qualify consistently in the Master category at APD; we had shot in competitions with Steve so we were all on the same wavelength.

After the grip came the stance. He taught the Weaver, the modified Weaver and the Turret stance. Basic marksmanship principals followed: front sight focus and trigger squeeze; then he went through magazine changes showing that hands find hands. The last thing he went over were the immediate action malfunction drills: Tap, rack, bang for failure to fire, sweep and rack for stovepipes, and drop the magazine, clear the weapon, and insert a magazine for failure to feed.

Steve issued some dummy rounds and had everybody practice these drills for the rest of the day and told us we would be shooting some bullets tomorrow. The morale was improving rapidly and the fact that our new team room was in the only indoor range at the Camp led the guys to believe we would be shooting all the time.

This was the beginning of SOD's emphasis on weapons competence. I had advocated this from the beginning. My experience as a police officer taught me that if you were confident of your ability to use your weapon if needed, you would probably never need it in this kind of work. Of course, you would have to learn how to convince your adversary to comply with your commands while maintaining your readiness to shoot them if they threatened deadly force. Role playing scenarios were incorporated in all our training.

Captain Akers had a friend named Doug who was running a law enforcement training program at Texas A+M in College Station. He made a deal with Doug and sent the boys to the advanced Special Weapons and Tactics or 'SWAT' school. They had a great time mixing with the elite of the law enforcement agencies who sent their people there for training. Most of the cops in the class were from the San Antonio PD SWAT team and they were a rowdy bunch. George was a big hit with them and to show him some special love, on the last day, the San Antonio cops grabbed him and held him on the ground while they pulled his pants off and one of their number bit him hard on his butt. The guys got good marks at the course and this was the start of a long-term relationship with Doug and his school. Soon we were teaching some of the classes in a five-day course called Tactical Operations for Narcotics Enforcement or TONE.

James F had joined us during Unity 6. He was a good soldier and a good artist. Like most artists he had a liberal temperament. I was amused to hear his amazement when he encountered some of the down-home Texas cop things. His Goth friends would have been outraged with him had they known what he did for a living. He and Clayton were big buddies and shared an intense interest in the Teenage Mutant Ninja Turtles, some cartoon people I knew nothing about.

James brought his brand-new Glock to the team room one morning and someone managed to fire a round across the team room without hitting anybody. I was out of town at the time and didn't hear that he had been transferred out of the detachment until I got back. Because of my own recent failure to control my weapon, I wondered if I also should be sent away. I went to Steve and he told me that it had happened before SOD and that he, Captain Akers, and Bil were happy with my work and not to mention it to anyone.

Unity 10

We were alerted for another mission to the Big Bend. This time we would work with the SMU and the Marfa Sector SOG agents. JTF-6 was involved and an A-Team from the 1st SF Group was added to the operation. The plan was for the SF guys to occupy OPs in the Big Bend National Park with the SMU and we would operate on the east and west sides of the park with the Marfa SOG. I spent a lot of time going down to Marfa and updating our target folders for the various crossings on the river. The SF team would be using our target folders; which is always a bad idea, but, because of various bureaucratic decisions they were unable to send their own people to conduct site selection.

Traveling Men

I met Bil and Colonel K in El Paso. The colonel was an older man and Uncle Bennie sent him out to be 'in charge' of the operation. Since we were working with an active duty SF team, who were overseen by their company commander, a major, we needed a colonel. Bil took me and the colonel out to dinner one night. He wanted to go to some steak house on Franklin Mountain which overlooks Fort Bliss. We were driving along the road looking for the place when I saw a brightly lit building and said, "There it is, Bil." He laughed and kept on driving. "That's where they cut your heart out and eat it," he said.

Then he turned to Colonel K and asked, "Are you a traveling man?" The colonel immediately replied with, "No I'm not and I don't want to talk about it." The fear in his voice discouraged any more talk about 'it' and Bil found the restaurant a few minutes later. We ate and returned to the motel. It wasn't until after the colonel said goodnight that I was able to ask Bil about the place where they cut your heart out and ate it and what that had to do with being a traveling man which was the title of one of my favorite songs by Ricky Nelson.

Bil then explained to me the Masonic 'secret society' and that many of the senior officers in the Texas National Guard belonged to it. He said that they all took care of each other based on their oaths and secret rituals down at the lodge. I was disturbed by this and said that the military owed their allegiance to the constitution and army regulations and that those secret societies had no place in our ranks. Bil looked at me and laughed, "You've got a lot to learn, don't worry about it; they still need us to do the work."

Mission planning continued the next day. We, the soldiers, would be driving the infiltration vehicles this time, and it would be a long drive because it was decided that the isolation would take place at Fort Bliss and that there would be no presence in the Marfa area for security reasons. The FSE would be in Fort Stockton where we would keep our MEDEVAC helicopter at the local airport. Steve decided that he would remain in the CP at Fort Bliss and I would run the FSE at Fort Stockton.

The 1st Special Forces Group

We moved the detachment into some barracks in the Logan Heights section of Fort Bliss where we set up our operations base. The SF guys showed up and we went through the usual ball sniffing, which was made more difficult because the SF were all required to wear sterile BDUs and we complied,. This was someone's idea of OPSEC, but it made us stand out like sore thumbs at a regular army basic training installation. The SF team had been 'granted' relaxed grooming standards which served to make them even more unusual at the active duty consolidated mess hall that we were required to use if we wanted to eat at government expense.

There were so many conflicting rules for the active duty SF that we felt like the 'real' Counterdrug special forces because this was the only mission we had; our chain of command had supported us realistically and here these superb active duty SF were being treated like children.

The biggest problem then as it is now is the lack of flexibility in the Armed Forces Officer Corps. We as a force have painted ourselves into a corner with paint provided by the lawyers who have so insinuated themselves into the highest levels of the decision-making process that no one can make a decision without a legal review.

The SF major from their company who had come down from Fort Lewis for the initial planning had created a problem when he made some remark to the Border Patrol that pissed them off. When we got everybody together, Steve told me to give an in-briefing to the SF team. In doing so I referred to their major as a "dumb-ass." This got quite a few laughs from the SF guys who were also not impressed with him; and we started to get along a little better.

The operations sergeant for the SF team wore a Ranger Battalion combat patch because the CP members were told to wear regular uniforms. He used to give me a hard stare whenever we went over the target folders but after a while he got used to dealing with an old National Guardsman of undetermined rank and background.

Route Recon in the Big Bend

A few days after they arrived, I was told to take the SF guys who would be driving the infiltration vehicles, down to the park so that they could recon the routes they would take on the night of the infil. I led them into the Big Bend, but it was taking us a long time, so I leaned out the window and gave a double time signal to the SF guys in the other vehicle. They took my signal to heart and took off.

Around the next bend in the road was a DPS corporal named P and he pulled them over. They were not supposed to tell anyone about our real purpose there, but they managed to let him know that they were Special Forces who just happened to be passing through, hoping to appeal to the corporal's patriotic sentiments. Corporal P wasn't feeling very patriotic that day and gave them a ticket. This made them very angry with me and when we got back to El Paso I got a lot of flak from the SF, Captain Akers and Steve.

I told this story to Billy B, one of our National Guard pilots who also flew helicopters for DPS. He told me that he had Corporal P as his Field Training Officer after he graduated from the DPS academy. One day they were out patrolling in the Big Bend and Corporal P told Billy to do something that he didn't agree with and Billy refused. Corporal P told Billy to pull over which he did and then P told him, "I'm getting out and I'm coming around to your side and I'm going to kick your ass!" Billy said, "I told him that that was fine but that he better bring his gun because I had one." "P stayed in his seat," Billy said, "and I drove us back to the station and that was the end of my field training. I was assigned to my own station a long way away from the Big Bend."

Rambo

I came into the dayroom that was set up for the troops and watched 'Rambo Part 1' with about half of the SF contingent. I couldn't believe that they would watch a movie like this. At the end of the movie I told them that after watching the movie with them I felt like I really understood them much better. One of them said "Hooah" and I went to bed.

At the end of isolation, we had the briefbacks and I was impressed with the amount of work that the SF teams put into their presentations. They used a lot of butcher block paper to diagram their immediate action drills and other actions during the operation. Our teams did credible briefbacks but the SF guys who watched them were full of suggestions on how to make them better. After the briefbacks the teams did their final packing and I watched one of the SF men run into the barracks with a BFR, which he proceeded to stuff into the rucksack of one of his teammates. He and his teammates were having a great laugh as he hid the big rock in the bottom of the ruck.

Infil Time

I drove a team to the site that was the farthest from El Paso, an eight-hour infiltration. We went through Fort Stockton and one of the agents told me to stop at the local market. They knew that we were not supposed to stop anywhere but after the train wreck in Del Rio I didn't feel like antagonizing the BP. They were used to working the border and if they thought it was ok to stop, well I was ok with it too.

Jimmy K drove one of the other teams and when they went through the Sierra Blanca BP checkpoint, the agent in the front seat told Jimmy to let him do the talking. They were stopped and were asked the routine question, 'where are you going?' and he told the agent that they were going fishing. "Where are you going fishing?" asked the agent. "Monahan's," said our agent, naming the only town he could remember in Texas that was east of the checkpoint.

The agent looked at them and said, "Please pull into the secondary inspection area." Monahans is the home of the Sandhills State Park; a place noted for its sand dunes not its fishing. Our agent had to show his badge, but the agent still insisted on looking into the back of their vehicle where he was treated to the sight of some unusual automatic weapons and some large rucksacks.

I drove my team to their drop off and they proceeded to walk about two miles through the dark to their hide site, a movement that took about an hour. They walked down a hard dirt road and brushed out their tracks as they went. One of the agents, Dave, was not happy with this route and let me know during isolation. I told him that it had been selected by one of his supervisors because there were no viable alternatives that we could see. The men would be carrying rucksacks that could weigh a hundred pounds and five-gallon cans of water, another 40 pounds. We didn't want to have the guys risk serious injury walking over rough terrain in the dark, so we opted for the dirt road. He told me that if we ever went here again with him, he would pick a better route.

Back to the Future

Map recon is always a good idea and aerial photos are also a big help when you plan your route. Dave the BP agent, a good guy who meant well, must have brooded about that route for some time. Two years later when he was alerted for an operation with us, he insisted on coming back to the same place. This time he would bring the team in using a route that would get them in undetected.

He chose a steep sided gully that ran from the same drop-off point all the way down to the river. What he didn't know and couldn't know from his map and aerial photo recon was that the gully he had chosen was used as a dump by local ranchers for many years and the main thing they threw away here was old barbwire. By the time Jim Walker, the team leader, realized that they were in trouble; they were too far down the steep path to go back and the walls of the gully were too high and too steep for them to climb out; they had to press on. It took them 12 hours of hell to get to their OP on Dave's 'better' route and that infil was entered into SOD lore as 'one for the books.'

The Forward Support Element/MEDEVAC

When I finished my infiltration, I drove to Fort Stockton and checked into the motel we had chosen because it had outside doors that would allow us to go in and out of our rooms without going through the lobby. The MEDEVAC pilots, Scott and Jerry were staying here also. They were both Vietnam veteran pilots who had recently transitioned to the UH-60 Blackhawk helicopters and unlike other older pilots they liked the Blackhawk and said it had much better performance characteristics than the Huey.

There had been a number of crashes involving these helicopters, people called them 'crashhawks,' as they entered into service. Scott and Jerry told me that the one problem that had been identified was the automatic reaction that a pilot made when a certain problem arose. Instead of pulling back on the stick you had to push it forward and I was glad to hear that they were learning from the mistakes of others.

Two SF medics had been assigned as infil drivers and they checked into the motel. The funding problems rose up again, tax exempt forms, and the hotel staff now knew that there were two men from the US Army staying at the motel. We avoided each other in the restaurant and only talked in our rooms. The SF guys used to come to my room and hang out. One of the fellows sold me a Woodland Camouflage Gortex jacket that looked cool. I still have it but if I would have waited 6 more years, I could have been issued one for free.

Emergency Resupply

I made a big mistake when I packed the FSE box in El Paso and left behind my copies of the infil routes for the teams in the Big Bend Park. A few nights after the operation started, I got a call from the CP telling me with code words that one of the teams in the park had failed to make commo and for me to initiate emergency resupply. I couldn't find the infil sheets and I couldn't ask for the details on the motel phone. Since I had written them, I decided to wing it and drive down to the park in hopes that I would recognize the designated point rather than use the odometer reading from a road junction that would tell me exactly where to stop. Our far recognition signal was my vehicle in the right place at the right time and a near recognition signal that only they could see.

I drove 4 hours down into the park and when I got to the turn-off to the SF team's site I started praying. I drove to a spot that seemed familiar and stopped. I was in the middle of nowhere, close to the Mexican border, so I took out my .45 and put a round in the chamber, safed it and kept it in my right hand. I rolled down the window and looked at the stars, but I was unable to enjoy the view because of my uncertainty about the location given to the team for an emergency resupply point.

I sat there for what seemed to be a long time then I decided to drive forward about a half a mile and stop again in case I was not in the right spot. I drove forward, lights out, holding a pair of PVS 7 night-vision goggles to one eye. I stopped again and waited for about 15 minutes. I was starting to panic. I called on the Lord, started the vehicle and put it in drive. Just as I released the brake, I heard a loud knocking on the back window. I braked and grabbed my gun off the seat and aimed it at the open window to my left and waited.

Someone knocked on the side of the vehicle and then stuck his face into side window. He was breathing heavily like he had just finished a PT test. "Wait," he said, "wait! We have been trying to catch you since you first stopped back there." I got out of the vehicle and there were two SF men breathing hard but glad I didn't make them run another lap. I stayed while they set up the radio and made a commo check. Now, the whole incident was pretty funny, but I decided to let them think that it was all part of the plan rather than telling them the truth and risk an ass-kicking on the side of the road at midnight in the Big Bend.

The Special Forces in Big Bend

The other SF teams that were inserted into Big Bend Park had a few adventures of their own. One team observed a vehicle enter a secluded parking area after midnight and it remained there for an hour. The occupants were seen from a distance walking down toward the river and later walking back to their car carrying a bundle. The agents decided that this was suspicious and stopped the vehicle. The driver was a National Park Ranger and the female that was with him was also a National Park employee who wasn't the driver's wife. He was so indignant that they had stopped him that the agents finally let him go without searching his vehicle.

The stop caused a big uproar in the park and the BP got a flaming rocket from Washington telling them stop interdiction operations in the park unless the Park Rangers were involved. One of the SF guys was with the agents when they stopped the vehicle and when the CP asked for a physical description of the vehicle occupants, he hesitated which caused the CP to prod him for a description, now. "Well," he hemmed and hawed, "she's really ugly," he finally blurted out, thinking, I guess, that they were wanting a barroom report rather than a physical description for law enforcement purposes.

One of the other teams was spotted by a rancher and compromised. The BP agent on the team knew him and asked him to keep their presence a secret which the rancher agreed to do and so they stayed where they were. A third team assisted in the search for a team of pack horses that had crossed the river and set off a BP sensor. One of the horses was found carrying 292 pounds of marijuana and the BP gave the team credit for an assist. All the SF teams in the park observed a lot of suspicious activity and were happy that they were operational in such a lively place.

Weekend Drill

During the middle of the operation we had to pull out all the LRSD troops. Captain Akers drove with me because he would stay behind and pick up the rest of the team while I was gone to drill. We all flew back to Austin. The next night they did a parachute jump and spent the rest of the night and all the next day in a field training exercise. Monday morning, we all flew back to El Paso and the guys were reinserted into their OPs on Monday night. While we were gone Captain Akers drove back to the site but got lost, "It looks a lot different out there after dark," he said later. It does indeed. It took him an extra five hours to finally find the place and the guys were wondering if they were going to have to activate their E+E plan.

While we were gone the team that remained at that site spotted a suspicious vehicle pulling a horse trailer and reported it to the CP. The local BP agents stopped the vehicle when it got to the highway. The agents knew the man driving and let him go without searching the vehicle. Apparently, they rethought their actions and radioed to Fort Stockton and asked the DPS trooper to take a look at the vehicle when it came into town. The DPS trooper did take a look and his dog found 177 pounds of marijuana in the trailer.

Exfil, Debriefing and AAR

All the teams were pulled out by the same drivers that put them in and we all returned to El Paso for debriefing. The guys were really tired after a long time in the field, but they answered the questions to the best of their ability. The SF had their own internal AAR. They were indignant about the lack of support from their Group at Fort Lewis. At one point they used the word criminal negligence when describing this lack but the major had them take it out of the final document. They were angry with the way JTF-6 had supported them in El Paso. They enjoyed the actual operation, but they were mad they hadn't caught more dope and they blamed this on their higher HQ.

During the operation we had a lot of commo problems and the commo officer from JTF-6 went above and beyond to help us stay on the air. He saw that we were getting interference from the giant antennas on the mountains above Fort Bliss and suggested that we move our antenna to a spot where there would be a building between the antenna and the mountain. This solved our commo problem and was a big help.

The active duty military was just getting started with the drug war and it would take a while for them to adapt to the new environment. Budget problems, vehicle problems, license plates, credit cards, tax exempt forms for the motels, civilian clothes, relaxed grooming standards, rules of engagement, private property, Posse Comitatus, cover stories, intelligence oversight, etc. were all areas which would require common sense leadership. We had some in the Guard but the career considerations for the officers in the regular forces were a big impediment to adaptation.

As far as intelligence operations go, without complete operations security there is no way of knowing if you are getting the real picture and there is no way of determining whether you have achieved perfect OPSEC. Good OPSEC was fairly easy to maintain if it was an operation conducted by the military alone, but when you added law enforcement and the military together it was almost impossible to impose strict rules of behavior on the cops and it was impossible to run operations without the law enforcement supervisors knowing the details. Beside loose lips and moles was the problem of sending strange faces into areas where a footprint or a rock that was moved could cause alarm to the local inhabitants who were related to people on both sides of the river. In other words, it was almost impossible to run successful operations against the drug traffickers along the river. Not impossible, but almost.

Job Security

While we were on the operation, orders were issued that put us on active duty until 30 September 1991, the end of the fiscal year, the first time we had some assurance that the program might continue. Before this we had been getting orders for a month, then several months, nothing that would indicate that this was any more than a flash in the pan. Now the guys felt like they could make plans to do things like move their families to the Austin area, buy cars, and do all the things that people do when they are financially secure. It helped to stabilize the detachment, but it also changed the mind set of operating professionally to operating for our future. This was the beginning of bureaucracy in the program and in the detachment.

New Guys

We got the go-ahead to expand the detachment to 20 people. The word went out at G Company and LRSD that there was a high speed 'secret unit' and the 'born to be wild' guys started knocking on the door. A sergeant from G Company arrived with some fanfare. A tabbed veteran of an 82nd Airborne Division recon platoon, he soon entered into the constant struggle for domination among the team guys.

Sergeant Juan Hinojosa a former Air Force jet mechanic and Army medic came; he was now a member of G Company and would prove to be one of the steadiest members of the detachment. He became famous later for having 270 days of TDY in one year.

Sergeant Dennis was a former active-duty soldier who had been in the Guard for a while. He would make big contributions to the detachment and when he went to Ranger School in his thirties, he came back with his tab and continued to go to the field like nothing happened.

Another sergeant came from active duty with the 7th ID where he got his tab and an interest in martial arts. He had a knack for breaking things and soon became known as Z-Man after he inadvertently lost the encryption in a radio once too often. He also wanted to lead and would get an opportunity. Specialist JT was a young Guardsman from LRSD and the son of a retired Colonel. He too would have a role to play in the future of the detachment.

Sergeants Brian and PB came from active duty to LRSD from the Rangers. Brian had gotten his tab, and then screwed up and was sent to the 24th ID where he was punished for his sins, finishing his enlistment in leg-land. PB had participated in the invasion of Panama, where he was air landed with the vehicles, so he wasn't able to wear the 'mustard stain' as they called the gold star, on his jump wings.

Brian and PB were rah rah Rangers but Greg Scott had had enough of that. He wore a tab and did his job but told the others that he wasn't interested in turning SOD into a Ranger Battalion. They all got out of the Army to go to college but found it boring, so they came in the Guard to jump out of planes and be rangers. Brian and JT were destined to play big roles in our future.

Specialists Joe A and CL came from active duty where Joe had been a Ranger but didn't get a tab and CL had had something to do with the Rangers as a medic. Specialist Miro came from G Company. They called him Spock because he looked like a guy on Star Trek.

Steve and I were used to police work where everyone was expected to do their jobs without supervision. We were operational and building a unit from scratch. We did what we could to train the new guys for operations in the unique environment of the drug war but with all the testosterone there was bound to be problems.

The lack of mature supervision among the team guys would be a major factor in what was to come. This lack of maturity in the leadership was why we started SOD in the first place and now we had to find the real leaders among the leaders. Some of the older guys were getting burned out and the constant complaining of some of them made fertile soil for the ambitious ones to grow their little plots to gain leadership roles in the detachment. Eventually the detachment began to resemble Saigon with coup after coup and the men wondering to whom they owed their loyalty that day.

When we returned to Austin after the Big Bend operation, I tallied up my first complete year on the program. I had participated in five operations which resulted in 211 days of TDY. I spent 15 days at AT and 30 days at drill. For some reason I neglected to notice that I had accrued 30 days of leave and when I went on orders for the next year, I lost it. I had done the same thing the year before and when I finally realized what I had done I asked if I could get my leave and the personnel people in CD said no. I was busy at the time and soon forgot about it and the next year we were required to take block leave so I was able to take my family to a few places for vacation.

Laredo Part II-Site Selection

Ten days after I returned from El Paso, Steve told me to go back to Laredo and conduct planning and coordination for an operation in the Laredo Sector. This should be interesting I thought as I remembered my trip to Laredo with Tom Payne where we met Dangerous Don and went on a river patrol with his Deepolitos. I drove down with Tom Stephens and we met with Don at the Laredo Sector HQ. The A-Chief that was overseeing the operation welcomed us and told us that Don would make sure we had everything we needed.

Don asked about Tom Payne and I told him that he was no longer working with the program. He didn't seem curious as to why and I didn't say anything. He was more interested in regaling us with stories about his Marine Corps fantasy and he now let us in on his 'secret' which was that he was the XO of the 4th Recon Battalion of the Marine Corps Reserve in San Antonio. 'I'm a major," he said, "but I don't tell everyone because it's hush-hush and we don't want our enemies knowing about it if we go to war."

Don took us upriver and we did site selection on a number of crossings north of Laredo. There is a stretch of border where there is no paved road between Laredo and Eagle Pass. There is a Tick Rider HQ there. These are cowboys that patrol the river on horseback looking for stray cattle that may have crossed over from Mexico. Many of the Mexican cattle are infested with 'fever ticks' and it's the tick rider's job to catch any stray cattle and quarantine them to stop the infestation of U.S. cattle and destroy the cattle industry.

These cowboys stayed out of the drug law enforcement, but they would let the BP know if they saw something suspicious. Don was going to stop and have a talk but when we got to their compound there was no one there. We spent the next three days going from crossing to crossing and preparing target folders in the motel at night.

Before we came down, Don had invited a salesman from a company that was selling non-lethal munitions to law enforcement to come to Laredo and demonstrate his wares. Don had promised him that there would be a representative from every state and local agency in the area but only two of his BP agents showed up. Don begged us to go where he introduced us as DOD people who were so high speed that he really couldn't talk about us, but he was sure we would place a big order as soon as we saw the fireworks.

The guy knew he had been had but went along with the show just in case Don was telling the truth. He demonstrated several types of distraction devices, better known as flash bangs and he had one that sent out a bunch of rubber balls when it exploded, guaranteed to take your mind off whatever you were thinking about. Don was able to get several samples and use them later when boredom set in.

The next night he sent us out with one of his Deepolitos who brought along his MP-5 and took us to some crossing areas right in town. These places were too heavily populated for us to put teams in, but they did present interesting problems for the drug interdiction efforts in Laredo. We sat and watched the main bridge for a while.

It was still difficult for us to grasp the enormity of the problem because we were trying to put small teams into areas where we might catch a two hundred pound load of marijuana when the large trucks that were crossing the river right in front of us could and did carry multi-ton loads of cocaine and marijuana daily.

The last thing I did after flying the river for the aerial photos was to sit with Don and write up the OPLAN and isolation training plan. He decided to establish the CP next to the Webb County Detention Center in an abandoned radar site, part of the old Laredo Air Force Base that had been closed in the 70s. Everything was good he declared but as I drove back to Austin I had a bad feeling about this one.

Isolation

The day came and we began isolation at Camp Swift. We formed the teams and they moved into the tin huts and got to know each other before the mission briefing. The agents were not happy with this operation or with their boss. "According to Don, with 24 hours' notice, we can deploy anywhere in the world for up to 27 years."

We gave the mission briefing and isolation schedule that Don and I had agreed on and this caused a small revolt among the agents. They had no intention of staying locked up at Camp Swift when the bright lights of Austin were blazing 30 miles away. It was getting out of hand, so we compromised, and Don took his guys to town in Bastrop where they had dinner and went to the Super Wal-Mart.

As if things weren't crazy enough, there was a detachment of Army Reserve soldiers from New Orleans that had been assigned to this operation as radio operators and CP personnel. This was another attempt by JTF-6 to be helpful. Except for their NCOIC, they were all black, making their stay in Laredo, TX a subject of conversation.

The army had changed since I first was a soldier in the 1960s. Then the infantry and the airborne were full of young black men and many of them were the best soldiers in the unit. Their sense of humor was very refreshing, and I never felt uncomfortable in their company. I could never forget John Brantley and Carl Johnson, two young American soldiers with black skin in my squad who were killed in action while feeding ammunition to a machine gun manned by a third black skinned American soldier named Robert Pointer. They were part of the defense of the left flank of the 1-16 Infantry at the battle of Xom Bo II, 17 June 1967, before most of my comrades in SOD were born.

I gathered the New Orleans soldiers together and welcomed them to Texas, a fact that didn't escape their attention, and they listened to every word I said. I told them that I was happy that they had come to help us with the mission and after impressing on them the need for accurate communications, I told them that we would be staying at a motel in Laredo Texas and that I wanted them to follow a few simple rules; one, no drinking, two, don't go out by yourself and three, let me know if you have any problems. They rogered my transmission and did a good job supporting the mission.

When the Deepolitos got back from their exciting trip to Bastrop, we had to readjust the training schedule. Don told his guys that he had double printed the schedule so that's why he thought they had four days instead of two. They were still not happy but there were bigger storms brewing half-way across the world that were beginning to compete with our guy's attention.

Saddam Hussein's invasion of Kuwait had led to a large-scale military response by the U.S. and right in the middle of isolation, the U.S. led coalition began air strikes on the Iraqi Army.

Our guys were freaking out. They were sure that G Company would be called up along with III Corps from Fort Hood. Unfortunately for them the Army decided to have a Corps command from Europe run the armored attack into Iraq. The III Corps units went but the Corps HQ and G Company were left behind. Someone had a small portable TV set which they set up in the CP and whenever there was a break in the training the guys would be watching CNN and moaning about how they were going to miss the war.

One news story came out saying that Saddam was sending agents to the U.S. to kill our leaders and sabotage our infrastructure. Shortly after that Don was told to send the two agents that were part of his SOG unit back to Laredo because they were needed for security at JFK airport in New York. Our guys came to me and demanded that they be issued more ammo because of the possibility that they would encounter some of Saddam's commandos on the river.

That seemed a little far-fetched to me but there was a war. I recommended to Captain Akers that we give the boys a basic load of 210 rounds for their M-16s and 50 rounds for their .45s. He said ok so we issued them the extra ammo and I could see the gleam in some of their eyes as they prepared to defend their country on its southern border.

We finally finished isolation and prepared to conduct the briefbacks. About an hour before they started, Don called his men together and issued them a 'patrol order' that we knew nothing about. His guys were pretty angry at everything by now and when they showed up for their briefbacks there was so much confusion that we had to stop and run through the operations order again.

Then Don got mad and took Captain Akers, Steve, and me outside and told us that they didn't work for the National Guard and that this was a Border Patrol operation. I was getting a little hostile myself because Don and I had gone over every aspect of the isolation training schedule and the OPLAN. He had assured me that he both understood and agreed with all of it and now he was acting like he never heard any of it which made me look pretty stupid in front of my boss.

Laredo-Infil

We finally got the teams on the same sheet of music, we thought, and headed for Laredo in a bus because the aircraft option had been scrubbed. In Laredo we went out to the CP site, driving passed all the prisoners in Webb County Detention Center. The BP infil vehicles were there and we loaded the teams into them and after dark, away we went.

I drove one of the teams in a brand-new BP pick-up truck with camper. There was a lot of banter with guys in the back and I managed to put a sizable dent in the side of the vehicle going out a narrow gap in the barb wire fence that surrounded the site. I drove the team in and dropped them off. Then I returned to the CP where I spent the next hour writing a memo to the BP deputy chief explaining how his brand-new pickup truck didn't look new anymore.

The New Orleans troops set up the CP under Tom Stephens and Steve's supervision and we were on the air. The first night passed without incident and I went to the motel for a nap. When I returned to the CP, I heard some of the New Orleans soldiers struggling with the pronunciation of Spanish names of towns in the area. My favorite was San Ignacio which quickly became Sam Obnoxious and I cracked up every time I heard it mentioned in the radio room.

Crazy Stuff in Laredo

After a few days, Don's number one Deepolito decided that he had had enough of where they were and contacted Don on his BP radio and demanded a change during the automatic resupply. The situation was discussed with Captain Akers, Steve, and I and we agreed with Don's proposal that the team be moved to the ____ site. Don told us that he would take care of the movement and brief the team on the new target when they were in the vehicle. Tom Stephens went with him.

They pulled the team out, relocated them to another location, then returned to the CP and Don told us what he had done. The original plan was for the team to be inserted on the south side of ____ but the team leader wanted to go to the north side of ____ so Don inserted them there. When he returned to the CP, he showed us on the map the place we had agreed on, not the place he had actually taken them.

About thirty minutes later they called in their initial entry report but didn't give us a grid. We asked them for one and they asked us to wait and they would get back with us. This was pretty strange and while I was musing on this Tom motioned me to follow him outside where he told me that when the team climbed out of the vehicle they threw the large laminated map sheet at him and said they didn't need it. "That might be why they can't give us a grid."

The team came back up on the air and told us that from a certain landmark on the map they were so many meters up and to the left of that object. Now I realized they were using their E+E map on which we had purposely left the grid reference numbers; now they really couldn't tell us where they were, but, where they said they were was dangerously close to another team and we didn't want to risk having a shootout between teams on the river. We asked the team to confirm their location and after a short pause they came back with a new location several hundred meters south of the first one.

We showed Don where the team said they were, and he said that they had made a map reading error. He pointed to a new position on the map several hundred meters south of where the team said they were and said, "They are right here." When I asked him how he could be sure of that he said, "Because that's where I dropped them off." This was really getting screwy.

I told Captain Akers that Don had put the team in a location known as____. Don immediately denied this and told us that he had put them into the area known as ____, a third location. It seemed that we had entered the Twilight Zone.

I had made target folders for all the locations that Don was claiming that he placed the team. When I showed him the three locations on the map, he then said that he had made a mistake and assured us that the team was at ____. Captain Akers, Steve, and I realized that no one knew where the team was, a bad situation for them if there were injuries or the need for law enforcement backup. Later that morning, after more discussion and before sunrise, I went with Don back to the actual drop-off point. I gave the team their gridded map and later they gave us their actual OP location on the SATCOM.

Don spent some quality time on the SATCOM during the rest of the operation; one time he was overheard arguing with his number one guy who had his own ideas about the extraction time, pizza, and the need to be in front of his TV for the super bowl. This call took twenty minutes and interfered with the mandatory SITREPs, and other traffic on the crowded SATCOM channel.

Don attempted to order a USMC unit in Arizona that was doing a Counterdrug operation to minimize their radio traffic by identifying himself as 'Major Don' when he called them on the telephone. The Marines complained to JTF-6 who was the 'owner' of the SATCOM channel and they called our HQ and demanded that we go through them for any problem solving regarding their channel.

Meanwhile, out in the field on the team whose location caused us so much consternation, one of the guys was becoming an irritant to the normally quiet Chris. After repeated attempts to get him to STFU, Chris took matters into his own hands by grabbing him by the neck and thigh, and lifting him high over his head, body slamming him to the ground while the other team members watched. Another sergeant, who had just joined SOD, told me later that this completely blew his mind and he wondered what kind of outfit he had wandered into. On another team, one of the Deepolitos got so angry with Don's double talk about extraction times that the SOD guys watched as he took off his shirt, built a big fire and spent the afternoon singing at the top of his lungs.

The CP got an X-Ray message about 11PM from the team at _____. Some Webb County Sheriff's deputies were in a running gunfight with some smugglers that was moving upriver toward their location. The BP agent with the team ID'd himself to the deputies, a precarious move on the river in the dark, and a potential blue on blue was avoided.

Back in the CP, we, the National Guard side of the command group kept a close eye on Don. Captain Akers had a long heart to heart with him. Don told him that, "Yes, there have been a few minor mistakes but let's keep them in-house, we don't air our dirty laundry." When he asked if Don's supervisor was aware of any of the problems, Don said that they knew that a few small problems had arisen but that 'we' had taken care of them. He then told Captain Akers that if any of these problems came up in the AAR then we would never work in the Laredo Sector again.

After the teams were extracted, we debriefed them back at the CP. Don cautioned each team that they shouldn't mention any of the problems that came up during the operation. I told Captain Akers and Steve that we needed to lay all the problems out at the AAR, but Captain Akers knew that would cause bigger problems than he was prepared to take on, so he called Bil.

When we conducted the AAR, it was like none of the problems that had caused us such heartburn, had ever happened. There were two Marine Officers in the back in civilian clothes who sat there listening and never said a word. I figured that they were there at Don's invitation and were checking us out as we had come onto their turf.

We never did go back to Laredo to work with the BP, but I got a report from one of the agents with whom I had made friends. He related the next chapter in the remarkable life of Don. The story was that Don got drunk one night and in a moment of exuberance, flash-banged the BP Checkpoint on IH-35 just outside of Laredo.

The flash blinded the agents temporarily, but the bang blew all the glass out of the windows permanently. Don went to the Sector HQ and after climbing onto a truck, fell off and injured himself. Because he was on the injury list, he couldn't be fired for the Checkpoint incident and what happened to him after that is unknown to me.

Personnel Changes

When we got back to Austin, none of us were sure about what the future held for SOD. After a good start in McAllen, we managed to alienate the BP there. We still had a good rapport with the Del Rio Sector, but we wouldn't be going back to Laredo anytime soon.

Chris decided that he had had enough and went back to the regular army. He had been an excellent operator and a real 'quiet' professional. He had done the research on this new thing called the internet and found all the information we needed to successfully request and receive 10 Motorola Saber Radios which greatly enhanced our ability to communicate with ourselves and the cops we would work with. The guys insisted on giving Chris a plaque which was one of the few given out in those days. Chris went to the Special Forces Q Course and was assigned to the 3rd SF Group. I saw him a few years later and he told me how he got crossways with his team sergeant when he told him that he had served with a National Guard unit that was more 'hooah' than his SF team. Later Chris went to OCS, branched artillery and was assigned as a fire support officer to the 75th Ranger Regiment. From there he went back to Special Forces and led an A-Team into Afghanistan after the war started. He was and is a great soldier and teaches a younger generation of soldiers as an ROTC instructor at a college in Texas.

George left not long after Chris and got a job with the Hospital Police where I had worked. He was accepted as a Warrant Officer pilot candidate with the aviation unit that supported us at LRSD and was our MEDEVAC support on the border operations. He completed flight school and became a Blackhawk pilot. He eventually became fixed-wing rated and flew for the regular Army with tours to Iraq during the war until he retired.

During this period, we moved from the indoor pistol range to a large room on the second floor of Building 15, built as a barracks during WWI. We occupied half of the eastern end of the building and CD HQ occupied the other half. We had an outside staircase which made it a little easier to haul up our equipment rather than having to go inside the building and use the other stairs. The Chief of Staff, my old CO, Colonel Danny Kohler, decided to get the Guard into shape and instituted a PT program for every military person on Camp Mabry. Monday, Wednesday, and Friday everyone was ordered to be on the parade field at 0700 in PT clothes. All the soldiers and airmen would be formed into one formation and led through the daily dozen, a series of calisthenics that were done in cadence with us shouting the repetitions. Then we would be divided into ability groups and go for a two-mile run, singing the old cadence songs. The guys hated this because our PT was much more intense than this so after the run we would reassemble down at the back forty and attempt to smoke each other with 'hard' PT.

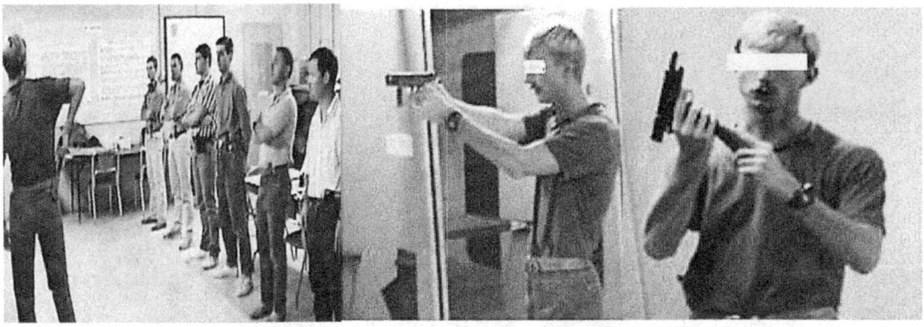

Pistolero Training with 'Steve'

SWAT School

SWAT School

Tom Stephens-Aerial Photos **Operator-Laredo Operation**

Operators-Laredo Operation

And do not be drunk with wine, in which is dissoluteness, but be filled in spirit-Ephesians 5:18

Chapter 12-April 1991

SE Texas I

In March we got a request for our first operation away from the border. It was from a TNCP Task Force. The Texas Narcotics Control Program (TNCP) was established in 1988 with the Byrne Federal Grant money that funded the establishment of multi-agency and multi-jurisdictional drug task forces across Texas. These were run in some cases by the Texas Department of Public Safety (DPS) Narcotics but most of them were run by local law enforcement agencies. The narcs that worked for these Task Forces were some of the most interesting cops I had ever met. There were some problems that arose with the freewheeling approach that these groups took but they caught a lot of dope and we were proud to work with them.

This request came from a SE Texas Task Force. The county is an undeveloped area east of Houston on the north side of Trinity Bay. I was sent down there about a month after we came back from Laredo to do initial planning and coordination. I met with Dave, the chief deputy of the Sheriff's Department and overseer of the Task Force. Dave was a part-time soldier with the Texas National Guard and a Vietnam Veteran. He introduced me to Jack, the TF Commander, a local narc.

They wanted to stake out a number of places that they believed were being used as landing strips for fixed wing aircraft from Colombia and Mexico. The TF wanted us to use encrypted military radios because they felt that all their radios were being monitored by drug traffickers. There were so many spots that they wanted watched that it required another Cecil B. DeMille. Together, we worked out the details. We would set up the CP in a large barn belonging to the judge in that area.

G is in Reno

The Air Guard added 'G' to Counterdrug. He was a targeteer and his job was to arrange for aerial photos for CD operations. G was attached to the Intel section and put under the supervision of SSG-R, one of the LRSD guys who had drifted into the Intel world.

SSG R had a history with the 82nd Airborne Division and he was not used to the work ethic or lack thereof of some of the blue suiters who worked with us. When G came on, SSG R was constantly hounding him for product and G was constantly flying to Reno, Nevada to 'work' it with an F-4 Phantom photo recon unit. He spent many days in Reno 'coordinating' these missions. It seemed kind of expensive to have an F-4 fly from Nevada to Texas to take pictures, those Phantoms being real beasts at the gas pump and all, but DOD insisted that the pilots had to fly anyway, and this mission would be useful training. I gave G the latitude and longitude coordinates for the areas in SE Texas that we wanted photographed and when I finally got them, I noticed the shadow of an F-4 in one of the pictures.

SE Texas-Isolation

We gathered the volunteers from LRSD and G Company and did isolation at the remote site where the CP was located. The teams were issued AN/PRC 77 radios with DMDG message burst devices that we used with our HF radios at LRS. There was a problem with our radio reception until Woodford, a G Company man, climbed a rickety old 75 foot antenna and installed one of our antennas at the top and after that we could talk to and receive from any location in the county.

'Gulf Wind'-Waiting for a Plane!

At the same time we were operating, there was a big JTF-6 operation called Gulf Wind going on nearby. I was sent over to their command post on the coast for liaison. After having my ID card scrutinized by a Navy chief, I was admitted to their briefing room. There I was introduced to the officer who was running the radar unit that was going to detect any inbound aircraft from the Gulf. He was very busy and told me so. They were conducting a test of their system and this radar information would be useful to us, so I was very polite and waited to show them the location of our teams.

They had an inbound target plane and it appeared to be further away than the radar was able to detect, so I went outside of their tent to look at the ocean. I saw a large four engine airplane about 5 miles out and heading right for us. It was the target plane and it was still evading the all-seeing eye of the radar as it flew right over us with a loud roar. When I reentered the tent, there were a lot of red faces and some serious naval expressions that I hadn't heard before. Nobody was interested about the location of our teams, so I gave the team list and their lat/lons to one of the chiefs and left. Our teams were inserted by vehicle into their target areas and we waited for the planes that never showed up. Later we found out that the smugglers used radar detectors on their airplanes just like we use them in our cars. Was Gulf Wind radar a contributing factor? Was it OPSEC or vacation time for the smugglers, only your hairdresser knows for sure?

CP Life

The boys in the CP kept themselves amused by shooting the large rats that had taken over the abandoned barn. They would put some bait out in a dark area. Then they would wait, watching with night vision goggles and when the rats showed up, they would shoot them with a .22 rifle provided by one of the cops. One night I was doing some paperwork in the small motor home when I heard someone shout, "No, don't shoot!' followed by a shot. There was some more shouting now, mainly statements of recrimination and a sudden migration of the shooters from one end of the barn to the other. I heard one of the cops shouting at the shooter, "Don't you know the difference between a pole cat and a rat?" The skunk in its death throes managed to spray a heavy stream of perfume around one side of the barn, reducing our working space by half and making the rest of the operation a smelly business.

The guys on the teams had their moments as well. One incident stood out; a cop on Brian's team fired a round from his M-16 into the ground behind an air guard guy that had been assigned to us as a token of 'jointness.' When the shot was fired, the airman crumpled into a fetal position and whimpered, which ruined any hope of his ever being accepted by them. They thought it was funny, but I wondered how they would have felt if the airman had turned around and shot the cop, believing that he had tried to shoot him in the back.

SE Texas-Eating Crawfish with Jack

At the end of the operation the cops decided to put on a feed for everyone, with crawfish and gumbo and all the things they eat down there. They also brought a case of Jack Daniel's which was a mistake. The cops all locked their guns in their cars and we started eating, drinking, and talking. Uncle Bennie showed up with Bil, George C, and Major K, who had been assigned to other duties with CD.

Bil and Uncle Bennie, Jack and the sheriff had a few drinks and then we had an awards ceremony where we gave the cops a plaque and they gave us some certificates and then the real drinking started. I was abstaining so I took some video and Tom took some group pictures. Things started to get out of hand when a few of the younger Guardsmen began to feel the effects of the whiskey they had chug-a-lugged in a show of bravado and now they had to be put in a place where they could throw up without hitting the rest of the drinkers.

Bil decided that he was going to show his stuff and picked out Brian as the one who he would beat on; bad choice, Bil. Brian was about 6'4" and weighed 200 pounds. When Bil came up to him and shoved him as hard as he could; Brian picked him up and threw him about 10 feet where he landed on his ass. The cops and soldiers jumped in to separate them and Steve declared that we were going back to the motel.

We loaded the vans and got on IH 10 and headed for our motel. The van I was in was driven by a G Company guy who appeared to be sober and Brian was sitting in the front seat, really hammered by now and wondering what he was going to do after he got fired. He started to get sick and opened the door of the van to throw up and almost fell out while we were going about 70 down the freeway. Juan reached over from between the seats and grabbed his belt and held him in the vehicle while he puked his guts out. When we got to the motel there were multiple vomit sessions and some of the guys gave IV's to the worse cases. One group took turns keeping one guy awake because they thought he would die if he went to sleep.

Bil's adventures didn't end there. George told me later that Bil began to verbally abuse Major K so badly as they were driving home that Major K was crying. Then Bil demanded that they stop at a hotel in Houston where the cargo inspectors were staying, and he insisted that they all come down to the lobby so he could 'inspect' them. After some wild moments in the hotel, Uncle Bennie finally got Bil under control and they returned to Austin.

When we got back Bil chewed Captain Akers out for changing the name of the operation from Unity to Overwatch I. I knew something was in the wind so when I saw Bil I asked him what was up. He was incensed with the captain but listened to my explanation of the name change. Bil decided that he had learned his lesson and said he would keep him around a while longer. So ended our first foray into SE Texas; there would be more.

The C-26

The Air Guard had found a number of C-26 twin engine turbo props. One of them was issued to Texas and our fighter pilots were being trained to fly it as an aerial interdiction asset by U.S. Customs. It also had an aerial photograph capability because in the rear of the fuselage there was a large window in the floor, kind of like a glass bottom boat. There was a door on the bottom of the aircraft that slid open after the plane took off. When it was on the ground and the door was closed you would never know about the window.

It was a spy plane. The rumor was that it had belonged to the CIA and was used in Central America during the Contra struggle in Nicaragua and the civil war in El Salvador. The plane was painted in civilian colors and there was nothing that would connect it to the Guard. I wrote a memo to Uncle Bennie giving him my opinion about the futility of the aerial interdiction mission envisioned by Customs and advocated using the C-26 for aerial photo missions. He red inked a few spelling mistakes and said that it was a good memo but too long.

Burleson County

The next mission was done in Burleson County with a few of our guys. They helped some local narcs stake out a marijuana patch that was guarded by booby traps using fishhooks on mono-filament line stretched across the path leading into the grow area. The grower showed up and the Narc ID'd himself and arrested him while our guys had his back. The grower had a pistol in his belt behind his back and the guys were pretty excited about the whole deal. Brian was one of the guys on the operation. He was still worried about his job after the fight with Bil but I told him, "See, I told you Bil wouldn't fire you, he thinks a lot of you because you had the balls to stand up to him."

Annual Training-LRSD

The 1991 Annual Training for LRSD was coming up in June. We had a new CO, 1st LT Rick Noriega, another General's aide. He was a politician from Houston and when he replaced Captain C, we hoped he would turn the morale around and he did. I introduced myself to him as the operations sergeant and we worked well together.

This year we were hosting a Mobile Training Team from the Ranger Department at Fort Benning. They would put our entire detachment through the Long Range Surveillance Leaders Course. This was an intensive two weeks where we were taught all the latest doctrine and tactics for conducting LRS missions. We were all enrolled in a correspondence course and if we passed this training, we would receive a Q6 identifier to our MOS. I had finished my courses and hoped I would be able to meet the standards that the Rangers held us to.

When we arrived at North Fort Hood we were met by Captain Hunter, the course commander, who gave us a run down on what he expected from us. The first thing we did was take a PT test followed by the combat water survival test. The LRSD had set the unit standard for all members of the detachment at the passing score for 17 to 21 year olds. I was recovering from knee surgery and although I could do the required push-ups and sit-ups, I really had to push myself on the run and barely made the minimum time for my age group which was the Ranger Department standard for this course. We did the swim test in the pool and I passed it with no problems.

Then the classes started, and we had long lectures on operations orders, commo plans, immediate action drills and various other things deemed important by the course creators. The big-ticket item was combat vehicle recognition. We had to memorize the distinctive characteristics of the entire inventory of both the Warsaw Pact and NATO. They gave us handouts and we watched slide after slide of tanks, armored personnel carriers, artillery pieces and various and sundry combat vehicles. We were tested on vehicle recognition and I got a go.

Next, we were taken out to a range near the airfield where we did a day and night land navigation course. I passed the day course with a survival compass that I always carried in my pocket. I used it again for the night phase which was a mistake and I failed. They allowed one retest and this time I signed for one of the lensatic compasses that the Army issues with a glow in the dark capability and passed.

The next event was SPIES or special purpose insertion extraction systems. This was a hundred-foot rope attached to a helicopter and dropped to a team in the woods where there was no place to land and after they attached themselves to it the helicopter could pull them out. It was a good idea that was used during the Vietnam War to extract SF teams from the jungles of Laos and Cambodia while under fire.

When it was my turn, I took my video camera and taped the guys below me as we flew 500 feet above the ground and a hundred feet below the Blackhawk. I called on the Lord several times and one of the other guys thought I was freaking out and told me to shut up. I told him I was calling on Jesus because I was excited, and he left me alone.

We spent a few days practicing commo procedures and patrolling. The course concluded with a four-day FTX or field training exercise. I was the operations sergeant, so I worked with Captain Hunter in the TOC. I told him about how we were utilizing the two LRS units to conduct Counterdrug operations on the border. He didn't seem very interested, but I persisted and was able to tell some of the lessons we had learned the hard way.

The FTX consisted of a parachute insertion, a three-day surveillance mission, and a helicopter extraction. Because I had been doing lots of operations this was a fairly easy task and we successfully provided command and control for the detachment under the watchful eyes of the Ranger Cadre. The commo section had a base radio team next to the TOC and the other BRS was flown to Camp Swift by Chinook helicopter where they listened for messages by the teams. On the last day we pulled the teams out by helicopter and debriefed them. The scores were tabulated and I was happy as my roster number was called as one of the ones who passed.

Operation Unity 15-The Dollar Ride

As soon as we got back from AT, we were alerted for another operation in SE Texas. This would be a waterborne operation on Trinity Bay watching for drug smuggling boats coming into Texas from the Gulf of Mexico. I went down to the Task Force HQ and met with Jack. Several other team guys came along to assist with site selection.

After the target areas were selected, I drove over to Ellington Field in Houston and found the C-26 hanger and introduced myself to Colonel Wilkerson, the commanding officer of the C-26 program. They had just finished an intensive air intercept training program with U.S. Customs, and they were ready to fight the war on drugs.

This was the first time I worked closely with Air Guard aviators, so I tried to establish rapport. Most of the air crew members were F-16 pilots from the 147th Fighter Wing of the Texas Air National Guard. This was still the Cold War and the Guard kept several F-16s on strip alert with full armament. The pilots had a war time consciousness, but they weren't used to working for an Army sergeant in civilian clothes.

After a little small talk, I briefed them on the purpose of the mission which was to take photographs and video of the sites that were selected for the surveillance operation. I gave them the lat/lon coordinates for each site and the systems officer or 'back seater' loaded them into the on-board navigation system and we took off.

I took a seat behind the systems officer; a young Captain Russ Malesky and watched as he worked an array of buttons and a joy stick while he watched the screen for the FLIR or forward looking infrared sensor that was in a pod attached to the bottom of the aircraft. He also controlled a gyro-stabilized Wescam video camera which was able to zoom in on a target a long way from the airplane and provide a steady picture of whatever you wanted to video tape.

After we took off, I walked to the back of the plane and lay on the floor facing forward. I was looking through an 18 inch by 24 inch piece of glass in the floor of the plane but I couldn't see a thing. I plugged in the headset they gave me and did a test of the intercom with the pilot. When we established communications, I heard him say that he was opening the door and as I looked down the door over the bottom of the plane opened from back to front and I saw the ground slowly going by as we gained altitude and flew toward Trinity Bay.

We flew the coastline of the bay and I took pictures of all the sites. The pilots coached me on how to get them lined up on the target. It took an hour to get the photos and video before returning to Ellington Field. This was the first operational flight of the C-26 and the Air Force calls this occasion 'The Dollar Ride.' I gave them a dollar bill and we all signed it and it's still framed in the Counterdrug Commander's office.

SE Texas-The 5 SFG Boat Team

Since we didn't have any watercraft, Bil got a boat team from the 5th Special Forces Group at Fort Campbell, Kentucky to come down and join the fun. The Gulf War was over, and these SF guys were still coming down from the war. They showed up with several pallets of gear to include five Zodiac boats with the quiet motors.

We took everything to Camp Swift for isolation where we were met by the SE Texas TF and the 'Jump out Boys'; members of the Harris County Narcotics Task Force. The Houston Narcs were a wild bunch and they were excited to be working with the Special Forces veterans of the Gulf War.

The teams were formed with two SF guys, two of our guys and a cop on each team. We gave the mission briefing at Camp Swift and then moved over to Lake Bastrop to conduct water operations training. We were given classes on the Zodiacs and we learned the capsize drills. They loaned us BCs or buoyancy compensators and fins and taught us drown proofing and how to swim as a team, towing rucksacks behind us. When the training was over, we went out into the lake and practiced casting and recovery like the frogmen in the old war movies.

I was on a boat with the SF captain and several members of his team. Since we were all in civilian clothes or sterile BDUs they didn't know what to make of me. After everyone took a turn steering the boat for a cast and recovery one of them asked me if I wanted to try it, saying "You probably know all about this anyway, right?" I just smiled at him as I took my place at the rear of the boat. "Ready," I asked. They nodded so I twisted the throttle on the outboard engine and we took off.

The SF guys positioned themselves on the pontoon sides of the boat in readiness for my command. "Go," I shouted and the first two threw themselves sideways landing on their backs in the water as we rushed by. "Go" the next two went. "Go" the last two went and I was left in the boat with one other SF guy. I turned the boat around as the SF man positioned himself on the left side of the boat toward the front. He had a big horse collar rope that was used to pull the guys into the boat. I lined the boat up with the men in the water who had formed a straight line and sped toward them. We got to the first one and the man in front held the horse collar out and the man in the water grabbed it and held on while the man in the boat pulled and flung him into the boat. A few seconds later we reached the next one and he repeated his actions until all six men were back in the boat.

One of the guys had dropped his hat in the water so we went back and got it. He jumped over the side to get it and I turned the boat so he wouldn't have to swim so far. As he was climbing back on the boat one of the SF guys said that I should have turned off the motor when he was climbing into the boat. "It's one of our safety rules in Group," he said apologetically. "Sorry about that," I told him, "it's the first time I've run one of these things." That got everybody's attention.

I gave up my coxswain position to one of the other SF guys and we continued to cast and recover until we noticed the other teams were racing around the lake. The training exercise transitioned to a giant water fight and I watched these battle-hardened desert warriors playing like a bunch of kids, really having fun.

We had a BBQ beside the lake that evening, and I was impressed with the quiet professionalism of one of the SF men. I heard him tell about how some civilian girls had asked his boat crew if they were Navy SEALS, "And I wanted to tell them that we were the Special Forces, so much better than the Navy SEALs, but I remembered OPSEC so I told them we were the National Guard."

The training on Lake Bastrop really blended the teams together. Camping by the lake was a pleasant time, the memory of which was marred when a colonel from logistics at Camp Mabry showed up sometime later and demanded money to give to the Lake Bastrop administration for a large phone bill that was incurred during the time we were there. Someone had spent hours on some 1-900 calls, talking to some lonely girls and trying to cheer them up. We denied any knowledge and the SF guys were back at Fort Campbell, so the colonel went back to the logistics section and we never heard any more from him.

With the boat training completed, we went back to Camp Swift and completed the mission preparation with time on the range. The briefbacks were given and the level of professionalism was never higher. When it came time to move to SE Texas, we loaded all the gear into a large rental truck; and the teams were loaded into our new Suburbans and some rented vehicles and we drove south.

SE Texas-Infil Hell

The teams and equipment were taken to a remote area north of I-10. The cops had found a large building next to the Trinity River that went into Trinity Bay. The plan was to launch the Zodiacs from there around 2AM and move to various sites along the banks of the bay. The SF team brought a new radio called SINCGARS that would be almost impossible for the drug smugglers to monitor but there were some problems with the new system that would cause big problems for us.

The teams launched their boats and headed down the canal at 15-minute intervals. They were camoed up commandos as they moved quietly through this backwater. Imagine their surprise when they passed one fishing boat after another with south Texas fishermen staring at them as they cruised by. That was just the beginning of the trouble.

The teams moved across the bay to their surveillance sites. One of them had a site on the east side of the bay that required them to stay in an ill-defined channel with shallow mud flats on either side. Someone made a wrong turn and the team started a four-hour ordeal that was reminiscent of Humphrey Bogart and the African Queen. The guys were wading through thick mud pulling their Zodiac behind them. The SF team leader lost his 9mm pistol in the mud and after numerous attempts to find it, they had to push on because the sun was about to come up and they were trying to get to a small inlet that was overgrown with vegetation where they could hide until night.

Their radio quit working, so they missed their mandatory commo checks which activated the emergency resupply plan. When the FSE people went to the resupply point there was no one there. This raised the alarm and we asked some game wardens that were supporting the operation to search in the area where the team should have been. The team was at the inlet by this time and hiding so they never attempted to signal the game wardens when they came near their location. More law enforcement personnel were called in to search for the missing team.

At this point, the atmosphere was very tense in the CP. Greg, one of the cops, walked into the CP where the SF sergeant major and the company commander were waiting for some news about the missing boat team. He stood in the doorway for a moment and then declared with a loud voice, "Well, now I know what SF stands for!" He paused for a moment, the SF soldiers leaning toward him to hear what he would say. "Specially Fouled up!" or words to that effect. There was silence in the CP. Nobody moved, nobody said anything. I felt laughter coming up my neck and I tensed the muscles in my mouth to suppress it.

The SGM's neck was swollen and his face was beet red. The major quickly turned and walked into the other room. Nobody was saying any-thing. Greg walked over to the refrigerator and got a coke. Then he went over and sat down in the relaxation room next to the CP and started to read a magazine. Slowly, things returned to normal.

I told the guy on radio watch that I was going to the store and asked if he wanted something. He said no so I left the CP and went out to my car. When I got inside and was driving down the street, I released the laughter that had been contained below my voice box. Later that afternoon, the narc on the missing team came up on his cop radio and told us where they were, and they were resupplied with a radio that night.

The command element had a discussion about the all-night fishermen that we had not considered when the operation was planned, and we all agreed that we needed to change our tactics. The teams were directed to return to their base the next morning before dawn. When they got there, they stored the Zodiacs and were infiltrated the next night into areas where they could watch for drug smugglers from hide-sites on the shore.

One team watched a bridge across the inter-coastal waterway and observed a suspicious vehicle and reported the description and license plate which came back to a wanted criminal who the local officers arrested. Another team observed what appeared to be a drug dealing operation out of one of the houses near their hide site. They requested an interdiction of one of the vehicles that was suspected of carrying drugs away from the house, but the cops were unable to find the vehicle after the team lost visual contact. At the end of the operation, we said goodbye to our south Texas friends and returned to Austin.

The SF Association

Bil arranged for the Special Forces Association to host a luncheon for SOD and the men from the 5th SFG after the boat mission on Trinity Bay. JT was in the latrine standing at the urinal when Bil stepped up to the one next to him. JT was a little worried just because Bil was so close to him and no doubt his activity made him feel a little more vulnerable. He became more alarmed however when Bil turned his body to face him. Bil stared at him for a moment and JT began to wilt.

Then Bil opened his mouth and said, "Mother blanker, I saw you eye blanking my daughter." JT by now had stopped breathing and was trying to figure out whether to run while peeing all over the latrine or finish what he was doing and let Bil attack him. He was unable to speak and after a moment Bil leaned a little closer and in a low voice said, "Mother blanker, I'm going to be watching you." He finished and walked out leaving a shaken JT standing at the urinal.

Scuba Class

There were several more consequences coming out of our time with the SF boat team. The initial plan after the operation was to go to Corpus Christi and have some recreation at the beach. Steve made arrangements for a scuba diving trip for all those who had a scuba certificate. The SF guys were a scuba and boat team so they were all going and Steve was the only one among us who had scuba qualifications.

This caused an enormous wave of anxiety to sweep through the unit. Eventually the recreation trip was scrubbed for the SF guys by the bean counters at Fort Campbell so none of us went to Corpus but the guys were so consumed with the thought that the SF guys could do something that they couldn't, that one of them found and enrolled us in a scuba course. We all took the course from an instructor who was curious about who we were. I told him that we were clerks at Camp Mabry, but he didn't believe it. He even asked some of the guys if they were interested in doing some kind of gun running in Belize. We all passed the basic scuba certification in anticipation of a scuba trip to Corpus with the SF, but it never happened.

Counseling Statement

I was getting some flak from a certain sergeant and some of his disciples. He was spending a lot of time around Captain Akers; talking, laughing; kissing ass is what soldiers call it. He had been made an admin team leader and he was eager to make changes that would show the others that he had the power and was being listened to by Steve and the captain. He came to my cubicle one morning, while I was wrestling with one of many projects, and began giving me what I felt was an undeserved ration of shit. I jumped up and told him that we could go outside right now, and he could make changes to my face to start with. He went to Captain Akers and told him that I had threatened him.

The captain who was already unhappy about my complaint's regarding the screwing of Tom Payne, decided that he had enough and gave me a written counseling statement that said I would be fired if another such incident occurred. Worried that I would lose my job, and convicted in my conscience, I went to the sergeant and apologized but he didn't seem to be receptive, so I went about my business. This kind of atmosphere would come and go in the detachment.

G Goes to Law School

Airman G, our man in Reno, had driven the Intel guys crazy. With the C-26 on-line, his skills were no longer needed. He drove a rent car into an immovable object and was injured. The Air Guard had a medical system with mysterious ways and before it was all over, G got a settlement that allowed him to go to law school; and an Air Guard Captain, who was standing when the music stopped, was fired. The captain had just married one of the female Army members of CD and the whole affair caused a lot of heartburn in the task force. This was another scapegoat execution like Tom Payne's, and it didn't renew any confidence in the Task Force justice system. Since the C-26 was now available for aerial photos, G was soon forgotten.

LRS Conference-Fort Benning

Soon after we returned from Chambers County, I got a letter from Captain Hunter, the Ranger officer who had brought the Ranger training team to our annual training and conducted the Long Range Surveillance Leaders Course. The Ranger Department was having their annual LRS conference at Fort Benning and he invited us to attend and make a presentation of our Counterdrug Operations using soldiers from the two LRS units in Texas.

When I went to drill, I showed the letter to Captain Noriega and he discussed it in a training meeting. Some of the other members of the staff were pretty jealous about our being asked to go when they weren't. Thanks to Captain C, there were a lot of divisions in the detachment and I was starting to feel some resentment from some of the others in the operations section.

We contacted Captain Hunter and told him that we would be happy to make a presentation and suggested that we bring some of the SF team with us to discuss the boat operation. He agreed so we contacted the SF people and they agreed to send their XO and several other team members to Fort Benning. They invited us to come to Fort Campbell and visit them so we ended up with a one-week trip to Fort Campbell and Fort Benning where we would stand in front of the Ranger Department and the Infantry School and told them what we knew. The captain and Steve began working on the presentation. Captain Akers would give the introduction, Steve would give the main briefing, I would cover operational issues and Joe A, would cover training the cops.

Fort Campbell-5th Special Forces Group HQ

We flew to Nashville and drove a rental car to Fort Campbell. We found the 5th Special Forces Group area and as we were driving down the street, we saw Terry, one of the SF sergeants from the boat team. I yelled at him and after he got over his surprise at seeing us so far from home he ran over to the car and shook our hands and made us follow him to his company.

We followed him inside and were walking to his team room when I saw the sergeant major who had been so agitated by the cop yelling what he thought about SF. He was sitting at a desk in the company office and he greeted us warmly. I noticed a large plaque on his wall that had a serious military symbol on it and asked Terry about it after we got to his team room. He told me that the SGM had been an Intel guy in Delta Force for 7 years. Now I could understand why he was so angry at the cop's remarks. SF was his life and his religion, and he didn't like anyone bad-mouthing it.

Some of the team guys were there and we sat around with them reminiscing about the boat op and some of the funny things that happened there. Chief ---- the XO showed up and Captain Akers and Steve went into his office to discuss the presentation. When they came out the XO invited us to visit their ISOFAC or isolation facility. This was where SF teams went to prepare for missions. The SF guys told us that when the Army proposed to move the 5th Group from Fort Bragg to Fort Campbell the SF commander demanded a new ISOFAC and so they got a state-of-the-art facility that they were proud of.

They took us to the place where their team was preparing to conduct a jumpmaster course and showed us a video of a female jumpmaster who took someone's static line and went flying off the back ramp of the airplane. As she flew through the air, she assumed a good body position and pulled the rip cord on her air force type parachute which opened and brought her to the ground safely. "She got fired but that was a good move on her part." Everyone agreed and then we were shown around the place. This ISOFAC would be used 10 years later by a new generation of SF soldiers to isolate for mission prep prior to being inserted into Afghanistan.

Fort Benning

The next day we all loaded up and drove to Nashville where we flew to Columbus, Georgia the home of Fort Benning which is the home of the 'Infant.' Huh? As we drove through the main gate, we passed a sign that had been modified by some disgruntled soul to declare the obvious; the infantry was made up of infants. We checked into a motel and Steve had us rehearse our parts for the presentation which would be on the last day of the conference.

I went out and scouted the used bookstores and ended up at Ranger Joe's, a combat gear emporium, where I met Steve. There were a bunch of Hispanic soldiers there who were attending the School of the Americas. Steve decided to try out his Spanish by telling them that I was an American congressman who was there to make sure that they, the Latin American students, were happy there. Several of the foreign soldiers came up to me and assured me, "Si meester Senador, we are happy." I just smiled and shook their hands saying, "muy bien, muy bien."

The LRS Conference

The next morning, we showed up at the conference in the Infantry Hall. This place was full of history and I felt very much at home. Captain Hunter spotted us as we came through the door and welcomed us. I introduced Captain Akers, Captain Noriega, Steve, Joe and the SF XO, and he told us to get with the A/V section for help with any slides or video that we might use. Steve had asked me to bring some of the video from the first SOD operation and I took it to the A/V office where they helped me copy a few clips for the presentation.

The conference had a presentation by the 24th ID LRS in the Gulf War. They started with a fast-moving video edited from CNN clips and accompanied by a fast-moving rock song, 'I need a hero.' The LRSD Commander made the presentation. The 24th ID managed to get a few of their teams into the field which was remarkable because the ground war only lasted a hundred hours and LRS planning can take that long.

After this the people who make the manning tables for Army units got up and proposed making the commo man the lowest ranking man on the team. Everybody sat there at the end of that presentation so when questions were solicited, I stood up and denounced the whole scheme. I told them that commo was the critical skill that validated even sending LRS teams out and they were making the position that required the most training and experience the lowest ranking position on the team. They gave their reasoning and I stood up and argued back. Eventually they said they would reconsider, and the conference moved on.

We got a NOFORN brief by an SF Major, which meant that the Brits and Aussies had to leave. He told us about the special reconnaissance teams that went deep into Iraq and were eventually compromised. The main point of his briefing was that in each case the SF teams kept the Iraqi security forces at bay by taking single well aimed shots that killed anyone who took the initiative to attack them. Rifle marksmanship saved them. This was especially nice to hear at the infantry school. NATICK told us about their new MREs and there was a display of night vision equipment. Finally, it was our turn.

Captain Akers got up and gave an introduction that set the stage and then he introduced Steve. Steve gave a very professional presentation of Unity 6, ending by showing the photograph that Clayton had taken. Then he ran the debriefing tape where the BP agent with George's team assured everyone that our guys had done "real well, I don't know anyone who could have done any better." Steve finished up by saying, "bragging, maybe so, but planning and preparation resulted in the seizure of 269 pounds, one pistol, and three arrests with no injuries."

The SF chief gave a rundown of the boat operation and at the end said that we, the SOD, were their kind of people. Then it was my turn and I went over the OPLAN but took the opportunity to emphasize the need for careful scrutiny of personnel that would be sent to the field because, "just because he can jump out of a plane and has been to Ranger school doesn't mean he can lay still under a bush for twelve hours a day in the heat or cold and stay focused on the target for ten days at a time." I was followed by Joe who gave an overview of our MTT training missions with the cops and we were finished.

Captain Hunter complemented us for our presentation and asked for questions. We were besieged by many in the room as everyone wanted to know something about the CD mission. When the captain finally stopped the questions and had the chaplain give a closing prayer, I noticed that a brigadier general was standing next to me and had put his arm on my shoulder. I never found out why he felt such a bond with me, a buck sergeant, that he would do that except that he was probably a Vietnam vet and we were starting to thin out in the ranks.

As soon as the formalities were finished, I had two Marine Corps sergeants in my face asking many questions about working with the BP. I answered as many as I could until Captain Hunter came up and thanked me for coming. "You probably thought I wasn't listening to you when you were telling me about your CD operations." I smiled and said, 'that's exactly what I thought but I guess I was wrong." I shook his hand and walked over to Steve who told me that my presentation was OK but that I was talking down to the audience, probably so.

A Major came up to me and he was mad. "We have a story to tell, why didn't they ask us to make a presentation?" I didn't know who he was or what his story was and told him so. He was from the California National Guard and had something to do with their LRSD and their Counterdrug program. I tried to sooth him with the suggestion that he talk to Captain Hunter about making a presentation at the next conference but he wandered away muttering to himself. I never went to another LRS conference nor did I follow events in the LRS community because I soon had other things to do. I spent the last evening driving around the Jump School area and videotaping everything. I still hoped to go here someday and earn my silver wings and if I did, I probably wouldn't be able to take video. As it turned out I never got to the Airborne school, but my son Richard did in 2018 and reminds me occasionally that I am still a 'leg.'

The Word Gets Out

Bil had been busy getting the word out to the TNCP Task Forces that we could do surveillance for an extended period of time and we wouldn't cost them anything. Criminal investigations of drug trafficking are mostly accomplished by the use of informants and undercover police work. The last resort for an investigation is surveillance.

It takes a lot of time and that means overtime which is always causing problems for police administrators who try and hold down any overtime due to budget restraints. When they have to give overtime assignments and they have no money, they give Compensatory or COMP time. This means that a cop can build up extra vacation time but usually the department can't give him the time off if they have to give comp time in the first place. This usually results in anonymous complaints to the Federal Government who sends people from the Wage and Hour Division of the Department of Labor to investigate the police agency. This happened once when I was working as a police officer and we all got a fat paycheck when the city was forced to pay us for all the comp time we could never use. After that a memo came down from the chief stating that there would be no more comp time but after six months another memo came down authorizing limited comp time for emergencies and the cycle began again.

Air Smuggling

The next operation was in a small town south of Dallas where we watched a rural airfield that the Task Force had information that it was being used by drug smugglers. We watched the place for 10 days, but

nothing came in. The narcs took the opportunity to inspect some of the planes that were hangered there and discovered that some of them had been landing and taking off from unpaved strips.
One of the guys wore a typical 6th Street t-shirt to the little party we had at the end of the operation. Once again, our guys failed to understand the environment they were operating in because the commander of the Task Force was offended by this and let Captain Akers know. On the front was a picture of two dogs copulating and, on the back, in big red letters was the exhortation to 'bury the bone on 6th Street.' Other than that we had made them happy. The Task Force told us they liked the way we worked, and they would call us again.

Far Southeast Texas

A sheriff asked us if we could support his task force's investigation involving drug traffickers from Colombia. The DEA had hidden an informant in their jail to protect him. He was lonely and told the task force that in exchange for some conjugal visits with his girlfriend, he would lure a plane load of cocaine, maybe 3 tons, to a small local airport but he cautioned them not to mention this to the DEA.

The initial request called for us to bring several machine guns and shoot down the smuggler's plane if it tried to escape. The sheriff was disappointed to find out that this was outside our scope of practice, but Captain Akers talked to him and they modified the plan.

Now we would go in and hide at the end of the airstrip and at night we would move to the half-way point of the runway and hide in a ditch. When the plane landed, the cops, posing as the ground crew, would seize the coke and arrest anyone on the plane. When the bust went down, they wanted us to fire hand-held illumination parachute flares to light up the area. They were afraid that there might be some sicarios protecting the load and if they tried to escape, they would run toward us. If that happened, we might be forced to defend ourselves with the shotguns we were all carrying.

Captain Akers gave us our assignments and I was chosen to lead the operators in the ditch. Brian and another operator were assigned the task of firing the flares from a position behind us. This really pissed Brian off because he was a Ranger, and I was this old leg. I was sure that he would give a good account of himself if we got in a fight, but the captain told him that I was a combat infantry veteran of Vietnam.

We worked out the details and the night of the infil we were dropped off about a mile and a half from our MSS. Brian was in charge and he charged down a path at a very fast pace through some rice paddies. We were all carrying heavy rucks and soon many of us were breathing hard. I had had knee surgery about a month before and this was kicking my butt. When we finally reached the vicinity of the MSS, Brian told us to find a place to lay down and sleep until dawn. I stepped of the trail and laid down in the mud and went to sleep.

At BMNT we moved into a wooded area about two feet above the swamp that surround it. The mosquitoes were fierce and could stab you right through your BDUs. We had some repellant but that was at the top of our first resupply request. I had a mosquito head net that kept them off my face but they buzzed around my ears constantly.

We settled in for what turned out to be a long wait. There were two cops with us and they had a new piece of technology which they used to keep in contact with our bosses, a cell phone. Several times a day they would check in and their boss would bring them up to date. 'Soon' was the bottom line of these conversations, 'soon.'

We stayed the first time for a week. Each night after dark we left our camp and followed Brian up the runway. I convinced him to use bounding overwatch like we used in Vietnam, in this case, half of our element would watch the area ahead while the others moved up to a place about fifty meters in front of us and they watched until we joined up and repeated the process. When we arrived at our OP, we settled in and fed the mosquitos. It was very hot and humid, there was water all around us in the rice paddies, and suddenly I realized that it was just like the many nights I had spent in similar circumstances in Vietnam.

We were pulled out for drill on Friday and returned on Monday. The TF had found a road that led to a canal near our MSS. Instead of walking in, we waterproofed our gear and swam about 30 meters to the opposite bank and walked another 50 meters to the MSS. I brought my camera and large camcorder and took photos and video of our activities while we waited for the plane.

On our next resupply, Clayton and Jimmy K decided to swim down a jungle lined canal that was overgrown with water weeds. Beside the many cottonmouth snakes, the area was known to have large caimans, a creature that is a close relative of an alligator. I waited for Brian to object but he must have thought it was hooah and let them do it. I prayed for their safety.

When they returned Jimmy K told me they both expressed second thoughts as they floated through that mess in the dark. When they reached the resupply point, they picked up the stuff and took the short swim across the canal and returned to the MSS. When we got back to Austin I found a saying that Bil liked, "A Professional Takes The Minimum Amount Of Risk Necessary To Accomplish The Mission."

Meanwhile back at the TF, the crook's girlfriend kept up her visits and he showed his appreciation by spinning more and more elaborate explanations of why the plane was delayed and when it would be coming. This went on for six weeks. My Vietnam jungle rot returned to my shins and I poured tetracycline powder into the open sores like I had when this happened in South America 20 years before. It worked.

Finally, Customs got wind of the sting operation and told the task force that they were going to be indicted for drug smuggling because of the unusual nature of our activity. The TF contacted DEA and fessed up. Then DEA set Customs straight and took over the case. This didn't set well with Customs so they began flying an AWAC off the Colombian coast in an attempt to catch the cocaine plane in flight. According to the TF, the plane had a automobile type radar detector and stayed on the ground. According to 'intelligence' the coke got through later.

Camp Bowie-Selection

Captain Akers and Steve decided, that with all the new personnel, we would run an assessment course as a kind of gut check and weather gauge. They sent me out to Big Bend to coordinate the next operation and on the way back I was directed to secretly go to Camp Bowie and set up a land navigation course. I took the young airman named Ron, and when we finished in Big Bend we drove to Brownwood and set up a long-distance land nav course in the shape of a star.

A few days after we returned, Captain Akers and Steve called a meeting of the entire detachment and told us that we would be leaving in 8 hours for an operation in West Texas. They gave everyone a packing list and told us that there would be a full field layout before we left, "so make sure you have everything packed according to the list."

As soon as they left the team room, the place erupted. The guys were arguing with each other about what was going on. There was a lot of complaining as one by one the guys left to get their things. I got my gear and was back about an hour before we left. I had a meeting with Steve. He was happy about the way things were going and now he wanted to add a few more spices to this volatile stew he created. He told me to spread certain rumors to certain individuals to see if they could keep a secret.

The guys showed up and they were freaked out. They kept asking Steve what was going on, but he had them lay out their gear for inspection. After inspection, everyone was given vehicle assignments and we headed west toward the Big Bend. About midnight we turned north to Camp Bowie. Most of the guys were asleep.

When we got to their drop off points and woke them up, we read each of them a set of directions for the course, gave them a map sheet, and sent them on their way. For the next 48 hours they would move from one point to another. When they reached a point, they would be tested on a skill like weapons maintenance, blindfolded. There were a few other Jedi Mind Screw activities going on that increased the tension as the men moved from point to point that night.

The intense atmosphere got to some of the guys who had been plotting against the leadership and three of them decided that they had had enough. A certain sergeant walked to the CP and told Captain Akers that they needed to talk. Steve had told the captain that he was becoming a problem because of his subversion of the troops and now he was exposed when he tried to convince Captain Akers that my idea for this exercise was a bad one and it was ruining the morale of the detachment. Since it was the captain and Steve's idea, he suddenly found himself in an untenable position and resigned. Two others had come under his influence and they too decided that they had had enough.

The rest of the guys took the challenge and tried to do the course faster than the others. Greg Scott came across a goat on his route and when he racked his shotgun the goat fell over like he had a heart attack. Greg thought he might have killed the goat and took off running toward his next point. Another man decided to take a nap and missed several commo checks. He finally came up on the air and received a rebuke from Steve.

I set up my video camera on a tripod to videotape the guys when they came to my point where they had to disassemble and assemble a M 1911A1 pistol while blind folded. PB came to the point and as I was reading him the instructions, the tripod leaned over, and my camcorder hit the ground. Later, when I looked at the tape, it was like a TV station that is suddenly hit by a bomb, all of a sudden there's a noise then lines on an empty screen. We returned to Austin and when I saw Bil and he heard about the three guys quitting, he said, "I guess selection works," with a smile.

Equal Opportunity Training

One of our men was working in supply. He had made a name for himself when he answered a telephone in the team room with the accepted format of declaring who was speaking and 'sir' at the end. The voice of an enraged female officer on the other end began to berate him and said, "Don't you address me as sir, what would you say if I were a female general?" He paused for a moment before saying, "Well, what would be the odds of that ever happening?" Before it was over he was getting a unique counseling statement which made him an honored member of the 'manly man female haters club' that the single guys in the unit called themselves when they went out to the bars.

Steve Leaves

Camp Bowie was Steve's last operation. He told me that when he was in an OP across the river from Piedras Negras watching a huge aluminum plant on the Mexican side at night with its orange red flames reflected in large pools of water in front of the plant, he felt like he was looking at hell. Tom Payne's departure had delayed things but now Steve was going to Houston to be an Intel analyst with a High Intensity Drug Trafficking Area (HIDTA) major drug squad.

Our next 'benefit' from our mission with the 5th Group boat team now came to fruition. Captain Akers and Steve had made some deals with the guys at the 5th Special Forces Group and we got a hard charging active duty SFC with the tower of power to transfer from the 5th Group to the National Guard without missing a day of work. Captain Akers wanted a hard ass to square us away after the fiasco at Camp

Bowie with his fair-haired boy leading a rebellion against him so that's what he got.

Sergeant Ralph showed up at the team room on a Monday morning and was introduced to the assembled men as the new detachment sergeant. He wasted no time in letting us know which way the wind was blowing when he announced unit PT the next morning at 0700. Anyone who was a second late by his watch would fill 50 sandbags.

Special Operations Detachment

SE Texas-Woodford Greg Brian AN/PRC 77 with DMDG

George C Sherriff Bil 'Steve' Time to Go!

C-26 'Dollar Ride.'

LRS: Foreign Weapons

Lt Noriega

SPIES

LRSLC: SPIES-Riding the Strings

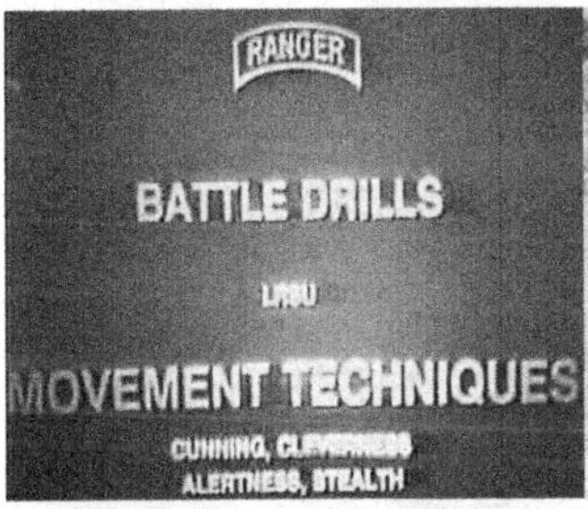

LRSLC: Tactics Class

Tom Reynolds

Juan Hinojosa

LRSLC: CPT Hunter

AN/PRC 70/DMDG

Hand Crank Generator

LRSLC: Honor Grads

CPT Hunter

Appreciations

Boat Operation

Waitin for the Plane in the Swamp

Jehovah will strike you with madness, blindness, and bewilderment of heart; Deuteronomy 28:28

Chapter 13-January 1992

The Golden Years

Our new detachment sergeant, Ralph, was clueless about our operations. He tried to push us during PT but we were so busy he didn't have much effect on what ever Captain Akers thought we were doing wrong. At the same time that they were trying to 'square us away,' Tom Stephens had been busy working with George C to get cameras and one day they came in. George had worked his logistics miracle and we now had six complete Nikon F-4s 35mm cameras with a 300mm and 600mm telephoto lens for each camera. We were also issued 6 four-wheel drive Suburbans and a lot of other gear was in the pipeline. This was the start of what I referred to as the golden years of SOD.

We were now established as a unit. Personnel turnover was high. The younger men would burn out quickly for various reasons. Part of the problem was that we were not doing what the typical Airborne Ranger had ever done in the past. We had entered into the 'secret' world of covert and clandestine type intelligence operations. They had a new name, "low visibility," which meant that if we were discovered we couldn't kill our discoverers and if we couldn't escape, we would have to face the music. There were plenty of organizations doing similar things, but their existence was classified, and we were not in the loop. If we had been, we might have saved a lot of time and effort but as it was, we had to reinvent enough wheels to drive around the world. We carried pistols and wore civilian clothes. We were issued ID cards that stated that we could carry concealed handguns and automatic weapons. It was high-speed but it was spying, not fighting.

Many of the young men were not suited temperamentally for this kind of work. Hiding in the brush along the Rio Grande was enough like an LRS mission that most of the LRS soldiers could hang. Hiding in a small house in San Antonio and watching a driveway for hours on end was something else. Other units doing similar work spend a lot of money selecting and training the right person. We had to take the ones who volunteered and those who were not suited for the work eventually left, but not before they expressed their discontent by rebelling against the leadership and causing dissension in the ranks. Our first rebellion took place at Camp Bowie and three operators quit.

Just before the Camp Bowie outing, Captain Akers decided that we needed to expand the operations section. I was given Airman Ron and SGT CB. Ron was a young Airman and happy in his work; CB was an infantry sergeant from active duty and he wasn't. Later Lt. Steve, the XO of LRSD was hired and assigned to work with me. It was an unusual arrangement because he was an officer and I was a sergeant, but we were able to get along.

The OPTEMPO began to increase. We had hired more men and we were getting more equipment. Bil was wheeling and dealing with all the Law Enforcement agencies: Feds, State, and local. We were one of the assets that he could bring to the table and when an agency requested free personnel for Intel and other types of support Bil would remind them that he could also give them 24/7 surveillance support. This resulted in more agencies requesting us for both surveillance and training.

Texas A+M

Captain Akers's relationship with Doug at Texas A+M was really paying off. We began to attend other law enforcement schools and soon we were teaching some. In January we went to Texas A+M to participate in their Tactical Operations for Narcotics Enforcement or TONE course. We were students but they wanted us teach some military subjects. We planned our classes, did rehearsals, shined our boots, and starched our BDUs; this began our detachment's involvement with police training. I taught a map reading course using the USGS maps and a lesson plan that covered a few simple techniques that I thought the cops could use when operating in a rural environment. We also taught first aid, small unit tactics and patrolling.

The class was a kind of SWAT school for narcs. There were a lot of dynamic entry procedures for clearing a room or a building. One of the techniques that one of the instructors had invented was a peculiar way of carrying a shotgun when one crossed in front of a doorway. It required some contortions and was difficult to master, even if you were interested, which most of the cops were not. Everyone was polite however and did their best so we could make the instructors happy and move on; all of us but one of our guys.

He decided to call BS on the whole concept which started no little turmoil in the class. The instructor was angry because we were military and were expected to follow orders without comment. I tried to get his attention by giving the hand signal for 'cease fire' but he was on a roll and ignored me. Finally, the instructor told him to leave the classroom and after a conference with Captain Akers and the instructor, he returned to the classroom in a more compliant mood. I agreed with him about the tactic, but he forgot the primary rule about rapport, so we had to apologize profusely to Doug, explaining that he had landed on his head a few too many times during parachute operations. We briefed the class on our capabilities and gave them the information they needed to request us to support their investigations. We met law enforcement officers from all over the state and they began requesting our services based on relationships established at these schools.

The National Guard Bureau-Paperwork

At the same time that our OPTEMPO increased, the bureaucratic demands increased. The NGB Counterdrug Directorate was feeling the need to remind us that even though they were not in our chain of command they did control the budget and that gave them the leverage they needed to demand more and more reports about our activities so that they could justify their existence to each other. They did need data to brief upstream on the money river that came out of the Pentagon and originated at the U.S. Capitol across the Potomac and they became more and more demanding each year.

They did streamline our AARs down to a one-page form that we had to get the cops to fill out and sign and we had a similar form to do the same. These two forms provided little intelligible information, but we tried to meet their demands for paper as expeditiously as we could.

Their lawyers created a Risk Assessment checklist that seemed to be designed to regard most of our operations as High Risk and not appropriate for us to conduct. We knew that NGB hated the fact that we were armed and they lived in daily fear that the cowboys west of the Mississippi were going to shoot some children at a Sunday School picnic on the Rio Grande and cause the demise of the National Guard

Counterdrug program which was the easiest and most money that most of them had ever made. The demands from NGB CD increased every year and our staff for the program bore the brunt of the work required to keep those people happy. Occasionally there would be extraordinary demands for complete explanations of what ever came to the minds of the beltway bandits.

Congress

There was a Congressman who was also a LTC in the Army Reserve and he took exception to the large budget that the National Guard was getting for CD work and decided that an investigation should be made to determine if the Guard's Counterdrug activities were degrading its member's ability to fight and win this nation's wars. Of course, that was a hard question because the Guard hadn't been called upon to perform such service in any great numbers since 1953 so how could they tell? This study took hundreds of hours of our time because a Reservist was jealous of the Guard having something to do.

The Army Reserve plays an important role in something and I'm sure that someone in the know could enlighten me as to what that is. A few years later, I had nothing but bad experiences dealing with the leadership of the reservists that worked for me. The reservists themselves were like most other soldiers. Give them purpose, direction and motivation and they'll do the work.

Laredo

Bil made some deal with the US Marshal's and some of us were sent down to Laredo to assist them in searching all the vehicles that had been seized by the BP and Customs and were stored in a giant impound facility. Vehicles involved with drugs were seized and eventually auctioned off by the Marshal's Service. There was the possibility that there were drugs and/or cash still concealed in the vehicles, so they needed help searching them one more time before they were sold. We spent 5 nights in the HoJo near the bridge and drove out to the lot each morning and searched cars. It was a drag, but we did the best we could and Bil never sent us on another mission like that again.

SWAT School

As soon as we got back from Laredo, I was told to be ready to go to A+M Monday morning to attend a basic SWAT school. I went with JT, Juan, Jimmy, Nick, and Spock. The first thing they did at the school was to give us a PT test which I barely passed. The rest of the detachment showed up for one day and taught a class on rappelling. One of the SWAT officers refused to rappel off the tower and his boss sent him home right then, he was off the team.

We had classroom instruction and practical exercises culminating in a scenario-based exercise that the whole class participated in. It turned into a pretty hilarious scene in the HQ where I had been assigned as the negotiator. The student commander was a police lieutenant and SWAT commander for a small department near Waco.

The school psychologist was the criminal who had taken some hostages and created a barricade situation. While I was negotiating with him, the SWAT team was planning an immediate action plan if he started killing the hostages and a more deliberate plan that they would put together if I could buy them some time. We had to have the exercise completed by a certain time because the course would be over, so the instructors told the student commander to execute a dynamic entry before he felt he was ready. He refused and now we had another scenario being played out in the HQ. I sided with the student leader and told him to tell the instructors that he was civil service and wouldn't order his team to go until he was ready. The instructors fired him anyway and told the student assault leader to execute whatever plan they had, and they did, killing the bad guy. We had a quick debrief and the school psychologist told me that I had been very persuasive as a negotiator and asked me where I got my training. "From my wife," I said and that got a good laugh from the class.

The Valley

Steve was now an intel analyst in Houston with a DEA run multi-agency multi-jurisdictional task force that was created when Houston was designated as a high intensity drug trafficking area or HIDTA. With his police and SF background he quickly became the go-to guy for their investigation of a drug trafficking ring based in a South Texas city. The investigation had stalled, and he recommended that they request SOD to do video surveillance on the main crook's house in a residential neighborhood of the city. They did and the request was approved. I met Steve at a nearby airport. First we went to the target neighborhood and did site selection. Then Steve and I flew over the neighborhood in the C-26 and took photos and video. This was the type of surveillance I had been envisioning so I was disappointed that I was not tapped for the operation.

Brian, JT, and LT Steve, went down to South Texas to get the goods on the crooks. They set up a newly released Canon video camera, the LX-1, which would accept other lenses and began watching the place. Steve stayed around to get them up and running.

They soon realized that there was a stop sign that blocked their view of the crook's front door, so their first task was to wait until 4AM when Steve and LT Steve went out with some screw drivers and raised the sign on the pole. No one seemed to notice, so they settled in and began to video tape the comings and goings of a heroin trafficker.

The Organized Crime Drug Enforcement Task Force (OCDETF)

The next job started off with a mysterious phone call to Jimmy K our admin NCO. I watched as Jimmy listened for a moment then said, "What did you say." He turned me and said, "That was strange." I asked him what he meant, and he said, "I just got a phone call from someone and all he said was'77.' Do you think I should have said Sunset Strip," he said, referring to an old TV show.

"Sounds to me like one of Bil's SF squirrels," I said and went to Bil's office and told him what happened. He told me that it was an FBI agent who had been in the 77th Special Forces Group back in the 50s. "They were always big on 'bona fides," he said, "I'll give him a call." I went back to the team room and a few minutes later Jimmy K was transferring a call to my phone.

The man on the other end introduced himself as an FBI agent. He said Bil told him to talk to me about setting up an operation. I asked him when he wanted to get together, and he thought we should eat lunch in Houston the next day so I got the address of where to meet him. This was the first time we were going to work for a Federal Agency besides the BP, and I was excited. I went to Bil's office to get a copy of the request and there it was, the 'Big Time.' The fax cover sheet for this request was from an Organized Crime Drug Enforcement Task Force or OCDETF. These were Task Forces established to go after major drug trafficking organizations, in this case, the Gulf Cartel.

I went to Houston the next day and met with the FBI agent in charge. We agreed to provide some operators with cameras to ride with his people and take photos while conducting mobile surveillance. He also wanted some administrative support in his listening post where they were monitoring a wiretap. We had a plan and we established a date to start. When I got back to Austin, I got with Captain Akers and Ralph and we worked out the personnel and logistics.

The Saga of Sergeant CB

When Captain Akers told us that he was going to hire a guy from one of the infantry battalions to work in SOD, he looked at me as he was talking and said that the new guy would be assigned to operations and would be working for me. We usually interviewed people at that time and except for the Air Guard people that were forced on us, we hired only people who were recommended by our own guys.

The new guy, Sergeant CB, showed up a few days later. Captain Akers introduced us, and I introduced him to the other members of the detachment. I told him what his duties would be and began training him on how to order maps and how to put together a target folder. At lunch, CB would go back to the operators' area and talk; his relationship with them soon became adversarial.

CB wanted to do what they did, the hooah things, and they re minded him that they were airborne, and he wasn't. "Just because I don't have those 'mosquito wings' doesn't mean I can't do what you guys do," he would say. They would argue back and forth until I would go back and tell him that we had work to do.

Captain Akers assigned Ron W, Jim K, and CB to work in the FBI listening post and we assigned some other operators to work with their FBI counterparts. They all moved to Houston and were staying in local hotels. Shifts were established for the guys in the listening post and I went back to Austin to prepare for other operations coming up.

About a month later I got a call one morning from Jim K telling me that CB had not shown up for work. Jim had gone down to the listening post and found CB's mission related materials and his .45 in a briefcase. I told Jim to check CB's hotel room and he found that CB had checked out. A van was missing and none of the other guys had seen CB that day. We told Bil about the situation and notified the FBI agent that one of our guys was missing. Then we continued work on our projects and waited to hear something.

That afternoon we got a call from the MPs at Fort Sam Houston in San Antonio. Ralph talked to them. They told him that they had found our van parked in the PX parking lot and in it they found a note from CB. The message was a fairly innocuous statement that he was quitting his job, but certain words had been lightly underlined, and they said that he had been kidnapped by the Drug Traffickers and was being held hostage. This was a WTF moment for Ralph and he said as much to the MP. His time in the Rangers and SF had given him a lot of experiences but this was far beyond those things.

Ralph sent me and Jimmy K down to Ft Sam the next day to get the van and to go by CB's house and talk to his wife. When we talked to her she said that CB told her he was on a secret mission and that she hadn't heard from him in a long time. We took the van back to Austin and continued to work on other projects. We waited and speculated. At one point the MPs called Ralph and I heard him say, "like, WTF!"

We went home that night and the next day when we came to work, the CD Task Force officers were in our office asking questions about the situation. At about nine o'clock we got the next piece of the puzzle. One of the Task Force members in San Antonio faxed us the front page of a San Antonio newspaper with the headline, 'National Guardsman in shootout with Bexar County Deputies.'

The story named CB and related that he had gone into a 7-11 and robbed it. He left on foot and was walking down the road near the store when two Bexar County Sheriff deputies recognized him from the dispatchers clothing description of the robbery suspect and pulled up behind him. When they hollered at him to turn around, he did and began shooting at them. They returned fire and they all continued shooting until CB ran out of ammunition and surrendered.

Captain Akers was as dismayed as the rest of us and asked me "What do you think we should to do to prevent something like this happening in the future?" I gave it a moment's thought and said, "I think we need to review our firearms training program." He didn't think that was very funny.

The story continued and we found out that CB had been a busy man since he dropped the van off at the Fort Sam PX. He took a cab down to Broadway which is near Fort Sam and checked into one of the many cheap motels there. He ordered a pizza and had it delivered to his motel room. When the pizza delivery man showed up at CB's door, he pulled a gun on him and made him come inside the motel room where he tied him up, took his money and the keys to his delivery car, leaving the delivery man in the motel room. He drove away in the pizza delivery car and began his short crime spree. Of all the things that happened, his ordering a pizza captured all our imaginations and for several years after, whenever someone was mad about something, or they wanted to convey their frustrations, they would say something like, 'I'm going to order a pizza if something doesn't change!'

The FBI visited CB in jail and debriefed him. Apparently, CB was spending a lot of time in a topless bar near the hotel. He had developed a taste for cocaine which he was able to buy from the usual suspects that hang around those places. He must have run out of money and any good sense, so he ended up in jail charged with robbery and attempted capital murder of a police officer. Our big concern was that he would reveal the existence of SOD and our operations. Whatever the FBI told him worked because he went to trial and was convicted, sentenced, and put in prison without any further publicity.

Texas A+M-Psych Eval

Captain Akers was alarmed by the whole business. He talked to Doug at A+M, seeking some solution to the possibility that we were all on the edge of 'ordering a pizza.' Doug offered to have us all evaluated by the psychologist at his law enforcement training center. The psychologist taught hostage negotiation and considered himself an expert in understanding the psychology of criminals. I don't know if Captain Akers believed that this guy could actually help us, but it would show the colonel that we had tried to determine whether there were some among us who were about to run amok.

We all drove out to A+M for some 'training.' When we got there the 'doc' assembled us in a classroom and told us that we were going to take an examination, purely for 'academic' purposes. He assured us that it had nothing to do with us personally, but we had been volunteered to help him practice using this test. I looked around the classroom and the faces of the guys seemed to indicate that they were not buying one bit of what he was presenting. Like deer during the hunting season in the Hill Country, we had become very sensitive to any threat to our existence which included our jobs.

The test was called the California Personality Inventory which is similar to the Minnesota Multi-Phasic Personality Inventory (MMPI). The MMPI, I was told later, was expensive so we used the California test which was free. The doc asked us to be completely honest about our feelings as we answered the questions, "purely for academic purposes, you understand." We understood alright.

I started answering the questions. Questions like, 'Do you ever feel like hurting someone?' were easy, of course not. This was a long test with many questions that were similar and it was hard to remember how I had answered a similar one but my rule of thumb was to find the answer that I thought they would want to hear and give it to them.

We finished the test in about an hour and were released to go eat lunch. We all met at a local restaurant and the guys were pretty agitated. "What's going on?" was the question of the moment but after while the group reached a consensus; that Captain Akers was trying to find out if any of us were ticking bombs so that he could defuse us before he had to deal with another CB incident.

Soon the guys were laughing and asking each other about how they answered certain questions. I realized that we were all thinking the same thing which showed that we were all paranoid; but 'just because you're paranoid doesn't mean that they're not after you.'

After lunch the doc went over the test with us. The gist of his analysis was that we had all scored very low on the integrity scale. Captain Akers was mad but there was little that he could do because the doc wasn't going to give us the same test again. I tried to console him by telling him that only a crazy person would have answered the questions according to his feelings, so he didn't have to worry about us.

East Texas

Captain Akers grabbed me one morning and told me we were going to East Texas to do something with a TNCP Task Force. The rest of the detachment was spread around Texas and he and I were the only ones there. On the way I was stopped by a DPS trooper for speeding. Since we were on a 'secret' drug mission I didn't say anything but when he saw my police uniform on my driver's license photo, he told me to slow down and let us go.

We met with Jerry, the TF Commander, and his deputy, Kenny. They told us that they wanted to get photos and/or video of the people that were hanging out at two crack dealing places in town. They didn't really have any plan. This would be the continuation of my experiences on the border where we developed a plan with the cops, helping them to integrate our assets while focusing on their investigation.

We went out the next night and did some drive-bys of the various sites and I took a little video to help with target folder development. As it turned out, we didn't need to make target folders because Captain Akers was called back to Austin several days later leaving me with the cops and everything we did after that was done in typical cops' fashion, pulling it out of our forth point of contact, as they say in the airborne.

East Texas-The Van OP

There were two crack dealing sites and two homes where the leaders of the crack dealing groups lived. The first place we watched was an apartment in a line of single-story apartments. The cops had a van parked at a nearby restaurant and they were using it to watch the apartment. The apartment was about 200 yards from the van and the cops would watch for a while then they would go into the restaurant and get something to eat. Most of the cops on the TF were pretty overweight.

OPSEC is not always a major consideration in police work. The cops knew all the workers in the restaurant, and they would confide with the middle-aged women who was the manager. She had a daughter who worked there also. The daughter was meeting her boyfriend during her breaks, and they would have intimate interludes in her car that she parked next to the van The cops would observe these activities as part of their surveillance duties.

One day we were sitting around in the Task Force office BSing and one of the cops told us that he had been in the van the day before and observed the low rent rendezvous in the car next to the van. When the couple finished their activities, the girl went back to work. She and her mother were behind the counter when he came in for a chow break. The mother asked him how the surveillance was going.

As she and the cop talked the daughter began to understand that her lunch time activities had been observed by the cops over the past few weeks. The cop told us he was watching her out of the corner of his eye while he was talking to her mother and the girl kept getting redder until her mother noticed and asked her what was wrong. She told her mother that something she ate was hurting her stomach and excused herself to go to the restroom.

I went into the van the next day with my video camera and my 35mm film camera. The distance was too great to do much with the video camera, so I mounted my 35mm to my spotting scope using a t-mount adapter and waited for the dopers to show up. Several hours later it was lunch time and I wondered if the girl would be joining her boy-friend in the parking lot for a 'nooner' but she never showed up. I guess she learned about the all-seeing eyes of the cops and would be more careful with any public displays of affection.

The crooks showed up in the afternoon and there was a small crowd of them gathered around the main crook. They were all looking at his 'ride.' He had a small sedan that was tricked out with some special lights and a boom box sound system. They were pointing and laughing while I took their pictures. The distance and the light made it difficult to get good shots, but I did the best I could.

The crooks all left at around 4pm and I called the TF on the radio in the van and they told me to meet them at the office as soon as I had locked up the van. I told them I would grab a snack in the restaurant and be there soon. I went into the restaurant and sat at the counter. The girl waited on me, but she wouldn't look at me and seemed embar-rassed. She knew I had been in the van and assumed that I was a new Task Force officer. I got my snack and left.

When I got to the Task Force office, I had a meeting with Jerry and he introduced me to Benny and Hutch, two of the TF officers. Hutch had really long hair in a ponytail. It was dark with a lot of gray. He looked like a Hell's Angel and so did Benny; later they told me that they operated undercover with biker gangs in another state. They had spent up to six months at a time riding with some of the nastiest criminals in this country. They both had real nice Harley Davidson motorcycles which they liked to ride. Hutch was a Vietnam vet, so he and I had some instant rapport.

East Texas-The Tree line OP

Jerry told us that he wanted us to do surveillance on the other residence the next night. He said he had information that there would be a party there and he wanted us to video tape everyone who went in and out. He gave me a Canon LX-1 Hi-8 camcorder and a night-vision device that could be attached to the camcorder. This was the same camera that the guys were using down south. I took the camcorder outside after the meeting and began practicing with it.

It took me a long time to figure out how everything worked. You had to find the setting on the camcorder that gave the most sensitivity to light. Then you had to focus the night vision device and then you had to focus the lens attached to the device. There was very little depth of field so every time you aimed at another target you had to refocus the device and the lens. I worked with the system until about10 that night and then I returned to the motel and went to sleep.

The next day at noon I went to the TF office and met with Benny and Hutch. We worked on our plan to videotape the crook's party that night. The crook's house was also near the restaurant but there was a line of trees blocking the house from the van. We would have to hide in the tree line and shoot video from there.

There were a number of problems associated with this plan. The crook had a lot of lookouts. He paid the kids in the apartment complex to keep an eye out for the cops. There was no ground cover so we would have to cross a large open area to get to the tree line.

Hutch and I agreed that we would need a diversion. The tree line itself was open at ground level and there was very little vegetation to hide in. There was one bush in the tree line that would give us a little concealment but if we lay under it our legs would stick out and if anybody walked by, they might see us.

I went to the local Wal-Mart and bought two olive drab bed sheets and a can of black spray paint. When I got back to the TF, I painted some black stripes on the sheets and hung them out to dry. I then cut a hole in each sheet at one end. The idea was to roll the sheet up to the hole. The hole was for the head. You put your head into the hole and draped the sheet-roll over your shoulders like a scarf.

We told Jerry that we would like to have a city police officer drive into the apartment parking lot with his overhead light on and stop some of the dopers for a talk. Hutch and I would be hanging around the dumpster of the restaurant posing as transients and when the cop drove into the apartment parking lot, we would walk to the bush we had identified and set up our OP. We would lower the sheet around our shoulders as we walked the last few feet to the OP, and we had a cape down to our feet by the time we laid on the ground under the bush. Jerry liked the plan and told us he would arrange everything.

That evening Hutch and I were dropped off on the highway near the restaurant and we walked slowly over to the dumpster. We got to the dumpster, expecting to see the cop car pull in and so we just stood there. After a minute, the cops had not shown up and we were starting to stand out, so I opened one side of the dumpster and started looking around for something to eat. Hutch got into it quickly and I handed several items to him while leaning into the dumpster.

After a few more minutes Hutch saw the police car approaching the apartment complex and told me to get ready. The cop turned on his overheads and we walked slowly toward the bush. We lowered our capes down across our backs and ducked into the bush.

We just lay there and waited. We watched the cop talking to the people in the parking lot as the lights flashed. I looked at everyone I could see, and it didn't appear that anyone had noticed us so I opened my bowling ball bag where I kept my camera and night vision equipment and began setting up the LX-1. I was up and taping in a short time. The city cop left amid the hoots of the young dopers standing around and little by little they settled into their routines.

It was completely dark now and I was adjusting the two focus points and the gain on the night vision device. I began to get a good picture and every time someone went in or came out of the apartment; I would tape them. Their faces were mainly in profile but when they turned a little in my direction, I would get a good picture of them. There are about thirty frames a second in video tape and you need only one frame to get a picture. I noticed however that the zoom lens was not getting me as close as I wanted to be. It is good to be able to put the subject from the shoulders up to the top of their head in the video frame. I was getting them from the knees to the top of their heads. With the light, night vision devices, and the zoom lens, the picture was pretty soft, and I felt we needed to get a closer shot.

We stayed out that night until midnight. It was springtime and we got a little chilly but not too bad. We waited until all the outside activity had ceased. The restaurant sign was off, and it was darker in the open area behind us. Hutch called the TF guys on his handheld radio and we got up a walked back towards the dumpster.

The TF car met us on the highway, so we walked down the road looking like two transients down on their luck. We got a call telling us that they were coming up behind us and that there were no other cars between us and them. We turned around and started hitchhiking. When they pulled up, we talked to the driver through the window for a few seconds and then we got in and drove to the TF office.

We went back to the tree line the next night so that Hutch could retrieve his back-up pistol which had fallen out of his pocket during the first night's surveillance. While we were hiding in the bush looking for the gun, I saw a young woman walking on a path to our right that passed right in front of our location. I nudged Hutch and froze. She was acting like she was just walking along but her eyes were going everywhere. I had seen her the night before hanging around with the other dopers at the apartment complex. She was doing a recon for the dealers, so I figured that we had been spotted walking in and she was there to confirm it. We waited until she passed us and because of the sparseness of the vegetation I had to believe she saw us as she walked past. Hutch and I agreed to leave as soon as he found his gun. He found it a few minutes later and we left the same way we came and were picked up on the road and taken back to the TF office.

East Texas-The 'Corner' OP

The next day we met with Jerry and Bennie. Jerry said he wanted to do the 'Corner' next. The Corner was a local gathering place where a crack dealing gang sold their goods to the locals. It was in the middle of town but there were several large vacant lots near the place, and we hoped to get close to an abandoned building that was near the gathering spot. There was a streetlight nearby which would help the videotaping.

We decided that we to continue with our appearance as transients and approach the target from a road about 100 yards from there. We would be dropped off on the road after dark, cross a barbwire fence and move through thick brush to the target. The distance was so short it looked like no problem.

That night we were dropped off along the highway and we walked down into the bar ditch. There was a 4-strand barb-wire fence and it took us a little while to cross it. Fortunately, it was pretty dark and we got on the other side without being observed. Once on the other side we were in thick brush. The ground was covered with leaves and other debris. I had calculated that we would have to move about a hundred yards. Each step we took made a sound. We tried to move very slowly but it was pretty much impossible to keep from making noise and soon there were several dogs barking furiously near the target. We stopped and waited until they stopped barking and then we would take a careful step. The dogs would go off again and it became apparent that they were alerting on our movement. We didn't know what effect this was having on the dopers, but it was making us paranoid.

We returned to the fence to consider our next move. After a half hour we gave it another try. We moved even slower this time but about two thirds of the way to the target, the dogs went off again and wouldn't stop. We turned around again and returned to the fence and called the TF guys who were waiting nearby. We requested extraction and they picked us up along the highway ten minutes later.

We returned to the TF office where we met with Jerry and Bennie. They gave us a ration of abuse. 'Vietnam vets who can't sneak up on some dopers, no wonder we lost the war, etc.' I was thinking hard. Hutch didn't seem to have any solutions and I didn't want to fail here. We were just becoming known to law enforcement agencies away from the border and I didn't want these cops to think that using us was a waste of time. I called on the Lord under my breath a few times and waited until they all stopped talking. They were looking at me and I wasn't sure what to say but I opened my mouth and started speaking. What came out surprised me but it made sense, so I just kept talking. I told them that we should go home and go to bed. Jerry and Bennie were looking intently into my eyes at this point. I pushed on.

"We should meet back here at 4:30 in the morning." Now they were really staring. "The dopers should all be asleep by then and the dogs will be sleeping too. We will leave the cameras at home and go in with garden clippers. After we cut a path through the brush, we'll remove the leaves and other debris. We should be able to get right up to the edge of the wood line and make a good OP. When we're done, we will come out; it shouldn't take more than two hours."

Jerry and Bennie looked at Hutch and asked him what he thought of the idea. "It might work; I think we should give it a try." Jerry and Bennie agreed and arranged for one of the TF officers to drive us in the next morning. I went to Wal-Mart and bought Hutch and me each a pair of hand clippers and a small whisk broom for brushing the trail. I returned to my motel room and went to sleep.

The next morning, I drove to the TF office where I linked up with Hutch and we were driven back to the drop-off point. There were few vehicles out on the road, and we were soon out of the vehicle and over the fence into the brush. I told Hutch that I thought we should move about 10 feet into the brush and establish a base there. We would clip out a small area that we could move around in. He agreed.

He watched and listened while I worked. I got down on my knees and clipped out a small area. I took the clippings and bundled them together. I enlarged the area, putting the small bundles of clippings into a space in the brush away from the entrance to our base from the highway and away from the direction we would be going towards the OP. Hutch was listening a few feet away and apart from my snipping of the clippers it was very quiet; so far, so good.

After I cleared a circular space about ten feet across, I stopped and took a break. Hutch was amazed that no dogs were barking. He volunteered for a turn on the clippers. I pulled out the new pair that I had bought for him at the Wal-Mart and gave them to him. He got down and started to clip his way towards toward the OP. I swept the trail of leaves and took some video during this project because I wanted the guys in the unit to see what they should be doing when they encountered a situation where the book solution wasn't working. We traded clipping duties and made it to a good OP site in about 45 minutes.

We developed the OP, camouflaging it and making sure we had good fields of view of the areas where the crack heads hung out. We moved back and forth on the trail, widening it here and there and making sure that the floor of the trail was clean. We agreed that we were done and called for extraction. The TF guy picked us up and took us back to the office. At about 8 o'clock that morning Jerry and Bennie showed up. We told them that we were done and that we thought the plan would work that night.

I told them that I was concerned that the lens wasn't strong enough to get the pictures we needed. Jerry gave me the address of a camera store and told me to go there and buy whatever I needed. As soon as we were done at the office, I drove to the camera store and found a 35mm lens that would give us good close-ups. I charged it to the TF and returned to my motel to practice with it for a few hours.

I returned to the TF office just before sundown. Hutch and I were dropped off after dark and we found the fence post with the beer bottle that Hutch had put there as a marker. We crossed the fence quickly; by now we had had some practice, and found our base area. I set up the camera and night vision equipment and then we moved up the trail toward the OP. I was waiting for the dogs to go off, but we made to the OP site without a peep from them.

Hutch and I set up our equipment. The abandoned building was about 30 yards to our right and the main gathering area was about 50 yards to our right front. It was 7:30 in the evening and there was a steady stream of people walking up the road from our left to our right. They would gather under a streetlight and talk and would pass what looked to be marijuana cigarettes among themselves.

I started shooting video and Hutch knelt next to me with a pair of binoculars. He was getting the license plate numbers of the vehicles that stopped there while the drivers talked with the dealers. Hutch was giving the numbers to other TF guys using his hand-held radio. I tried the first camera lens system with the night vision scope. I started taking video with it and saw that I still needed more magnification, so I told Hutch I was going back to the base and change camera lenses.

I moved quickly down our trail to the base and changed lenses. I returned to the OP and started shooting video again. The new lens brought the faces to fill the view finder. I had to learn to focus the new lens whenever a person moved a few feet toward me or away from me. I was totally involved, and I was counting on Hutch to watch for anyone coming up on us.

Surveillance Video-R+D

Taking video like this was what I had been working for since we started SOD. I used to sit around with Tom after a long day working on the current operation and we would speculate on what SOD would be doing in the future if congress continued to fund the counterdrug program. I told him that we should try and get video cameras so that we could learn to use them with night vision equipment.

I thought that if we could become experts at taking video from hidden positions, the law enforcement agencies who were conducting investigations of drug traffickers away from the border, would want us to assist them because cops didn't like to live in the woods and the overtime issues prevent them from conducting lengthy surveillance even if they could find one of their own to do it.

Because our equipment procurement procedures at that time were so slow and cumbersome, I started buying video equipment with my own money. I would eat cheaply while on the road and with my per diem that I saved; I bought a Hi-8 Video Camcorder. It was a small device and I thought it was perfect for surveillance operations. It was expensive but it was the perfect test platform for me.

I used my camcorder that night to take video of Hutch while he was looking through his binos and I got video of the dopers as they walked past us on their way to and from the dealer's area. I wanted to take this video back to the program and show Bil, the colonels, and the boys what we could do if we got some equipment. I didn't know at the time that the guys in south Texas were using the same type of video camera that the Task Force was letting me use and that they had reached the same conclusions.

I shot video using several configurations. I used the LX-1 with the TF night vision scope and the new lens. I used my Hi-8 Video Camcorder with a night vision device that was designed to be mounted on a .50 caliber Machine Gun. It had good light amplification but not much zoom capability. I could get good video of the cars and people, but I couldn't get good facial recognition. I used night vision goggles with my camcorder but had to hold the goggles with one hand and the camera with the other while trying to focus and adjust the lighting.

With both the weapons sight and the goggles, I had to align the end of the night vision device that you look through with the lens of the camcorder and hold them together while I tried to find my target. When I found it, I had to focus each device and adjust the shutter-speed and exposure to get the light levels right. I tried zooming in with my camcorder lens. You could only zoom in about ¼ of the way. All you were doing was magnifying the screen of the night vision devices. After everything was set, I would hit the record button and shoot my targets. This was a cumbersome way of doing things and I was constantly thinking about how to shoot low light video in a way that the average operator would be able to do it successfully.

East Texas-The 'Corner' Part II

We continued to shoot video of everyone who came to the dealer's circle. Hutch would call out LPs and I would shoot. Hutch wanted to smoke a cigarette every hour so we would pack up the stuff and move down our path, which we had begun to call the Ho Chi Minh trail. We were delighted with our ability to be able to move into the doper's backyard so to speak and the dogs never alerted on us.

Hutch confided to me that the TF had tried to go through this same area sometime before and the dogs had foiled them. He said that the TF guys had laughed at my suggestion that we cut a trail. Hutch told me that he had been willing to give it a try because I was a Vietnam vet, like him and he wanted to give us a chance to help them.

Jerry told us to come out at 1:00 AM and so when the time came, we called for extraction and went home. We did this for the next few nights, working through the weekend. Jerry and Benny were looking at the video and eventually they decided we had enough. I was in the TF office one afternoon and I saw the Intelligence Analyst and a cop showing the video to an informant. Every time a new face came on the screen, the informant would give information such as the person's name, where he or she lived, with whom they lived, the type of car they drove, what they did in the dope dealing world, etc.

Tx A+M TONE Course: Map Reading and Tactics

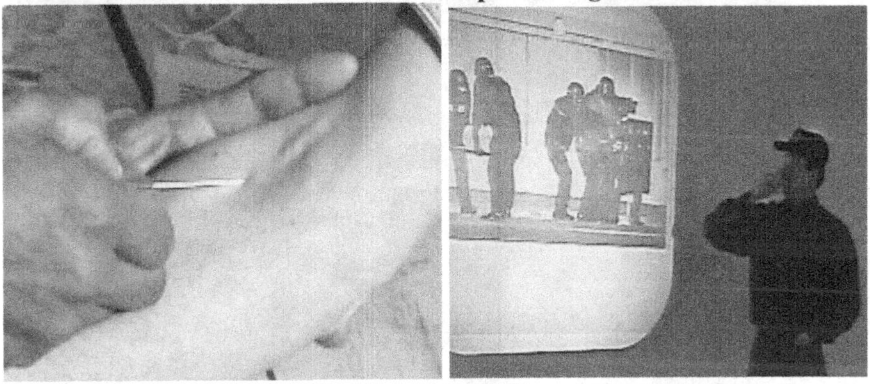

A+M TONE Medical Training (IV) and Dynamic Entry

E. Texas Van OP

E TX Corner OP-Clearing Path, Prepping OP, Call for Extraction

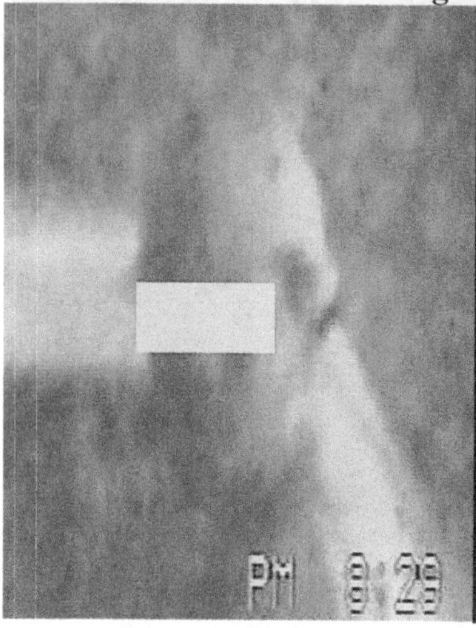

Suspect #1

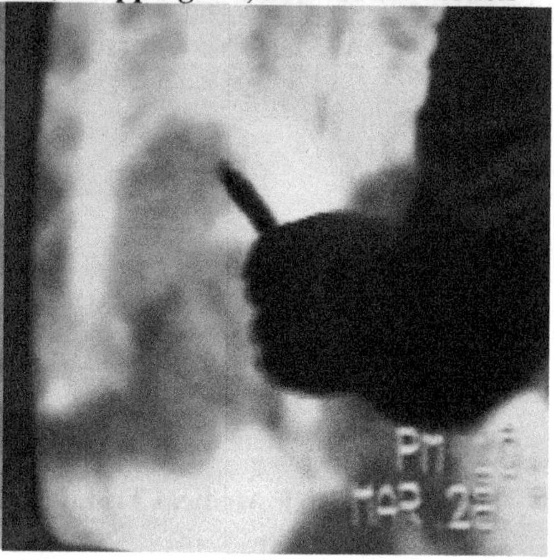

Analyst at work

Therefore putting away all malice and all guile and hypocrisies and envyings and all evil speakings, 1 Peter 2:1

Chapter 14-April 1992

Surveillance Video-The Rail System

When I came back from East Texas, I talked to George C about what we could get on our limited budget to enhance our ability to take evidentiary photo and video. I was sure now that we would have all the work we ever wanted if we could provide the cops with visual evidence of criminal associations and activities.

I had been following the John Gotti trial in New York. The video surveillance footage of Gotti and Sonny Gravano together established Gravano's credibility when he rolled over on Gotti as part of his own plea bargain for a pass on the 19 murders he had committed while a member of Gotti's mafia.

Federal Drug Law Enforcement was now using many of the tools developed for prosecuting the mafia like the RICO or Racketeer Influenced and Corrupt Organizations Act. The Texas state agencies were using similar laws so I advocated that our contribution to their efforts in the future should be photo/video products and detailed descriptions of all activity we observed.

George C hooked me up with a colonel who ran the maintenance shop at the camp. He and George C were members of the old airborne mafia and the colonel introduced me to his best machinist. I explained my ideas to him, and he walked over to a stack of aluminum pieces and selected a piece of channel stock. He began making measurements using my Hi-8 Camcorder and my spotting scope. While we were tossing ideas around, the colonel came in and handed me a tube from a set of night vision goggles saying, "don't ask" and walked out.

The machinist took the tube and measured it. He was taking notes and making sketches while I tried to figure out what he was doing. Finally, he told me that he had enough information. "Come back on Friday and I'll show you what I've got." On Friday he gave me a rail mount for my camera and the spotting scope. There was a tripod mounting screw hole and some machined rings of aluminum that would fit over the ends of the night vision scope and attach it to the video camera and the spotting scope. This was the beginning of a beautiful friendship. I was worried that he might not be around for long because his hobby was building and drag-racing powerful Harley Davidson motorcycles. He managed to stay safe and made us a number of rail systems that the guys used for several years until we acquired remote camera systems.

Epiphany

I took the rail with me to our next drill in San Antonio. I mounted my spotting scope to my video camera using the rail and I was looking through the view finder trying to figure out how to make the picture larger. The spotting scope was making objects far away a little closer but there was only a small circle of a picture. I was looking at a tower on the top of a nearby hill and I accidentally hit the zoom rocker on the top of the camera and, whoa! The picture filled the viewfinder and I was looking at a piece of electrical equipment on the tower that was invisible before I zoomed in.

It was an epiphany. By zooming into the spotting-scope, I was increasing the magnification by mathematical factor that was huge; way beyond my high school drop-out math capability. This was a poor man's surveillance system that would give us evidentiary video from several hundred meters away during daylight hours, enough stand-off to deal with every surveillance situation.

South Texas

Soon after we returned to Austin, we were heading to south central Texas to work for a TNCP Task Force. They wanted us to watch a horse ranch. We set up in an abandoned house across the county road from the target. The first morning of surveillance while the guys were still setting up the equipment a pick-up truck and another car drove to the gate of the ranch. The drivers got out and met at the front of the pick-up. The driver of the pick-up took a vial out of his vest pocket and poured several lines of white powder on the hood of the truck. The other driver, meanwhile, had rolled a bill into a tight tube, leaned over the hood, stuck the tube in his nostril and began to hoover an entire line off the hood with one long snort.

The guys were going nuts trying to get the camcorder going but the action was over too quickly, and they missed the money shot. Later they observed two individuals load packages in the trunk of a car and reported the activity. They gave a good description of the vehicle and its LP. The Task Force pulled it over 20 miles away in what appeared to be a routine traffic stop. They got 27 pounds of weed, 6 handguns, $6000.00 and two arrests.

Night Vision Heaven

In June, Brian and JT and I were sent to the U.S. Army's Night Vision Laboratory at Fort Belvoir, Virginia. We were to be trained on a piece of night vision equipment that had been developed for the Gulf War and now was being given to the National Guard Counterdrug program. We got a tour of the lab and we were drooling. They were the leading edge for research in how to see in the dark. This was something that the Armed Forces had been working on since WWII.

It started with the use of almost invisible infrared light being used to illuminate the battlefield for anyone using special equipment that could see this light. By the time of Vietnam, they had developed passive systems that didn't require any direct illumination but amplified the ambient light from the moon and stars. One of the first of these devices that was fielded was called a Starlight Scope and I had used one several times in Vietnam. It would fog up; the image was blurry and there were no instructions on how to use it, so we ignored it as another extra weight to carry for no good reason.

The Army had come a long way since then and there were some units like the Rangers that issued Night Vision Goggles to every man. The newest goggles were a huge improvement over the Vietnam era equipment, but they didn't have much magnification for seeing long distances and that's where the LEROS or Long Range Electro-Optical System was supposed to come in. It had been thrown together in a hurry when the Army was camped out on the border of Iraq in Saudi Arabia during Desert Shield. The forward OPs were trying to see into the desert for long distances at night. The LEROS was basically a large Cassegrain telescope with a night vision device and a series of magnification lenses and filters on wheels that would allow it to be used day and night. It was mounted on a tripod that was designed for a TOW missile, so it was steady and it was powered by a twelve volt battery. The whole system was packed in large cases that were much too large to fit in a car, much less a rucksack.

We went to the Fort each morning where a civilian who had been involved with the development of the LEROS instructed us on the intricacies of his system and there were quite a few. While he was demonstrating its features and something went wrong, he would have to tinker with it for a while until it was working again. We tried to pay attention at least as much to his troubleshooting as we did to his instructions because the LEROS was starting to look like a lot of trouble.

We wore civilian clothes and Brian and JT, who were very fit young studs, enjoyed strutting around wearing their multi-colored Oakley sunglasses like Delta Force operators in for the day on some secret assignment. As we walked through the Night Vision Lab's area people would actually stop and stare at us like we were some kind of secret squirrels who came to their lab trying to find better ways to see in the dark.

I had my new rail with me, and I showed it to the civilian scientist and let him compare the daytime image of the LEROS with my Hi-8 Camcorder and rail-mounted $50.00 spotting-scope. I actually had a better picture with more magnification than his device. That was because his night vision device stayed on all the time and used filters to keep it from flaring out in the daylight. His device did a lot better than mine for distance in the dark but I would be in R+D competition with the lab for several years trying to find a way to get evidentiary photo/video in the dark from far enough away from the target that we wouldn't be discovered.

At the end of each day I got on the Metro and went into DC. I visited my buddies on the wall. I went out to the cemetery at Arlington and visited Orde Wingate's resting place. I took photos and video of the changing of the guard at the Unknown Soldier's tomb.

Meanwhile the boys were taking in the local sights. Brian and JT hung out together and used the rental car to explore new places to continue their passion for exotic dancing and wine tasting. In a moment of youthful exuberance, they managed to charge drinks on their credit cards. When the training was finished, we arranged for our LEROS to be shipped to Austin and we returned.

When we got back, about a month later when the credit card report came in, they were called into the COs office and read the riot act. We all had a good laugh about that, and Brian's face would get red when we reminded him about it. The LEROS was stored in our supply room and there it sits to this day. It was one of the National Guard Bureau's many 'good ideas' back then but even a blind pig can find an acorn once and a while and they surprised us with a doozy later that year.

Central Texas

Many of the procedures we had developed had been done away with when Ralph took over as Detachment Sergeant. Since Ralph was a Special Forces guy, everything appeared to be fine. We got an urgent mission request to put a team on a large marijuana patch.

Ralph made up a team on the fly with mostly new men who did not go through the minimal OTC that we had at that time. He sent them out to watch a patch that had been discovered by a DPS Helicopter looking for domestic marijuana cultivation activities. They had found the patch northwest of San Antonio and a local Task Force decided to stake it out and asked for our help.

The new guys were all LRS soldiers so the idea of hiding while doing surveillance was part of their training. Using the cameras though, was new to them and they studied the equipment while driving to the target area. The narcs had determined that the cultivators were not around, so the guys were able to establish a hidden OP near the patch and they waited. Two days later the growers showed up. It was a husband and wife team and they worked quickly to irrigate their crop and do a little trimming. The team was holding their breath trying not to be discovered as they videotaped and photographed the couple.

They were using a video camera that someone had found at a Law Enforcement trade show. The camera was modified to work at night using an infrared laser light to illuminate the target. This activity was taking place in the daylight, but the video had a red tint to it from the modification. The guys used the auto focus feature because they didn't know how to use the manual focus. They videotaped some very clear shots of a tuft of grass that was growing right in front of their OP but much of the target video was out of focus. However, they did get some usable video and they never were discovered.

The narcs decided to take the growers down after several hours and they swooped in and made the arrests without incident. It could have gone the other way, they realized, when they discovered the arsenal that the growers had with them, including a Mini 14 with scope and a 9mm pistol with plenty of ammunition for both weapons.

The narcs pulled up 4000 marijuana plants, carefully counting each individual root system for evidence. Later they raided the suspects' house and found enough weapons for a good-sized Texas gun show. They seized a Corvette, and a nice travel van along with the house, the guns and some cash. After the plants were pulled up, they were taken to the sheriff office compound and burned. The guys took photos and video of all this activity and were quite the heroes when they showed us their adventures. Ralph's unilateral decisions about training requirements were annoying; his decisions on operational requirements, though made in ignorance, would border on stupid. Once when we were in east Texas, the cops wanted us to shut down the operation because nothing was happening and they had overtime issues. Ralph refused to stop it and they let us know they didn't appreciate our lack of flexibility and never asked us back.

Out West

In July I went down to Alpine with one of the team guys to meet with a local sheriff that I had met at the A+M TONE course. He wanted us to work with a game warden who had discovered some suspicious tracks in a remote area to the east of the Big Bend Park. The first thing we had to do was go to the sheriff's house that was way out in the desert for a planning session. The Sheriff was about 60 years old and he had just bought a water evaporation cooler for his house and asked us if we could lift it into the windowsill for him. It was pretty heavy, but we managed to get it in and that made him happy.

He wanted us to run the operation from a ranch which was owned by the state. We would stage our MEDEVAC helicopter there and drive the team to where they could watch a canyon where the Game Warden thought the smugglers where coming through after they crossed the river. The sheriff wanted to somehow follow the smugglers if we spotted them so I mentioned the C-26 and told him he could request it.

We went back to Austin and I told the captain and Ralph what was needed. He gave me the names of the men to take and we worked up an equipment list. I contacted the aviators and gave them the details of when and where we needed them. When the C-26 commander called me with a politely worded WTF, I filled him in on the concept and he agreed to meet me at the Airport on the day after the operation started.

We launched on a Monday morning and drove to the ranch where the sheriff had arranged with the ranch manager for us to stay in the bunk house; it had everything we needed. On one of the walls of the living room was a plaque from the Texas National Guard thanking the ranch for supporting Operation Gatorade four years before.

I set up the CP while the team prepped their gear. Then I prepared my video camera to tape the helicopter when it flew through the pass. At exactly the right time a Black Hawk helicopter flew through the pass, landed in front of the bunk house, and out stepped Scott and Jerry, our pilots. They brought Tina, a third pilot with them and two crew chiefs. The helicopter had the extended fuel tanks, giving it a high-speed look as the sunset glow reflected off its windshield.

The sheriff, the Game Warden and I went over the OPLAN. When they were satisfied, we signed the MOU and started the mission briefing. The team had 2 hours to wait before the Game Warden drove them to their drop-off point so they went over their contingency plan for a law enforcement situation with the young sheriff's deputy who would be going in with them.

Scott and Jerry volunteered to help me with radio watch along with Tina and the crew chiefs. All was quiet the first night after the Game Warden drove the team in and the operation started. The next morning, I drove to Alpine and met the C-26 crew at the airport. They were in civilian clothes and this was one of their first operational deployments. They didn't have access to the KY-57 encryption so I filled theirs from a device I brought with me.

None of them knew how to program their satellite radio so I had a few moments of panic. I prayed and the Lord led me through the procedures. I was able to get the frequencies in and the encryption installed, and we made a commo check with the ranch. I was relieved to hear Jerry tell us "Loud and clear."

I told the C-26 commander to expect the sheriff at about 10:00PM that night. The sheriff wanted to fly over the area, and I told him that we could conduct a signature check at the same time. He would fly back and forth at different altitudes and we would advise him about the noise and visibility of his airplane. I returned to the ranch by way of MacDonald's where I picked up some burgers for the helicopter crew.

That night the sheriff had a great time laying on the floor in the back of the C-26 and flying around his county, the largest one in Texas, in the dark. The sky was clear, and the moon was almost full so he could see pretty well and he thanked us all later for the great time he had flying that night. When they flew over the CP, it was obvious that they were not a Stealth asset. The team never saw a soul during the operation. On the last day we pulled them out, had a BBQ, sent the helicopter on its way and we drove back to Austin.

Austin—A Season of Discontent

Ralph had been the detachment sergeant for 6 months and things were getting bad. I spoke to him about his ignoring everything we had learned about rapport with the cops and operational techniques, but he ignored me. I wrote a five-page memo to Captain Akers detailing the problems I had observed in the detachment, most of which were caused by Ralph's autocratic infantry platoon sergeant way of leadership. I didn't get any satisfaction from the captain regarding my observations and so I wrote him a letter of resignation several days later. I didn't quit, I just asked for reassignment. There were no positions open, so I continued to march but it was definitely a season of discontent.

Green Thunder

Green Thunder was one of the CD operations we knew of but had no contact with; that was about to change. The Texas DPS had been working with the DEA for about 15 years in an effort to stop domestic marijuana cultivation in our state. According to one DEA narc I talked to, the program was driven by our attempts to eradicate marijuana in Colombia. The Colombians claimed that our eradication proposals in Colombia were designed to protect our own domestic marijuana crop from being undersold by the fine Santa Marta Gold and other potent smoke from Colombia.

We got involved when there were radio communication problems between the National Guard helicopters and the cops on the ground. Bil directed us to send some 'Go-rillas' out to where the aerial spotting campaign was being conducted each week. The guys would take several PRC-77 FM radios with which they could talk to our pilots. They would wait with the cops in a staging area for the aerial spotters to announce that they had found a marijuana patch. When they did, our guys would ride with the narcs as they broke all speed limits and some speed records getting to the spot that the helicopter was circling. They would go in like gang busters and arrest anyone in the proximity of the patch that couldn't explain their innocence.

This was a nice break for the guys and when the narcs took them for a ride, they had enough of an adrenalin fix to keep them happy. Our participation began to evolve when the DPS Narcs wondered if we could help them watch a patch when they found one. They wanted us to hide and catch the cultivator. Two of our guys did this one time on the fly so we developed a whole system for rapid deployment of a team and C+C element anytime there was a Green Thunder operation.

Pagers were just coming into use with our guys and most of us were paying for our pagers out of our pocket. If you were on a team that was in a stand-by status for Green Thunder and you didn't have a pager you had to call in to the office every 30 minutes when you were away from your phone, that's why most of us brought pagers. Later we would be issued them and cell phones, but this was new technology for us.

Hanging out with DPS Narcs was a trip. I spent a day with one who told me about the time he was working undercover in Texas. Someone in a nearby state found out that he was a cop and called the crooks he was investigating from a pay phone at a gas station and told them to kill him. There was a person using another pay phone and overheard the conversation. That person overheard enough details to call the police and tell them what he heard. The police used that information to call around to the surrounding states. Through a process of elimination, it was determined that the one telling the story was in danger and DPS was able to alert the police in the town where the undercover operation was taking place. The local police spotted the undercover officer's car that he was using to drive himself and his executioners to a remote place so that they could kill him. The local cops pulled him over on a routine traffic stop and got him out of the car before they arrested the others. It's hard to believe that they were able to save this man's life, with such flimsy information, a divine intervention, obviously.

The spotters found some weed that day. We raced through the back roads of a rural county to the place a Blackhawk was circling. The carload of narcs just ahead of us were piling out of their car, guns drawn. A typical hippie stepped out of his trailer with his hands up. I was videotaping the scene as the narcs put him face down on the ground and handcuffed him. He was a real stoner and he was also a horticulturist. There were five hundred plants growing around his house. When the narcs searched his house, they found several pounds of buds and a few joints. He had a cannabis growing library and High Times magazines.

When the narcs started asking him questions about his plants, he began questioning their suitability for even talking to him about pot growing. "Do you have the knowledge?" he would ask them. The narcs thought this was kind of funny, "What do you mean, do I have the knowledge?" one asked. "I mean, do you have the knowledge, because I can't talk to you if you don't have the knowledge." This went on for a long time until we were finished pulling and loading the weed on the trailer and then they took him to jail.

One of the biggest Green Thunder operations we went on was out in East Texas. We got the call early in the afternoon and drove to a place near Texarkana. We got there about 11:00PM and got a briefing from the cops. I put the guys in a motel so they could get a few hours' sleep and we inserted them before dawn to watch a patch. They stayed in for three days, but the growers never showed up.

I was parked under a bridge a few miles away maintaining radio contact with the team and hanging out with the DPS Narcs. The DPS Lieutenant was a fun guy and he amused us all with his feats of strength such as hanging from the bridge's I-beam and swinging hand over hand for long distances.

One of his sergeants told us the story of how he enlisted the aid of some good citizens. After he told them that he was trying to do surveillance on some criminals nearby, they invited him to use their house that had a good view of what he was trying to watch. "It turned out to be a dry hole," he told us, "and after while I broke off the surveillance, and after thanking them I moved on. It wasn't till much later that I found out that they were the main crooks I was looking for."

Change of Command

Captain Akers had received a promotion to major and was moved into the CD XO position under a female officer, LTC Gladys. We had a small party at Captain Smith's house and the new major gave a short speech and was given a plaque. He moved his things to his new office in the headquarters.

When Ralph found out about the promotion and impending move, he began cultivating Captain Smith for the position and he was cautiously on-board if Uncle Bennie gave him the nod. Captain Smith was a career Guardsman and a member of the 124th Cavalry which, before the airborne, was the elite branch of the Army. They had a tradition of gallantry during the Civil War and were the subject of many of my favorite western movies with John Wayne leading the charge across the west. Captain Smith was a tanker, wearing tanker boots to work which made some of the stomachs turn in SOD.

Central Texas-X-Ray

During this transition we got a request for surveillance support in a town near Austin. Ralph sent me down to do site selection with some of the guys who would actually be on the operation. After discussing the plan with the cops, we did a drive-by and took photos and video of both the target and the area where the team would hide.

A few days later, the team went in. They set up their OP in some sparse vegetation and began their surveillance. After a few hours a man carrying a rifle with a scope came out of the house they were watching. He crossed the road and disappeared from their sight. About a half hour later they saw him walking toward them, looking into every bush and pile of brush. This happened several times in the next few days. The team had to move their location several times and they observed people on the target watching the area they were watching from. Eventually a man walked right up to the bushes they were hiding in, "You couldn't have driven a thin needle into my anus with a sledgehammer," one of them told me later. He bent over and looked into their OP, "Hey," he whispered. The guys stood up as one, racked their shotguns, and answered, "What!" I don't know what he expected to happen when he whispered into the bushes but what he got was three green faced commandos pointing 12-gauge shotguns in his face.

The guys handcuffed and searched him. He had a 9mm pistol with a laser pointer in the back of his pants. They called 'X-Ray' and asked the cops to meet them where they had been inserted. The cops took the guy to the county jail and booked him for marijuana cultivation based on the marijuana plants that the guys found growing near their hide site. Later, the cops told us, this guy was beating his head against the wall of the jail and demanding his right to make a phone call.

Meanwhile, the Task Force commander decided that they had probable cause to get a search warrant for the place, and they got it. The DPS SWAT team was asked to help out with the search. Our guys recommended that they sneak some cops around the back before they went in the front door but DPS said that since there would be a U.S. Customs helicopter with a FLIR overhead, they were sure that no one would get away. They went in the front door and the crook went out the back door and wasn't seen again for a long time.

A search of the house yielded a pleasant surprise. There was about 1,000 pounds of marijuana and $250,000.00 in cash. What was really interesting were the countermeasures that this guy had employed. He had radio scanners all over the house and a signal strength meter that could tell him how much power a radio was putting out. That would tell him how close to his house a transmitter might be. There were guns everywhere. What we later surmised was that he had been picking up our radio transmissions, but he couldn't hear what was being said because of our encryption. He knew that someone was transmitting close by, so he sent his buddy out to patrol the area around the house.

WTF!

We were really happy with the way the guys had handled the situation. They had done everything we had trained them to do and they had done it well. Therefore, I was astounded when Colonel 'Buckwheat,' our CD Coordinator/Commander came to my office to talk about some officer's concern about our horrible failure.

The officer was very concerned that we were about to be shut down by the National Guard Bureau for arresting people when our role was surveillance only. He told Buckwheat that there were a bunch of loose cannons in SOD and unless he did something, we were likely to destroy the Counterdrug Program.

This was insane. It appeared to me that whoever said this was totally ignorant of our training and procedures. I was furious that someone who wasn't there would say that our guys had done something wrong. I just told Buckwheat that this had been a perfect operation and that the men should get a medal. "They did everything we trained them to do and when they had no choice, they restrained an armed man and took him to the cops so that they could determine his status."

Buckwheat took a few notes and thanked me for my time. The guys got letters of commendation. I had a much higher regard for Buckwheat after that. I knew he was a flyboy and we figured that because of politics a blue suiter had been put as a figurehead over a largely Army operation. That may have been the intention of someone in the higher echelons of the Adjutant General's Department, I don't know, but Buckwheat had my deepest respect from that moment on and I was proud to serve under a man who would take the word of a low ranking sergeant over that of an officer.

Out West

Jimmy K and I went out west to coordinate for yet another foray into the Big Bend. We decided to eat lunch in Terlingua, known for its annual chili making championship and festival. Jimmy loved this place and insisted we stop here. There was only one restaurant/bar in town and so we went in. It had about 15 customers and there were only a few empty tables. Most of the people stopped talking and looked at us intently when we walked in, kind of like a bar scene in a western movie. Except for the lack of cameras, I felt like we were in one of those movies. We ordered at the bar and the bartender tried to elicit some information about who we were and what we were doing here.

I hadn't eaten all day and my blood sugar must have been pretty low because I sought to reassure him by telling him, "We're from the government and we're here to make sure that everyone is paying their taxes." We walked back to our table and Jimmy decided to use the restroom. When he came back, I noticed that his shirt had caught on his pistol which was now in full view of the bartender and everyone between us and the door. I told him to fix his shirt which he did, and we sat there waiting for our food.

The atmosphere was thick enough with bad vibes that you would have needed a chainsaw to cut through it. I kept scanning the place with rapid eye movements while trying to appear uninterested with my surroundings. The bartender walked past our table with someone's order in the back of the restaurant. He lingered to talk to the men at that table for a few minutes and then walked back to the bar. I saw one of those guys get up and walk to a window that opened up toward the open desert behind the place. He stood there looking out the window like he was trying to spot some Indians who might be sneaking up. He turned around and stared at us before returning to his seat.

This was getting really interesting, but I realized that things might get out of hand. We were there on a low-visibility operation and if we stayed here much longer our low-visibility status might be upgraded so I told Jimmy that we needed to get our food to go. He was disappointed that we wouldn't be able to have a relaxing afternoon in Terlingua. We went up to the bar and I told the bartender that we wanted our food to go. He seemed relieved and went into the kitchen and brought our burgers out in a bag. We paid him and took off.

New Detachment Leadership

Ralph decided to go back to the regular Army and returned to the 5th Special Forces Group at Fort Campbell. Captain Smith chose Sergeant Brian to be the detachment sergeant and the two of them worked together to bring SOD up to the next level.

Brian had been a rock steady operator, determined to be the best at anything he put his mind to. When I put him through the pistol qualification course, he missed a few targets and qualified as a sharpshooter instead of an expert. He went home and bought a .45 caliber 1911A1 like the ones we carried and practiced at a local range on his own time. The next time we qualified he was a high score expert.

We were running multiple operations at the same time under what we called the 'mission commander' concept. The mission commanders were me, the CO or the detachment sergeant. I would usually go and set up the operation and then one of us would take a team in to conduct the operation. I had been doing most of the coordination for our operations up until now and we needed to decentralize. One of the first things they did was to direct me to create a checklist that could be used by anyone they put in charge of a mission to make sure that all aspects of mission planning were completed correctly. This checklist made for a much more efficient planning and coordination process.

The next thing they did was to develop a better method of keeping track of where everyone was at any given time. They brought some large dry erase boards and created a matrix where everyone in the detachment was named and there were thirty one spaces after each name where a letter could be placed showing at a glance who was on what operation and when they were on leave or at drill. At this point we were using the pool of operators to form ad hoc teams for a particular operation. The CO and Brian would choose one of the more experienced operators to be the team leader and assign him and his team to me or one of them. The mission commander and the team leader would work out the details of the operation, request the equipment from supply, brief the team and execute the mission.

Permanent teams were something we considered from time to time but the turnover of personnel and the requirement that everyone had to attend drill and AT with their unit made that a distant goal. We assumed that the men would embrace the professional ethic; but that would be unrealistic in a SF unit, and why we thought it would work here is something that I would attribute to our own burnout.

The CO and Brian both had an interest in technology, but Brian would become consumed with the quest for night vision video and eventually remote video systems. Captain Smith was content to let the NCOs do most of the heavy lifting, but he wasn't afraid to intervene when we were all wrapped around the axle.

One place where he and Brian had different views was toward PT. Brian came from the Rangers and in his view, physical fitness was as important as anything we did. If an operator was enthusiastic in the PT program, Brian could overlook a multitude of failures in other aspects of the job.

Cultivating

Busted

Green Thunder

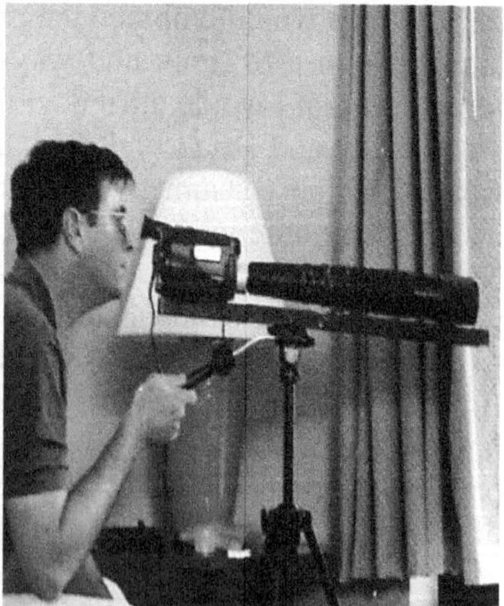

The Rail

County Sherriff SSG PB Game Warden

West Texas: Scott and Jerry Cops, Air Crew and SOD

CPT William (Len) Smith

SSG Brian Stevens

SOD Command Team

And the tongue is a fire; the very world of unrighteousness, the tongue is set among our members as that which contaminates the whole body and sets on fire the course of life and is set on fire by Gehenna. James 3:6

Chapter 15-January 1993

Central West Texas-Buggin the Plane

We got a request from a TNCP Task Force out to the west of Austin. They wanted us to watch a suspected smuggler who had his own airplane and landing strip at his house. The cops arranged for us to run the CP from a nice bunkhouse on a large ranch nearby. There was a brush patch across the road from the suspect's house in which the operators would be able to set up their OP. I did a commo check from the CP to the OP when we did site selection and although the signal was weak, I thought we could communicate.

The operation was scheduled to start on Monday morning when we left town. All of the team including myself had drill that weekend. Sergeant ___, the designated team leader, had most of the rest of the guys in his LRS team. He made arrangements with the detachment sergeant to leave drill early for some crisis. Later the other guys said that the crisis was his girlfriend who was feeling neglected by his many absences. When the others found out that he had left early and didn't try and get them released as well there was a small-scale mutiny and Brian had to come down to the Camp and arbitrate between the sergeant and his team. We drove west the next morning and by the time we got to the CP the guys were speaking to ___again and they wasted no time getting their gear ready for infil that night while I set up the CP with the ones assigned to work in the rear.

We infiled the team at about 10PM and they made commo with the CP using one of the jungle antennas; and we were up. The next couple of days were routine with no activity at the target. Then, the cop running the operation showed up with a change. "We need to go into the suspect's hangar," he said, "The feds want to bug the plane."

The Federal technicians showed up the next day and we began detailed planning for an on-the-job training seminar which would involve 10 techs watching one of their own installing a transponder in the suspect's plane so they could track it the next time it flew a load. Now that air smuggling had gone into decline, they had a whole generation of techs who had never done this before.

It was arranged for our guys to perform security and give an early warning if someone was approaching. We also provide security around the hanger while the techs did their work. I videotaped the activity and gave them a copy for training purposes. The cop who was running the operation actually knew the suspect and arranged to meet him in a nearby town for a drinking session at a bar.

It took two Suburbans to get us and all the equipment to the suspect's house. The team across the road assured us that the coast was clear, so we were dropped off on the road. We crossed the fence and walked across a field to the back of the hanger. After assuring that the place was deserted, I placed our guys in the front and back and the feds went to work. It seemed to take an eternity. First, they had to make sure that they didn't disturb anything in the hanger. Then they had to decide which plane to bug because there were two. After a short conference, they flipped a coin and chose one of the planes. The lead tech got out a power screwdriver and went to work.

The rest of the techs gathered around on both sides of the plane. I managed to get a few shots through the crowd using a handheld set of NVGs and my Hi8 camcorder. The tech kept unscrewing screws which were kept in small trays until a piece of the airplane was freed and handed out to the other techs. Finally, he called for the 'bug' which he proceeded to attach to the plane. When he completed this task, he got out of the way so that the other techs could see what he had done. After that another tech went in and began to replace the piece of the airplane that had been removed.

We were in that hanger for an hour and a half before the lead tech announced that we were done. He made sure that everything was the way he found it before closing up the plane and telling me to call for a ride. I did and 15 minutes later we were waiting by the side of the road when our Suburbans showed up and we went back to the ranch.

The rest of the operation went on until the agreed upon end-date and we all bid farewell to our friends and went back to Austin. Like most of these operations, if there was no bust during the operation, then we never knew the outcome of the investigation until we saw one of the narcs we knew who was involved. In this case I saw the narc about

6 months later and asked him how things turned out. "Well, we still haven't got him, and we put the 'bug' on the wrong plane."

The SOD Tech Arrives

We had a hiring board for three vacancies in the detachment. We interviewed 5 people and after the board voted, we chose three men from G Company, as usual. The fourth man on the merit list was a quiet guy from an artillery unit. He looked slightly overweight and even though he had airborne and air assault wings the guys rejected him because he wasn't in G Company or LRSD. I was very impressed with his demeanor and after the board results were announced, I took him aside and told him to be patient and try again the next time there was a vacancy. The next day, one of the candidates that was chosen dropped out for some reason, so we scrambled to get a hold of Sergeant Pat, the one who I had talked to.

He was working for the CDTF in Del Rio, at the Border Patrol Air Operations, and when we told him he was hired he climbed over the fence of the storage facility where he kept his things and threw them over the fence where one of our operators, was waiting. They loaded his things into one of our Suburbans and Pat left Del Rio in his rear-view mirror and never looked back.

Pat was a genius. He was a perfect complement to Brian's relentless search for a bigger and better technical approach to our surveillance operations. Pat started out as a surveillance operator but when we had problems with programing our radios, he read the manual and became an expert. Over the next 10 years he would become the 'go-to guy' for all our hi-tech equipment and the techniques necessary to employ them. He was good in the woods as well and if I had to choose anyone to operate with in difficult circumstances Pat would be my first choice.

Houston

In February we got a call from Steve down in Houston. His HIDTA squad had a heroin buy set up with one of their informants that night and they wanted to videotape it. I was tapped and took my own video camera, the rail, the night vision tube, an Insta-scope adapter that turned a telephoto lens into a spotting scope and a 300 mm lens from our new camera systems. When I got to Houston, I met with an FBI agent on the squad who took me to a house where the deal was to go down later. I met their informant who was wearing a body microphone and transmitter and we talked about what was going to happen. The narcs wanted the deal done in the driveway and so I was put in a utility closet with a small window where I could video the transaction and the squad tech could run a cable from the body mic receiver to my camera for the audio. When it was showtime, I was set up and videotaped the buy that went down with no problems.

I ate dinner with the FBI agent afterward and he told me about his work in counterintelligence before he got into drugs. I spent the night in Houston and drove back the next morning. A week later, we were gearing up for yet another operation in Del Rio that was staged in San Antonio. I got a call from Steve who informed me that the house we had used for the video hide was owned by the informant and he had just been diagnosed with active TB and AIDS. "We are making everyone involved with this guy get a TB test and you should to."

I went to a nearby Air Force base and got the test. While I was in Del Rio, I went to get the results and was informed that I had tested positively for TB. I was put on a medical regimen run by the County Health Department and at the end of that time I was declared non-contagious and inactive. There was a lot of paperwork for the line of duty investigation. Our Task Force commander was quite concerned about my health and wellbeing so whenever she asked, I would reassure her with a gentle coughing spell.

Waco

Captain Smith told me that some ATF guys were coming to talk to us about borrowing some surveillance equipment for an investigation in Waco, TX. He told me to tell them that we couldn't loan equipment, but we could support them if there was a drug nexus. I met with the two guys who were from the ATF Austin office and they told me that they were trying to conduct surveillance on a place that was guarded by armed men. The distances were too great for their equipment and they had heard that we had some big camera lenses and night vision equipment that might work.

I told them what we had and that we couldn't loan our equipment, but we could support their investigation if there was a drug nexus. I was pretty interested in their investigation and asked them if they could give me any more details. The younger guy looked around to see if anyone was nearby and then he told me that they were investigating a religious cult that was stockpiling guns, including automatic weapons in their compound. "They have guard towers and night vision equipment," one of them said. This sounded like a big deal and I wanted to be in on it so I told them again that we would love to help but we had to have a letter from their agency requesting our support for an illegal drug investigation. One of them said there had been some talk about a meth lab at the compound but it was some time ago; they would get back to us.

Several weeks later we got a request to support ATF's investigation of a meth lab at a location in Waco, TX. ATF decided that they needed to search the place and asked CD to participate in the planning. As a result, our support included a UH-60 for command and control during the search, two RAID birds for recon during the event, and three SOD operators to act as pathfinders during the search operation.

Forward Looking Infrared (FLIR)

I was directed to take aerial photographs of the compound from the C-26. The C-26 guys were also making a video tape of the FLIR images of the compound. These FLIR images would be used by ATF to justify the possibility of a meth lab. Our C-26 was down for maintenance so another one from Alabama was used for the flight.

I didn't really think about the FLIR image as evidence of a meth lab until I went to a Thermal Image school being put on by the DEA in September. The lead instructor seemed a bit contemptuous of the Texas National Guard, so I talked to him in private and asked him what was up. "Waco" was all he said. The course lasted a week and one night we were down in the bar with some of the instructors and he was sitting at our table. He had had plenty to drink and was by now our friend. I sat next to him and he told me how mad the DEA was because of how the ATF had used FLIR images from the C-26 to provide them with probable cause for their search warrant.

The whole school had been about the use of thermal images to provide the 'icing on the cake' for probable cause, not the only evidence.

The DEA was very excited about the use of thermal imagers for investigating indoor marijuana grows. They had a whole checklist of tasks that needed to be accomplished before you used the thermal image: tasks like obtaining the electric bills, analyzing them, etc. According to their strong recommendation, the last step in an indoor grow investigation was the thermal imager.

As far as using FLIR to find a meth lab, the DEA said that it was ludicrous to think that a meth lab's heat signature and a kitchen stove's heat signature were different in such a way as to be detectable. The FLIR image that the ATF interpreted as a meth lab was later, I believe, found to be a water heater. The DEA was struggling to keep thermal image technology from being knocked out of the drug war by bad case law and the Waco operation was a landmark in bad case law, bad judgment, bad training and bad... The DEA guys were not impressed with their fellow federal law enforcement agents.

WACO

The day I flew the mission was about a week before the raid. We flew over some targets in Bryan and then we went up to Waco. I shot the photos using my own personal 35 mm camera and I rigged up my Hi 8 to video the targets. I was dropped off in Austin and gave my roll of film to the ATF guys and went home.

The next Sunday morning I woke up and while I was waiting to go to the morning meeting, I turned on the TV to look at the news. There was an amazing sight. Men dressed in SWAT clothes with MP-5s were carrying dead or injured people away from a building that looked just like the building I photographed. "Holy shit," I shouted. "Come here and look at this," I yelled to my wife.

Our guys told us later that they were in the dining room of a hotel in Waco that had been commandeered by ATF to use as a briefing room. "We were filling out the order forms for t-shirts that they were going to have made to commemorate the heroic moment," said Rob, one of our guys chosen to be a pathfinder for the helicopters. "All of a sudden a guy came running in shouting, 'they know we're coming; they know we're coming, we got to go;' so off they went."

Sergeant Tommy Reynolds, one of our sterling operators, was told to video the raid from the C+C helicopter. Years later when he was going to the FBI Academy, they showed the video to his training class. He told one of his fellow trainees that he took that video and the trainee reported this to one of the instructors. The FBI investigated Tommy and eventually called him into the agent in charge of the Academy and accused him of an integrity violation because he had lied to a fellow trainee about the video. He pointed them in the right direction where they soon discovered that it was him who took the video.

Thanks to our involvement with the ATF from the beginning of the fiasco, CD continued to support them by providing personnel to drive a 5 ton truck with a water buffalo used for providing water to the ATF and FBI during the siege that resulted when the Davidians resisted the search warrant service. We sent a few operators to do other work.

February 1993: Waco-HRT

I pulled a tour in Waco during the long siege. One day I was directed to meet two men from the FBI Hostage Rescue Team and take them to Fort Hood. We spent the day together and talked about many things. They were both senior members of HRT and had recently finished up an operation at Ruby Ridge in Idaho. One of their snipers had killed a woman holding a baby and there was a lot of controversy about their actions but the HRT was more concerned about the fact that they had put in a lot of overtime and hadn't been paid for it.

We arrived at Fort Hood and I dropped them off at a motor pool where they were taught to drive and operate the Combat Engineer Vehicles, armored vehicles that they would use to try and force the Branch Davidians from their compound. I brought a book with me, so I made myself comfortable and spent the afternoon reading. When they were done, I drove them back to Waco where we ate dinner. I insisted on buying, "I'm on per diem." I told them, knowing how sensitive they were about finance. As we ate, we talked about our police backgrounds. One of the agents told me that he had worked for a certain Police Department near Washington, DC. "Do you know an Officer Dodger," I asked him. He looked shocked and told me that he was his partner on the force. "How do you know Dodger," he asked. "He arrested me back in '68' when I was raising hell at a party. He came to the door and asked me step outside so he could talk to me and when I did, he arrested me." We all laughed about it and I was reminded that it was a small world.

The siege ended in tragedy. Seventy-six people died in the fire that consumed them and the building they refused to leave. I was very disappointed in my lack of discernment regarding the ATF's initial contact with us. I wanted for us to be involved in any important case to raise our profile among the law enforcement agencies because I thought we had something to offer; if they only knew about us, they could use us. Buckwheat would spend a lot of time in congressional committee inquiries about our participation years after the rest of the country forgot about it. One of the ATF agents, a man who had worked with Operation Alliance for a long time, couldn't forget about it and went to work helping children after he retired.

Spring 1993: Turmoil Again

Captain Smith put a sergeant in my section and made him the operations supervisor. That pissed me off for several reasons. One, I had been doing the job to everyone's satisfaction and two, the man was a spotlight Ranger and he had a serious drinking problem. This was creating problems among the guys who wondered why nobody was jacking him up. He had gone hunting with some of them and they complained that he had shot birdshot at them from a distance. Then he got arrested for DWI, and though he got a lesser charge for a technicality, it was clear he had been driving while drunk. In the latest incident, he was found passed out in a motel corridor during a school where we were teaching the cops about booby-traps. The captain told me that since the sergeant was in OCS and would soon be an officer, he wanted to 'develop' him as a leader. I wrote a scathing memo to the captain denouncing the man for the incidents that everyone knew about and the effect it would have on the unit if he remained as the operations supervisor and a while later the man left the detachment.

SOD CD Doctrine

That whole business left me feeling like we had to establish a doctrinal basis for how we operated in order to educate people like Captain Smith and the new people we would continue to train as the older guys got tired and left. I began reading the manuals to educate myself on

Army doctrine. I started by reading FM 100-1 (The Army) which gives the legal and moral foundation of this country's Army.

The Principals of War were there and FM 100-5 (Operations) helped me understand the doctrine for doctrine in the introduction to the new version: "Army operations doctrine builds on the collective knowledge and wisdom gained through recent conduct of operations, combat as well as operations other than war, numerous exercises, and the deliberate process of informed reasoning throughout the Army. It is rooted in time-tested principles and fundamentals, while accommodating new technologies and diverse threats to national security."

All the conventional Army manuals were too vague to make a case for what I felt was right in the way we had been instructed by Bil in the beginning so I went over to the manual library and snooped around until I found some doctrinal manuals for special operations. In FM 31-20 (Doctrine for Special Forces Operations) I found what we needed. It was a brilliantly constructed treatise on how to think, not what to think.

In the Special Operations Imperatives, I found the framework for our doctrinal approach to our operations. They provided the legitimacy needed to prove to the more conventionally minded members of our detachment like the captain, and our superiors in the CDTF, that what we were doing was grounded in accepted military doctrine.

Understand the Operational Environment, the first imperative, spoke volumes to me and substantiated what Steve, and Tom Stephens, and I had done when we introduced the soldiers to copworld. Recognize Political Implications, Facilitate Interagency Activities, and Ensure the Legitimacy and Credibility of Special Operations were some of the other imperatives whose titles alone tell you that this is essential brain food for people doing this kind of work.

I wrote up some lesson plans for teaching the Special Operations Principals and Imperatives and showed them to the captain who was not feeling very friendly toward me. I was surprised when he told me that he had read them and that he thought they were very good. From that time on I felt we had a better professional relationship.

The operators paid little attention to the classes. That was because the HQ element was still doing most of their mission planning and coordination. That would change as we began to decentralize; doing more and more missions with smaller teams, and by necessity, using the team guys to run operations without the core leadership being present.

New Detachment Sergeant-More Turmoil

The CD program got involved with manning a state-wide 24/7 Deconfliction Center that meant having someone sit in a room across the hall from the team room and wait for a telephone call from a narcotics agent who was about to do an undercover buy or bust some drug dealer. Our people would check to see if another narc was working in the area on the same person or was undercover at the same location. Although it was a great idea, the cops didn't use it for a multitude of reasons; but for six months it gave a few of our guys a nice break.

Brian was one of the ones who volunteered for this job and that meant we would need a new detachment sergeant. For reasons known only to him and the captain, he recommended one of the young sergeants who had been there for a while. The sergeant was a good man, but he wasn't capable of ramrodding the pack of wild dogs that were running around in the detachment at that time.

The Team Supervisor-More Turmoil

Another sergeant became the team supervisor when we got the new detachment sergeant. He had some ambitions and had spent some quality time with the new detachment sergeant convincing him that he could take care of the team guys and free him up to handle other things. The first thing that he did was to reinstate the 50-sandbag policy for being late to PT. The next day he was late and filled the sandbags. The guys were impressed with his honesty and integrity. Two days later he was late again and again he filled the sandbags. This really impressed some of the guys, but the wind had shifted and now instead of 'hard-ass,' 'dumb-ass' began to be used when referring to the team supervisor. This little comedy plowed up the ground and when the seeds of discord were sown by the malcontents, there was plenty of cultivated soil and rebellion broke out like weeds in garden in the spring. Cabals were formed, plots were hatched and the 'Team Supes' days were numbered as his buddies schemed to do him in.

The troops were fried from the OPTEMPO and certain individuals were spreading discontentment among the weaker ones and it was starting to seem more like a low morale infantry unit than a special ops environment. Whenever this happened it usually meant that the guys were not happy with the leadership and the focus of their discontent was usually the team sergeant. This kind of atmosphere in the team room could cause some attitudes to degrade on operations, which is when I noticed that there was a problem.

My office was in the team room area and I could hear the men complaining about the team supe like a Democrat complains about a Republican president. After a while I went into Captain Smith and asked him if he had consulted with Brian before he made the man the team supervisor. In a display of anger, the like of which I never saw before or after, he verbally threw me out of his office. I did what I could, but I had to watch the whole drama play out like the ones before.

A month later, when anarchy was reigning, the CO finally accepted the fact that the men were not responding to the team supe's leadership and replaced him. Several months later, the former supe decided to try something new and began doing a lot of swimming during PT. After he left the program, he joined the Air Force to be a PJ. He was in his thirties when he started, and we were proud to hear that he made it through the grueling selection and training for the Pararescue Jumpers and has served as a PJ all over the world in the war on terror.

'Bienvenidos a los Estados Unidos'

I was sent down to the valley to meet with the PAIC of a BP station which ran a checkpoint on one of the highways. The chief had a MACV-SOG plaque on the wall, so I figured that this was another Bil deal. The chief wanted us to set up some OPs about a half a mile from the checkpoint on each side of the highway. He assured me that a lot of dope was coming around the checkpoint every night.

He sent me out with one of his agents to look the place over and after I completed site-selection, I went back to the station chief and we went over our rules and agreed on some dates. I went back to Austin and briefed the command on the operation and recommended that we borrow some TOW missile night sights from one of the infantry units in the valley. The CO thought we should borrow some Hummers so we could mount the sights up high enough to see over the scrub brush in that area. We had just been given some AN/PVS 13 Thermal Image Scopes and I decided to take them along also.

We sent some of the guys to a unit in San Antonio to sign for the Hummers and I went down to the valley and picked up the night sights. We staged our equipment and met there for equipment familiarization and rehearsals. I had used the TOW sight before, and several others had also. When we set up the sight on a tripod, I noticed that mounting the sight system onto the tripod was difficult in the daylight so I had everyone mount it several times until they could do it quickly. One of the new guys kept objecting to the repetition, assuring us that he knew what he was doing so I grew weary of the rehearsals and called it a day. We were going in that night and we'd see if our training was effective.

We drove out to the checkpoint that evening and after dark we drove in with one Hummer to set up an OP. I took my personal video camcorder and a set of PVS-7s and stepped away from the Hummer to videotape the set up. The new guy took the lead in mounting the sight system and I started videotaping so I wouldn't miss this part of the operation. I needn't have worried because he made the same mistake made by many others who found their way into the detachment before they had matured in life. Skills are developed by repetition.

'Slow is smooth and smooth is fast' is the mantra for skillful hand gunnery and it can be applied to many hands-on tasks found in this type of work. He assumed that because he had been selected to serve here that he must be 'special' so rather than make himself special by hard work, he chose to throw in with the malcontents and now we would have to wait an extra 15 minutes while he was working and cursing as he learned how to mount the sight in the dark.

He finally got the sight up and we started spotting groups of people walking through the brush. When we called the checkpoint, they sent two agents out to the area we were watching and we vectored them until they found the group, who were now squatting on the ground when they realized that they were caught. This went on all night and by morning we had helped them catch over a hundred illegal aliens who were trying to walk around the checkpoint. There were no drugs.

I was a little surprised with the reception we got at the checkpoint when we stopped to let the agents know that we were leaving. I thought they would be happy with all the apprehensions that they made but they weren't. It would take a few days before one of them let me know their version of what was going on here.

According to this agent, the PAIC was trying to put some points up on the big board at sector by catching more illegal aliens than anyone else. I told him that we had been called in to catch drugs going around the checkpoint and he laughed. "The drugs are going around 5 miles out not a half a mile from the highway." He told me that the checkpoint was swamped that night trying to process all the illegals we had 'helped' them catch and the agents were far from happy about our being there to 'help.' A few nights later, I was standing on top of a Hummer watching a group of about 20 people slowly approaching our location. I called the agents at the checkpoint to keep them abreast of the situation. They acknowledged my reports, but I didn't hear any further traffic indicating a response unit was on the way.

The group got closer and closer, so I called again and this time the senior agent at the checkpoint came back and told me that they were too busy to respond. I came back and asked him what we should do now because they were all squatting in front of the Hummer, having walked upon us while I was talking to the agent.

The BP agent told me with an impatient tone, "Let em go." That was interesting. I looked down at the 20 men squatting in front of me whose hopes had been dashed for a quick trip to San Antonio tonight. They were docilely accepting their fate and planning their next attempt when I addressed them, "Senores" I said with a loud voice.

They all looked at me with big eyes as I said, "Bienvenidoes a los Estados Unidos!" They didn't believe it the first time, so I used my California, Mexican, and Colombian Spanish to explain that tonight God had blessed them and they were free to continue on their way. They finally believed me and got up and thanking us profusely, they took off up the trail to a rest stop a mile or so north where they would meet their ride, marveling at the random happenings in Gringolandia.

North Texas

The operations were coming in like a flood. The detachment sergeant was fried and said that we needed to start turning down requests. I objected to this in an open meeting and the captain sided with me. This is probably what destroyed whatever hope he had in giving the guys a break and he left shortly after this. I was assigned to a mission with two other men, one was our personnel clerk and the other was an operator with nothing going on.

We drove north and met with some Federal agents who were investigating a crack distribution network in a large Texas city. One of the places they had identified was a business with public access. The Feds rented an apartment across the street from the target. It was a low-income neighborhood and most of the inhabitants of the apartment house were people of color. The agents took us there at four in the morning and we managed to get into our second-floor apartment with our equipment, unobserved.

The plan was for us to take photos and video of every person and vehicle that visited the target. They wanted us to closely observe every transaction in hopes of getting a money shot of the surreptitious transfer of contraband for money. Most of the action would take place during daylight hours but the agents wanted us to watch the place as long as any of the employees were present.

The guys set up the camera equipment. The main system was a Nikon F4s with a 600mm F/4 lens. We shot through a small slit in the window shades, invisible to the naked eye from across the street. We also had one of our VHS camcorders which were almost useless. I had my own video camera on a rail with one of our Nikon spotting scopes. If something interesting developed, I would use it to catch the action. We set up a schedule and settled in for a three day stay.

The first day passed uneventfully. We took photos of every person and vehicle that came onto the target, making sure we got the LP on film and in the activity log. I used my down time to make a sketch of the target area. I wanted to add this practice to our OTC, and I was still working on my own skills. That evening there continued to be activity on the target, so we continued watching. The traffic on the landing outside our front door picked up after dark. People were going to an apartment next door for short lengths of time and then leaving. We went on full alert when someone from next door came out and sat down on our doorstep, leaning against our front door.

There was a crack house next door and while we were watching one crack distribution operation across the street, the people next door were watching for the cops while they ran their own business. We were scarcely breathing when, 'clunk,' one of the guy's pistol fell out of the shoulder holster he was so proud of and hit the carpeted floor. I was at the other window trying to get a shot of the guy leaning against the door when the gun hit the floor and I almost shoved my camcorder through the front window. I gave him a short message expressing my disapproval for his choice of holsters and suggested that he put the pistol somewhere where he could better control it. He was a little offended with my rebuke but soon things returned to normal. I went back to the window. Nothing had changed. The guy leaning against the front door was still there and it didn't appear that he had heard anything. I repositioned my camcorder and began to take more footage of our friend.

'Clunk,' the pistol hit the floor again. I almost shoved the camcorder through the window again, but this time my remarks were less gentle in verbiage but just as quiet. After a few minutes of total silence, we continued our surveillance of the target while the watcher for the crack house next door watched the street below. This went on for two more days. The Feds showed up at 4:00AM in the morning to pick us up. I gave them our exposed film and activity log and we left town.

Credit Cards

Our funding for the program was still sporadic at times. The program hired sixty people for cargo inspection in the belief that there were funds to pay for them. After two month they found out that a mistake had been made and fired them. These people had quit their jobs to come to work and many of them had families. Suddenly they were unemployed. They received their per diem payments and when their credit card bills came, most of them in excess of $2000.00, instead of paying the credit card bill, most of them used the money to take care of their families while they looked for another job. When the bank who sponsored the Government Credit Cards notified the program that a large number of credit card holders had not paid their bill for two months and that hundreds of thousands of dollars were at stake, the head shed went through the roof. After the dust settled there were more rules about credit card use than there were about the use of deadly force. Everyone was required to bring their credit card bill to work as soon as we got it. There, it was checked by our own personnel clerk. Then it was checked by the Task Force XO. If something didn't look right, you were called down the hall to explain the questionable charges

Green Thunder

The next Green Thunder operation that I was on was in East Texas. The helicopter spotters had found what looked like a sea of marijuana in a rural county. Usually, the growers would try to hide their plants, but this was a large farm operation. We were requested to put a team on the ground to watch for the growers and we did.

I ran our CP from the Suburban while we waited. The DPS Narcs passed the time with yet another series of amusing war stories. These guys, who had once been Texas Highway Patrol officers with knife edged creases in their immaculate uniforms, spit shined boots, and high and tight haircuts that were covered by their Texas Cowboy hats; now looked like a bunch of white trash dopers with a bad attitude.

We got a report from the team that some people had showed up and we got ready to charge if they called X-Ray. The growers appeared to be Hispanic males; the team heard them speaking only Spanish. They spent the day cultivating, smoking their crop and eating snacks to stave off the munchies. Later in the afternoon, they pulled out a .22 rifle and had some target practice. They put their targets on a large tree that stood between them and the team so there was some tension but the team assured us that they were in defilade and not in danger.

The higher ups at DPS decided to try and follow these guys when they left and see who they were connected to in Dallas where their car was registered. There was a hand-off about halfway to Dallas and due to someone's screw-up they lost visual contact with the suspects and the operation was called off. We all went charging in with the DPS Narcs in the lead on their 4 wheelers. The rest of the day and the next day were spent pulling plants and counting root systems.

Eventually the cops determined that the growers had been smuggled in from Mexico to grow dope for some criminal organization here in Texas. I don't know why they decided to use the same techniques they used in Mexico, maybe they burned a lot of candles and said a lot of prayers to the patron saint of dope growers in hopes that their huge plantation would go unnoticed by the Green Thunder operation which was hardly a secret; there were news stories every week on TV using B-Roll of Guard helicopters and cops pulling weed. It was however an interesting case of reverse outsourcing though hardly the kind we need to restore our economy.

Things were getting pretty dull for me at LRSD. It was clear that I would never get to jump school. I had just received a reply for my latest request for an age waiver that was denied. Lt Steve was now a captain and our CO and while he and I had worked well together in SOD I was having a hard time working in the detachment.

Mac, the detachment sergeant had formed a group of NCOs around him who were not friendly towards the members of SOD in the same unit. Most of the SOD operators had left and gone to G Company. There was a former member of G Company who outranked me and wanted my job. He had attended a training school for Danish Jægerkorpset and had been given a shoulder tab for that school.

The tab was not authorized for wear on a U.S. Army uniform and when he wore it for a class A uniform inspection during our December drill one of the rangers made him take it off. That didn't set well with him and since the ranger was one of the SOD guys, he became a hater. He was moved into the O+I section in a vaguely defined position which caused me to waste time that I didn't have.

When I went to Captain Steve and asked him to clarify the duties of each of us, he got us together and said, "Now Greg you're the Intel guy and Sergeant Johnny you're the ops guy. Now I know that Greg likes to do ops stuff and you like Intel things, so I want you to work together." That didn't solve anything so I tried the direct approach with Johnny and asked him to tell me what he wanted me to do but he informed me that as an NCO I should already know what to do and he wasn't going to tell me anything.

Because I'm not the brightest light in the sky, I never really under-stood what they were saying; which was to go away. I did one more annual training and was more or less ignored. I finally got the message that I could either spend some quality time with Mac and work my way into his good graces or I could find another place to drill.

I basically despised him and his Navy ways. Since I had accidentally discovered his book by Alistair Crowley, which led me to believe he was a devil worshiper, I stayed as far away from him as I could. He had asked one of the younger guys in SOD to go the annual Halloween bacchanal on 6th Street in downtown Austin dressed as a pet on a leash for Mac to lead around while he was dressed in some kind of S+M costume; like I said, I despised his Navy ways. For some reason, once when I was stuck with him in the same vehicle, I asked him if he was a 'traveling man.' "Indeed, I am," he said in a voice that sounded like someone talking to me from the other side of the Gates of Hades, "why do you ask?"

I was completely flustered by his reply and began explaining how I had heard that Masons used this challenge and password when they were meeting someone for the first time. This started a short discussion on Masonry in the Guard. He told me that the Masons had never given him this job, but they helped him keep it. A few months later, when the entire detachment rose up and demanded that the CO get rid of him, the Masonic conspiracy couldn't keep him, but he did get another active duty position and finally retired as a sergeant major.

The LORIS

The National Guard Bureau finally found the acorn they had been sniffing around for and they sent us six LORIS systems to use. I went down to our supply room when I heard about them and set one up. LORIS stood for Long Range Infrared System. It was a tripod mounted Thermal Image camera with a battery and a control box that had a small monitor. It came with a tripod, pan and tilt mechanism, and was stored in two large carrying boxes. It was large and heavy, much bigger than our old TVS-4 and when I had it set up Captain Smith and Brian came down and looked at it. We all shook our heads and wondered when NGB was going to consult with the people in the field before they spent all the money on expensive crap like this.

I was very interested in thermal imagery for detection and even though these things were big and heavy I decided to come back that night and try one out. The CO and Brian decided that they too would take a look. That night I set one up. It took a while to figure out the controls. The device came with a manual but we, in true SOD fashion, ignored it after we made sure we had the power right, so it took us a little longer to figure out how to pan, tilt, and focus. It took even longer for the white hot/black hot and zoom settings.

Eventually we were looking at people across the large parade field. They were walking or running on the track that was open to the public in those days. We could see, very clearly, their facial features, from where we were. It would be difficult to use this image in a court of law for identification, but you could sure see a long way in the dark.

We were not doing anything on the border at that time, but we all agreed that it was a nice piece of useless equipment that worked pretty good. Unfortunately, we had no use for it at that time and stored the systems in the back of the supply room next to the LEROS.

FY 94 Roll Up

In FY 1994 we did 36 operations and we were credited with $3.5 million dollars in drug seizures. The drug seizure figure didn't impress me because of the convoluted methods used for determining those statistics. I was more impressed with the number of missions we conducted without any compromises or injuries.

October 1994: The Gulf Coast

We received a request from Federal Law Enforcement organization to help them with a drug trafficking investigation on the Gulf Coast. I went down to get things started and met with Agent Dan, a seasoned law enforcement professional who was the head of the investigation. He told me about the success they had had with catching drugs that were being floated across the Mansfield Cut and driven up the beach to Corpus Christi, avoiding the BP checkpoints on the roads coming up from the valley. From Corpus the road was clear to San Antonio and points north, west and east.

He wanted some help staking out the channel between North and South Padre Island. We went over all our rules and came up with a plan. We would drive 55 miles down the beach to the cut and camp in the dunes; watching for the smugglers. As I was driving back to Austin, it occurred to me that this might be a good opportunity to try out the LORIS. When I got back, I briefed the CO about the operation. Almost everyone else was deployed so they assigned one man, a former team leader, to go with me. I casually mentioned that I might take a LORIS for kicks. No one objected, they probably didn't even hear me, so I notified him and we worked on the equipment request.

We left Austin and drove to Corpus. In the back of our Suburban was a LORIS with four fully charged batteries. We talked on the long drive. Eventually he brought up the subject that no one was talking about and asked me what I thought about those backstabbing so and so's on his former team. I thought about how I could comfort him because I knew he had been deeply hurt by what had happened.

After some consideration I tried to use the episode as a lesson to be learned, since we were both sergeants and professionals. I thought that after I explained the principle, he might be able to apply it to future endeavors. "Generally speaking, there are no bad soldiers, there are only bad leaders." This didn't seem to have the desired effect. It did end any more talk about what happened; actually, it ended all talk. He didn't say a single word to me for the rest of the operation.

When we got to the beach we met with Dan, signed the MOU, and took off down the beach. He had us lower the air pressure in our tires, told us to avoid running over wood because of nails, and to keep moving in soft sand or we would get stuck. I was really enjoying the sea air as we chased Dan and his partner down the coast. We were going 70 miles an hour at some places which was pretty fast for driving on sand. Dan was an expert and as long as we did what he did we had no problems and it only took us 90 minutes to get within sight of the jetty that sticks out at the cut. Dan pulled off the beach into the dunes following a path wide enough for our vehicle that opened up into a flat area surrounded by dunes but invisible from the beach. His partner brushed out the tracks we made driving in and joined us. Dan said we would operate out of this area. He told us to walk through the scrub to a point on the channel and set up an OP and wait for the dope.

We unpacked our vehicles and threw camo nets over them to keep from being seen by aircraft. Dan suspected they were being used to scout the area before a load was crossed. I decided to try the LORIS that night, so I set it up on its tripod and kept it standing out of the way until later. Dan watched all this and asked me what I had. I told him it was a thermal image scope and that I had brought it down to try it out. He was interested to see how it worked.

After dark, I carried the LORIS on its tripod to the top of a dune near the vehicles. When I turned it on I was once again impressed with how far you could see in the dark. There were some people camping by the jetty and we could see all of them moving around their camp. It was a mile away and you could tell which was a girl and which was a boy.

Once Dan saw what the LORIS could do, he wanted us down there all the time. We had lots of work, but I assured him that we would get down there as often as we could. We went down there several times in the next year. Dan was getting cops from all over to provide our close proximity requirement and one time he got some Coast Guard Law Enforcement sailors from the Coast Guard Base in Corpus.

A Coast Guard Hero

We returned to Padre Island and made the drive down the beach. There was a part-time Park Ranger and a Coast Guard Law Enforcement sailor going with us. This time we set up an OP on the channel and I remained in the camp with the Park Ranger and the Coastie. Pat, the coastie, was a nice guy with a tale to tell when there was nothing to do all night but monitor the radio and stare at the LORIS monitor.

He was working on an anti-smuggling detail in this area when they spotted a Mexican shark-boat. They signaled it to heave-to, but the boat took off and the coasties took off after it. The chief managed to get them alongside and told Pat to jump in which he did. Just as he jumped, the other boat's operator swung the shark boat away and Pat made a desperate lunge for the gunnel of the shark-boat.

He grabbed it with both hands and now he was being dragged along side of the shark-boat which was moving at about 40 knots. The Mexican crewman nearest him reached over and tried to grab Pat's pistol while his partner changed course again and tried to crush Pat between the two boats. "I threw myself into the boat at the last minute and now I was fighting with three men on the shark-boat," he said.

"By now there was a Coast Guard Helo right above us and one of the aircrew was taking video of me as I threw the closest guy over the side and worked my way to the back where the other two were. I threw the second guy into the water and jerked the gas line from the engine and the motor died. The guy driving the boat was an old man and didn't give me much trouble."

Pat told the whole story so matter of factly that I had no doubt I was in the presence of a real American hero. I congratulated him on his excellent adventure and asked him what kind of medal the Coast Guard was going to give him. He said he didn't know if they were going to give him a medal because they were trying to kick him out of the Coast Guard for being a cowboy and getting injured in the line of duty. I was speechless. I did ask him if I heard right that the chief had ordered him to jump in the boat and he said yes.

I didn't see Pat again after that operation, but I did work with some other Coasties and when I asked them about Pat's story they assured me that it was all true, one of them had seen the video, but they weren't sure what had happened to Pat's case because he had been transferred. This was my first contact with the Coast Guard in the drug war, but it wouldn't be my last.

We did three more operations with Dan's guys. On one, Juan was down at the cut with one other cop when they spotted a smuggling operation. The cop decided to take them down and of course, Juan didn't back him up with his Remington 870 loaded with 12-gauge slugs. The cops were really happy with Juan's help and asked for him again.

On another operation we had to leave the beach because a hurricane was coming straight for us from out of the Gulf. The cop wanted to wait until the last group of people left so that they wouldn't see us when we left. The wind was really blowing, and the water was getting higher up the beach when we finally left.

I was driving and as we continued up the beach the visibility got worse as the rain started pouring down almost sideways from the wind. I was straining my eyes for any glimpse of the Park Ranger who was ahead of us. If he stopped for any reason, I didn't want to run into him going this speed. The waves were pushing the water closer and closer as we drove as close to the edge of the dunes as we could. The guys were starting to get worried and several of them wanted to drive; sure that they could do a better job. I wasn't going to stop and risk getting stuck, so I told them to chill out and let me drive.

The water was up to the dunes in front of us and I couldn't see the sand. I felt the SUV shudder and saw a wave breaking on the passenger side window. No way, I thought, we are going to have to abandon the vehicle and find some high ground inland if the engine drowns out. I told the guys to get their jackets and some food ready in case we had to bail and continued to drive. We finally got through the narrow part of the beach and now we had a little more room. I could see the cop ahead of us and it looked like the visibility was getting a little better. It took us a lot longer than the usual ride up the beach and we didn't get off the beach until after dark. The weather was a little better, so we spent the night and then drove to Austin.

I spent a lot of nights on that beach fooling with the LORIS. It needed to be moved by a vehicle and the batteries needed to be recharged every day, but an operator could detect a man 5 miles away in total darkness. It was an incredible piece of equipment and I hoped that we would be able to use it in a target rich environment.

G2 Intel Geek

I finally got the point at LRSD. It was time to move on. Captain Smith told me that he had arranged for me to transfer to the state headquarters to the unit that handled the National Guard's piece of the state emergency management system. He told me that it was a done deal, so I went to the full-time officer who ran the unit and presented myself, ready for duty. He looked at me like I was crazy and told me that he didn't know who I was, and that he didn't have a place for me in his unit. When I reported this information to the CO, he appeared shaken, so I believe he thought it was a done deal.

Something had occurred and though I was only vaguely aware of it at the time, the Lord moved his hand and I went in another direction that was to take me to distant lands and have unforeseen consequences in my life that are still on-going as I write this story.

I talked to Steve and Tom Stephens who were in the G2 section of the 49th AD. They seemed to think I might be able to transfer there and be trained as an Intelligence Analyst. Since I didn't have any other prospects as an infantryman in the Austin area, I went over to the G2 people and asked if I could transfer in. Captain Smith's wife was a full-time Intel captain there and she said OK. As soon as I completed the transfer paperwork they enrolled me in a reserve component Intel Analyst course that was taught in two phases; the first phase would be taught during weekend drills in Austin and the second would be two weeks the next summer in San Antonio. At the same time, they enrolled me in PLDC, a course I would need before I went to BNCOC which I would have to pass before I could be promoted. My life had gotten even busier now and the OPTEMPO in SOD was increasing also.

PLDC

I received orders for PLDC in December at Camp Shelby, Mississippi. I was 49 years old when I arrived at Camp Shelby. I processed in and was assigned to a squad with other National Guard and Army Reserve soldiers. The oldest one there was 30 and most of them were in their early 20s. There was one guy from Alabama who was a parachute rigger for a National Guard Special Forces Group. Full of himself, he began his quest for dominance of the squad. He showed us his large ALICE pack and told us that it was a 'Ranger' rucksack, only issued to the most elite forces of which he was a member.

I tried to ignore his continuous self-serving chatter, but his bunk was near mine, so it was hard to tune him out. We had several females in the squad. They were billeted in separate barracks and joined us each morning for PT. The rigger decided to harass the prettiest one; being the airborne stud and supposed informal leader of our squad.

The rigger was continually chiding this female soldier with sexual innuendo and she was a married woman. The female soldier in question was a good troop who minded her business and wasn't taking any crap from this guy, but he stayed on her more and more until one day she came to me in tears. This was before the Equal Opportunity and Sexual Harassment training came that would soon take priority over Rifle Marksmanship in NCO training.

I didn't know what was going on because I had never served with women in such close proximity. I had told the rigger to lay off several times because he was disrupting training, but I hadn't made an issue about it. The main point of equal opportunity and the fight against sexual harassment was to build cohesion in military units. It wasn't a feel good, do good, liberal love fest for all God's creatures; it was based on the grim reality that in combat soldiers fight for one another and if there is cohesion in the unit they will fight harder for one another at the risk of life and limb. If there is no cohesion you can expect the opposite. I really hadn't given this whole concept much thought, but I was offended that the rigger was harassing a married woman and causing her distress. He was a distraction to the squad's mission which was to take in the information that they needed to be leaders in the Army. I went to the rigger and demanded that he stop messing with the female. This led to a long verbal fencing session which I half hoped would lead to fisticuffs because I really didn't like this guy. He was young and fit so I was considering how I could hurt him quickly to gain the advantage. On top of everything else, he was displaying all the signs of being a racist and that just made me angrier.

At one point he was accusing me of not liking him because he was airborne and a parachute rigger. "That's not true," I told him, "some of my best friends are riggers." When he heard that, the double entendre stymied him, and he backed off so I told him again to stop bothering the females or I would take him to the class instructor. He a face-saving retort and walked away but he didn't say anything to the female again.

Like all Army leadership schools, PLDC required you to pass some written tests and some hands-on performance evaluations. If you failed the retest, they sent you home and you could only hope that your unit would send you back for another try. If not your potential for advancement was nil. We were fortunate to have an excellent instructor. He seemed to care about the troops and spent extra time with anyone who was having a hard time. A few years later I would be an instructor and I would use SSG Sweeney as my role model. I enjoyed the classes on leadership using the Maslow Theory of the Hierarchy of Needs which I thought was a great way to define human nature.

The PT was boring. It wasn't designed to maintain our physical fitness but to teach us how to establish a PT formation and lead a PT session. Drill and ceremony wasn't much fun either and I had to concentrate on every movement, which is what D+C is designed to do, to keep from making a mistake while marching in a formation or while commanding a formation. I did like calling cadence when I was in charge of marching or when I double timed the squad during PT.

Land navigation was emphasized and there was a make or break land nav course that wasn't difficult but if you didn't concentrate on using the principals that were taught you could get lost like the student a few classes before who got lost and stayed lost until they found his body out in the woods. I wanted to be in the top 20% of the class so I could get a distinction on my 1059 school completion form, but I didn't.

BG Daniel James III

COL 'Buckwheat' Pettit

LTC Gladys Tinsley and COL 'Buckwheat' Pettit

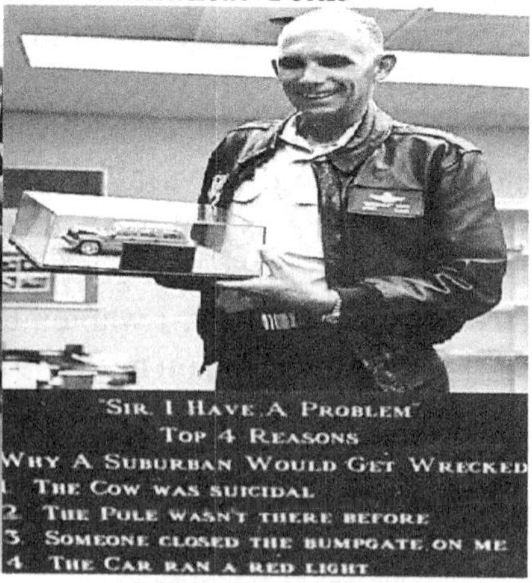

Colonel Pettit and author

SOD Momento

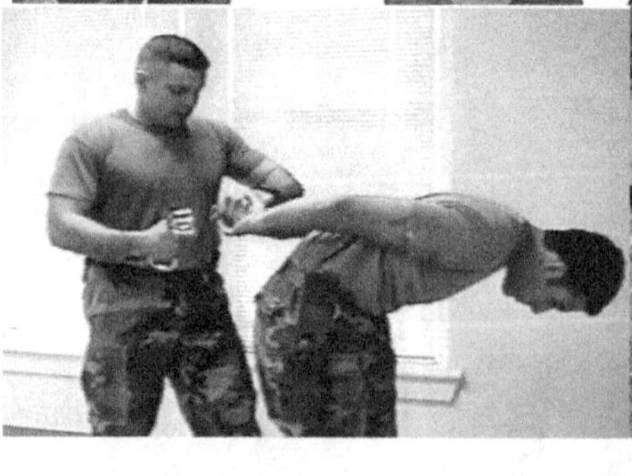

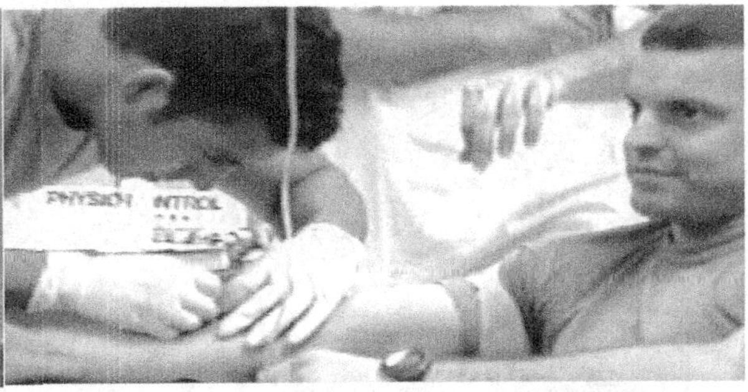

Let the elders who take the lead well be counted worthy of double honor
1 Timothy 5:23a

Chapter 16-January 1995

The Golden Years Continue

The Special Operations Detachment had arrived. During FY 94 we conducted 48 missions. 36 were surveillance operations and 12 were training operations where we trained cops how to rappel from helicopters, read maps, find booby traps, and use ranger tactics when they were in the woods. We had been given credit for 3.5 million dollars in drug seizures. The leadership of the Counterdrug Task Force and the Texas National Guard had seen that we were consistently able to operate safely and successfully.

One sign of their approval was our budget which was in the millions of dollars. We used it to buy the latest and greatest video, commo, medical, and personal equipment that came to our attention. We were getting catalogs every day from vendors who wanted to help us spend our money before the end of each fiscal year. This was crucial because common bureaucratic sense dictated that if you didn't spend your entire budget each year you wouldn't get as much next year.

We had come a long way since we first started with some binoculars and borrowed Vietnam era night vision equipment. We had excellent still camera and video camcorder equipment and a state-of-the-art communications system. Terms like 'one is none and two is one' and 'better to have and not need than need and not have' were heard often during mission planning sessions. We had plenty of everything.

Operations had become a logistical nightmare requiring two members of the detachment to run the supply room; the CD J4 Logistics section spent a lot of their time working on our equipment procurement requests. The detachment XO position was eagerly sought by several 2nd Lts but the ones that got the job soon wished they had never heard about it. Brian would give them endless demands for more and more high-tech equipment research and ordering. The military logistics system is cumbersome enough when you are dealing with items that have a federal stock number but most of the things we wanted had never been heard of in NGB or the Pentagon.

We developed a fill-in-the-blanks hand receipt that was 10 pages long. The mission commander would fill out what he wanted, and the supply guys would gather the equipment and stage it for a coordinated time when he would draw it. Every serial number or stock number would be recorded on the receipt and we would have to check each one as they handed the item to us and we checked it off. This could take several hours for a big operation.

After we drew the equipment and signed for it, we would stage it in the team room the night before leaving. The guys were all in the latter stages of advanced sleep deprivation so when we noticed that we were forgetting small but important items of equipment we started adding steps to our load-out procedures. This resulted in 7 full serial numbered inventories for each operation.

North Shore of Oahu

I took block leave at the end of December and took my family on vacation. We drove to Travis Air Force Base in California and flew Space A to Hawaii. I had just bought a Sony 3 CCD Hi8 camcorder and was determined to get some surfing footage at the Bonzai Pipeline. After landing at Hickam AFB, I rented a car. We'd taken a tent and rented a camp site at Barbers Point Naval Air Station. We left our camp site each morning at 4 AM and drove to the North Shore. There is a McDonald's in Haleiwa and we stopped there for breakfast. I was on the beach and ready to video when the sun came up.

The surf was good and most of the pros were still in town after the final event of the season, the Pipemaster contest. Every day I shot a two-hour video tape. The surf breaks a long way from the beach, and I was disappointed that I couldn't get better close-ups. Ever the surveillance freak, I spent most of my thought time working on a better camera system that would give me the close-ups and not put my family in bankruptcy. One reoccurring theme in the soundtrack on the tapes is the plaintive voice of either my wife or my son asking, "Can we go now?" and me replying, "just one more wave."

We did go swimming together at Makapu Beach and Hanauma Bay. We shopped for Aloha shirts at the Goodwill in Honolulu and watched a group of Hawaiian Christians baptize some new believers in the ocean water at Magic Island near Ala Moana. After a short wait in the Space

A terminal at Hickam, we returned to the mainland on a C-5a and drove down the coast of California before turning east and heading back to Texas. The trip requirements were intense enough that they took my mind off my work which is what leave is supposed to do.

G2

My first G2 drill in January was also my first class for my 96b MOS training. We met in a classroom for one weekend a month for phase one; then that summer we had a two-week session at Camp Bullis in San Antonio. The classroom work mainly consisted of map reading and posting information on a situation map that would be used in a headquarters. We studied the Warsaw Pact weapons and tactics even though the Cold War had been over for five years and there was no Warsaw Pact. The lack of synchronization between the schoolhouse and the real world was a source of frustration to some of the students but the instructors reminded them that they had to pass the course if they wanted to be Intel Analysts regardless of what their own analysis of the situation was telling them.

Back to the Beach

At the end of January, I was assigned to take a team back to Padre Island to watch the Mansfield Cut. This time we replaced a Marine Corps Reserve unit that had been watching the north side of the cut while they began watching the south side. I gave the dates to the two team leaders who rotated their teams in and out of the OP while I would remained on the beach a mile north with the cops. The team leaders had some problems with the math and one team decided that they were getting screwed. The continual complaining of some of the operators became like the hiss on an old record. I had everyone go down to the back-forty and rehearse the tasks that were required including the erection of a camouflage net for each vehicle. Rehearsals had become a critical event in pre-mission preparation. Much of the equipment we were using now was unfamiliar to the new guys and the some of the older men hadn't used it for a long time. The rehearsals also served as a final check of the equipment we were taking, and this allowed us to identify and replace any malfunctioning equipment before we left.

We had an enormous amount of equipment for this operation including portable generators for charging our radio and LORIS batteries. One LORIS would be used to watch the Cut and the other would be with me at the CP where we would watch the beach and the Gulf. We brought the solar panel that was part of our radio repeater system to see if it would charge a LORIS battery in a day. This was part of our ongoing quest for sustainable renewable energy.

Most of our equipment needed electricity and we didn't always have a place to plug in a battery charger, so a great deal of our research was on better ways to acquire electricity and store it.

I enjoyed the early mornings as the stars began to disappear and the sun rose out of the ocean. Large flights of pelicans flew along the shoreline and the lousy surf made a pleasant sound. I enjoyed these trips to the beach even though they were unproductive. We had become fairly competent with the LORIS equipment and although I didn't know it at the time, our skills would come in handy.

Bil Enney Departs

Things were changing at the headquarters. Bil had been put on a big salary and he bought a Corvette. His wife had died recently, and I knew he was hurting. My wife and I went to visit him at his house one Saturday and while we were there, we preached the gospel to him and he prayed to receive the Lord Jesus Christ as his savior. We gave him a small New Testament before we left. I didn't see Bil very much because he and we were always out of town. He was in DC at NGB and the Pentagon getting money for the program. He used to say he didn't care how much we got as long as it was a dollar more than California.

We came back from an operation and were told that Bil had resigned. I knew he had done that before to make a point and the Governor's Office would ask him to stay after giving him what he wanted. This time it had something to do with a public event in a nearby town where we were helping to tear down some abandoned houses that were being used by crack heads. I waited to hear that he was not leaving, but the word came down that he was out of there and wouldn't be coming back. He had made some enemies; they moved on him and that was it. We had a big going away party for Bil. I hugged and thanked him for all that he had taught me and done for me, and then he was gone.

Bil Enney was the man of the hour when the program started. He took the bull by the horns and led the way. The battles that he fought for us were often in faraway Washington D.C. He was the epitome of what a Special Forces soldier should be; able to operate in whatever environment he finds himself in and ready to fight.

CD Aviation

When we started running operations in the drug war, we always had a MEDEVAC helicopter standing by in case of an emergency. The crew was usually a couple of full-time pilots from the Austin Aviation Support Facility. They would get a crew chief from the full-time maintenance workers and a helicopter from the facility and fly wherever we needed them. In the early 90s, NGB started a Counterdrug aviation program which resulted in the creation of the Reconnaissance and Interdiction Detachment or RAID.

They were equipped with OH-58 helicopters with a lot of unique features. The 58 was very familiar to most of our pilots because it had been used by the Aero-Scout platoon of the Air Cav troop for years. The Cav used it with a pilot and an enlisted observer. I had always wanted to be an observer because the pilots would teach them how to fly in case the pilot was injured or killed while they were flying.

The RAID detachment had grown during the last few years and they now had 10 pilots and a number of crew chief/mechanics. The birds had a FLIR pod and a Midnight Sun searchlight with IR capability. The FLIR and a video camera were observable through monitors in the front and back seats where two passengers could ride. The aircraft had also been fitted with a Wolfsburg Radio system that would allow the aviators to talk to the cops' radios on the ground.

Many of the pilots were Vietnam veterans who had 8 to 10 thousand hours of flying time. They were the best. When a regular Army aviation brigade commander listened to a briefing about RAID and he saw the slide with each pilot's hours and qualifications, he almost fell out of his chair, interrupting the briefer by saying; "My God, every one of your pilots have more hours than the combined pilot hours in my brigade!"

My old boss from the Police Department, Sam Warren, had retired and came to work with RAID. Several pilots that I knew from the Air Cav were there as well. Scott and Jerry were the pilots that I worked with the most. Scott had flown a tour with the First Cav in Vietnam and was an officer who reverted to Warrant Officer Status when there were no longer any flying slots available and as a Major, he would have to leave the service if he didn't get promoted. Scott liked to fly so he accepted this gracefully and never looked back.

Jerry was a Texas rancher who went into the Army during Vietnam and flew as a pilot with the First Cav. He was always working toward the day when he would retire to his ranch up by Fort Hood; he must have liked flying though because both he and Scott stayed at it until they reached the maximum age limit and were sent home. They were both members of the Aviation Battalion that supported LRSD operations, so I got to see them now and then at AT. They would insert our LRS teams all over Fort Hood and do poncho parachute resupply missions in the crowded night skies.

We did one operation in the Big Bend where we all stayed in a bunk house on a ranch in the middle of nowhere. Scott and Jerry were the primary pilots and there was a female pilot as a backup and a couple of crew chiefs. I was running the CP and Scott and Jerry volunteered to pull radio watch. Scott liked to talk, and Jerry seemed to like to listen so there was never a quiet moment when these guys got together. The female pilot turned out to be a team player and when Scott made dinner each night because he fancied himself a chef, Tina washed the dishes and Jerry complemented both of them on a job well done, illustrating the cowboy philosophy of sitting in the middle seat of the old pickup truck so he wouldn't have to drive or open the gate.

East of Big Bend Again

In April, I was sent down to Marfa to coordinate yet another operation in the Big Bend. I met with Jerry A who was the Marfa Chief BP agent. I had known him for 5 years and we had developed a good rapport. He told me that he wanted to do something different this time. He wanted to do a helicopter insertion into some areas of the Big Bend. We had worked a lot of these places before, but because of the helicopter factor I was going to have to go back to these remote locations and look for LZs and check for landing zone hazards.

I didn't think that the insertion plan was a good idea because a helicopter landing in the dark could be heard for miles in the desert and no one would mistake it for anything else but a law enforcement operation. I didn't say anything though because he wasn't stupid, having worked this part of the border for many years; if he thought it was a good idea, who was I to question him.

We had a meeting with the sector SOG leader and some of the Intel people to include a captain from JTF-6 who had just been assigned to Marfa BP sector as an Intel support officer. Like many military intelligence officers, he was sure he was smarter than everyone else in the room because his branch of the military was 'Intelligence.' When Jerry briefed the helicopter insertion, this captain jumped up and interrupted a senior Border Patrol agent, with the equivalent rank of a Brigadier General, and said, "I don't think this is a good idea!"

There was dead silence in the room. The captain didn't have a follow-on statement and while he was searching for one, Jerry cut him off by telling him, "Sit down, the helicopter insertion will be handled by the National Guard. We've been working with them for years and we know what they can do." He was looking at me when he said it and when I turned and glared up at the captain who was standing next to me, he sat down. After the briefing he wanted to try and get me to hear all the reasons why he didn't think it was a good idea. I had had many of the same thoughts but all I told him was that he should apologize to Jerry before the day was over or he would be working with meaningless statistics in the Sector HQ until they found a way to get rid of him.

We worked out the time frame for the operation. Jerry sent me out to meet with an agent who would show me around the area we would be working. It took three days to visit all the places Jerry had in mind for the operation. I was mainly looking for places that the helicopter could land where they wouldn't be seen, where there were grassy areas that wouldn't produce a brown-out from the dust when the bird was touching down. The plan was for a night insertion without lights, meaning that the pilots would be flying blacked out and wearing night vision goggles. This was the riskiest flying there is but with our pilots I was sure we could pull this off.

When I returned to Austin, I briefed the CO and Brian. We would put four teams in so it was 'ladie-dadie everybody' as far as who was going. We prepped our gear and planned a detachment movement to El Paso, 500 miles away. Monday morning came and our convoy of Suburbans and a large truck rolled out the gates of the Camp and headed west. We were running on a strict schedule and our first stop was 3 hours later. One of the guys almost exploded before we got to the gas station on I -10; he hadn't brought an empty bottle, or he didn't know how to drive and void his bladder at the same time.

Ten hours later we pulled up in front of the SMU HQ where we would conduct isolation and were greeted by Jerry and his SOG guys from Marfa. Gracious Bill had retired and the new SMU commander was someone we had worked with before. There were still some bad feelings about what had happened in Del Rio 3 years before but those of us who remembered what happened acted like nothing had happened and went about accomplishing the current mission.

I spent some of my time during isolation dealing with the blue suiters at Biggs. The Infil plan was complicated. After isolation, we would move the troops over to the Biggs Army Airfield and load onto a C-130. We would also load one of our Suburbans on the plane with the troops and fly to the Air Force Base in Del Rio. From there we would do an ERO and board the Blackhawks which would depart after dark and fly the teams into the desert a hundred miles north of Del Rio.

We were dealing with the Air Force people who ran things at the airfield, but they were overwhelmed. There was a huge exercise going on called Roving Sands and there were planes from all over the world involved. When we told them that we would be loading a civilian vehicle on a C-130 they got worried. When they found out the troops would be carrying weapons, they got more worried. When they found out the troops were carrying live ammunition they freaked out and it took a few phone calls to the Texas Air Guard's unit that was flying the C-130 mission before they calmed down enough to deal with the next issue; civilian law enforcement people flying with us.

They told us all the things we had to do with the Suburban before they would allow us to load in on the plane. They wanted all the ammunition stored a certain way and they told us that they would inspect everything before they would allow us to board the plane. We knew that we weren't going to comply with all the restrictions that this peacetime military was trying to impose on us drug warriors but there was no sense in arguing, we would show up and get on our plane regardless what these neck bones said.

On the day of departure, we showed up at the apron as our plane was taxiing up. The pilots parked it with the rear-end facing us and lowered the ramp. The Air Force guys were looking through the Suburban to make sure that we had done everything they told us at the same time watching the guys struggling with their rucks and other equipment. It was time for a commo check and status report to our CP in Marfa so Flounder broke out a SATCOM and aligned the antenna right there on the ramp. The SATCOM display didn't go unnoticed by the Air Force. They were not given any details about who we were or what we were doing. Only the most high speed units in the military were using SATCOM at that time so we played out another Delta Force fantasy for all the people in the area who were watching this strange group of people with longer hair, mixed uniforms, heavy rucksacks, strange weapons, and a civilian SUV staging there.

The radio operator contacted the CP and gave them a single code word that told them that everything was on track. He had been in the Air Force in a former life, and after he finished folding the antenna and putting it back in his ruck, he stepped in and told the Air Force departure control boys that their careers would take a big hit if they didn't stop messing with us and let us load the plane.

The BP agents that were going in with the teams were a mixed lot. Some of them we had worked with in the past and there were some happy reunions with some of our older guys. One of the new agents was not really happy about this assignment and insisted on bringing a cot, a large skillet and a cook stove so that he could have his pancakes and bacon each morning. He was carrying a cloth bag partly full of flour and bacon and a can of white gas for his stove that we got past the Air Force guys on the ramp. When the loadmaster saw it, he told the BP agent to get rid of it. The BP agent stood there and looked at him for a moment and then said, "If my gas don't go, my ass don't go."

Here we were at a critical moment in the execution of this complex plan and everything was coming to a halt because the BP agent had to have his pancakes. The loadmaster must have realized that this was one of those problems that called for a National Guard solution and told him, "Put the can in that bag and get on the plane."

Brian and JT went with the Suburban onto the plane and the guys struggled up the ramp with their rucks which were weighing a hundred pounds or more. One of our guys was carrying his ruck and the BP agent's cot in one hand and his M 16 in the other. The loadmasters were stacking all the rucks on the ramp and sending the guys forward to find a seat. Most of them had spent a lot of time in a C-130 but this would be the first time for many not wearing a parachute and the first time that they had landed in one.

The plane took off on schedule and I left for Marfa with the rest of the CP team after we sterilized the isolation area at SMU. Four hours later I arrived at the CP behind sector HQ and found out that everything had gone according to plan. The teams were flown into their LZs after dark and all the initial entry reports had been made.

There was one glitch which was my fault. One of teams was landed in the wrong place and as soon as they were on the ground, the team leader verified this by checking his GPS. When I converted the UTM grid coordinates of their LZ to lat/lon I made an error. As soon as the team leader saw that they were too far away to walk to the correct location, he called the bird which was on its way back. Scott and Jerry were flying, and they immediately turned around and returned. They landed and picked up the team and flew them to the right LZ, landed, dropped them off and returned to base for a nice cup of hot chocolate and some crew rest. You gotta love those flyboys.

A Lifesaving Resupply

The teams saw and reported suspicious activity at every location so apparently the helicopter insertion did go unnoticed, but no big loads were caught. That weekend, most of us in the CP had to go to drill in the middle of the operation, so we left Ken F to do resupply along with two men to man the radios. He was coming off the river after a resupply run and came upon a bad car wreck on the highway. He was in a BP vehicle and the agent pulled over to see if he could help. A man traveling west on the highway to Alpine had fallen asleep and his car had gone off the road at a bridge across a large wadi.

The bridge was split with one span for each lane of a two-lane highway. The car steered to the left between the spans and launched out across the wadi going about 70. When it nosed into the bank on the other side the car stopped with the impact, but the driver kept going through the front windshield and onto the flat ground past the wadi.

The local paramedics were there trying to stabilize the driver, but he was in bad shape. Ken walked over to them and asked if they wanted a helicopter to fly the victim to a trauma facility in Midland. The paramedics looked at him like he was crazy and told him to stay out of their way. Ken watched for another few minutes and asked them again if they wanted a helicopter. This time the paramedic told him, "sure buddy, whatever" and continued to treat the driver before they loaded him up for a long trip to Alpine and the local doctor.

Ken went back to the BP vehicle, took out his SATCOM and called the CP. He told them that he needed a MEDEVAC for a seriously injured civilian. The CP called and got permission from Gladys and launched the MEDEVAC. Scott and Jerry got there 15 minutes after Ken called, loaded the victim and one of the paramedics, and flew them to a trauma hospital where they saved the drivers life.

As Ken was folding the SATCOM antenna a Marine who happened to pass by and had stopped to see if he could help walked up to Ken and said, "Who are you?" It was one of those perfect times when the old saw, "If I told you I'd have to kill you," seemed appropriate so he used it with a smile and drove away.

At the end of the operation, the helicopters went back and picked up the teams and took them back to the Air Force base. When they got off the helicopters, they started walking toward the C-130 that was parked nearby with its engines running. An Air Policeman stopped them and told them to stay where they were while he checked with his people to find out what to do; someone always fails to get the word. Tommy was leading the troops and after 30 seconds he told the young fellow that he had a plane to catch. The AP was trying to control the situation, but Tommy grew weary and pushed past him and the rest of the troops followed him and got on the plane.

Gifts From The Border Patrol

The next time I was in Marfa, Jerry A said he was really happy with the way the operation went even though we didn't catch any big loads. The regular Army Intel officer was gone, "We sent him back to JTF-6, he was too intelligent for us Marfa folk," Jerry said. "I have some things I would like to give you guys." He took me to the storage sheds behind sector HQ and opened the door to one of them. When he turned on the light, there sitting in the middle of the floor was a brand new F-470 Zodiac, two of the quiet 'MARS' outboard engines and a small rubber boat. “Somebody from JTF-6 gave these to us about 6 months ago. Nobody wants to use them on the river, so do you want them?”

Without thinking twice, I told him that we would love to have them. We didn’t do many water operations, but it would be nice to have the capability. This was the same kind of Zodiac that the 5th Group guys had when we did the training for the operation in Trinity Bay a few years before. Jerry was pleased with my answer so he turned off the light and we walked around to the back of the shed and stopped in front of a travel trailer and he said, “Could you use one of these?” pointing at the trailer. “Sure,” I said. “Great, I’ll get the paperwork ready and you can pick this stuff up the next time you’re here.”

96b MOSQ: Camp Bullis

At the end of May my wife dropped me off at Camp Bullis. When I reported in for phase two, I was assigned a bunk in a tin hut that was probably built to house the Rough Riders when they assembled in San Antonio before they went off with Teddy Roosevelt to Cuba. We had some of the same instructors for phase two that we had for phase one and we continued to study the ghost of the Soviet Union. A lot of army people were left bewildered by the sudden collapse of the Russian Empire. Some thought or hoped that it was a trick and that they would let us know the punch line soon.

I ran into Colonel B, who had been the G2 for the 49th AD. He had transferred over to the reserves and was now the Commandant of the 5th Army Reserve Intelligence School. I had a run-in with him a few years before at AT when some females accused the LRSD of peeking at them in the shower. There was a big brouhaha that came to nothing when the girls couldn’t identify the culprits.

Colonel B and I were laughing about it until I asked him if he wanted to see the photos. Of course, I didn’t have any photos nor did I know if LRS guys were involved in the alleged incident. The culture of the military was in the midst of some heavy contractions. Activities that could be construed as sexual harassment were a scary thing to a lot of people. Joking about such things for some people was like joking about bombs at the airline ticket counter.

The course itself was pretty easy, more Warsaw Pact stuff. I got the highest score on the final, but they wouldn’t give me a distinction, 'exceeded course standards' on my 1059. One of the instructors told me that they didn't like the paperwork involved so that's why they didn't do it. I did get a new MOS; 96b, Intelligence Analyst. I also got a Top-Secret Clearance and now I was ready for the 'secret world.'

OCS Graduation

Brian, JT, and several other SOD operators graduated from OCS. They were now officers and theoretically no longer eligible for service with us. The CO made Lt Brian his XO and others were given jobs as Intel people after they finished their officer basic course. We needed good officers in the program, but funding was pretty spotty at this time and some men lost their positions, found other work, and we never saw them again. Lt JT branched Intel and went off to Fort Huachuca to be initiated into the 'secret' secrets of the 'secret' world but he would be back. Lt Ken also branched Intel and was eventually hired by the DEA. Not long after this Captain Smith was promoted to major and went back to the operations section of the Task Force. Lt Brian became the SOD commander and Lt JT came back from his basic course to be the XO.

Another Operation in Marfa

We continued to have a good relationship with both the Del Rio and Marfa BP sectors. In July we got a request from Marfa to conduct an operation in the Presidio area. I went down to do site selection and took one of the guys with me. We finished up and went back to Marfa where we talked to Jerry A, the sector chief and Simon Garza, his deputy, who I had worked with in Del Rio on my first operation. They were two senior BP agents that we had worked with for the past 5 years.

JTF-6 had an SF A-team sitting around in El Paso with nothing to do. They sent them down to Marfa to take horse riding lessons. We were asked to include them in the operation, and we agreed. They would operate an OP in Presidio, and we would put our guys on the river with the sector SOG agents. We decided to have everyone meet in Alpine for the initial briefing with all elements to include the aviators who would be our MEDEVAC option.

I met with the SF team sergeant who was less than thrilled with having to get mission essential information from a National

Guardsman. I went through the secret rapport building techniques that I had learned at the feet of Wild Bil and soon we were getting along in a more pleasant manner. Before the operation was over, he would approach me and tell me that he was retiring soon and wanted to know, "What was it like out there?" like I was some fountain of comforting knowledge about the civilian world. Rapport is a beautiful thing.

A Ghost from the Past

The team sergeant said he would send one of his commo guys to meet with me and get the frequencies for the operation. I was in my motel room waiting to meet him when I heard a knock. I opened the door and almost had a heart attack. There stood Captain C, our former CO of LRSD. He had faded away after leaving LRS and I, like most of the troops there couldn't have cared less about what happened to him as long as he didn't come back.

There he stood, Sergeant C now. He was with another guy who gave me the code word that IDd him as the SF commo man. I invited them in and we small talked for a minute while the SF guy pulled his notes from his briefcase. C told me that he had gone back to active duty as a sergeant and had been assigned to the 5th Group at Fort Campbell.

He was assigned as a support element based on his Ranger Tab and was acting like an eager puppy as he told the other guy how he had worked with us before as a team leader on a mission in 1989. I guess he had forgotten about his less than stellar performance, but I hadn't, nor had I forgotten how he had fired Tom, brought in Mac, and single handedly destroyed the morale of the LRSD.

I wanted to throw him out of the room, but the other guy wouldn't have understood, and we were there to accomplish a mission, not work out unresolved conflicts from 6 years before. I gave the SF guy the frequencies and other details of the commo plan and answered his questions. C sat there looking at me and trying to be helpful, but he was still clueless about the nuts and bolts of this type of operation. Several times the SF commo guy told him not to worry about some aspect of the commo plan and we finished in about 30 minutes. As they were leaving, C told me that we would have a lot of time to catch up once the operation started because he would be working in the CP with me.

Marfa-The Special Forces

The next day I met the SF ODA captain and answered his questions about the procedures we followed for various contingencies. He was a bright young officer and was quite pleased with the training his team had gotten riding horses around the Big Bend area. He was sure that he knew the deal about Counterdrug operations because he had spent a lot of time with a particular BP agent who had been their escort for the last two weeks. When I heard who it was, I bit my tongue and continued to listen to what he thought was going on around here.

We all went into isolation at the Marfa sector HQ where we set up our CP and staged the teams for infiltration. The SOD teams were going in for a week and the SF team would run a nighttime only OP on some Federal land that overlooked the town of Presidio. C was there during isolation and tried to strike up conversations with all the guys he knew from his days in LRSD, but they were busy. Lt Brian told them to be polite and not start anything because of the mission.

The SF team was having trouble with their SATCOM, so I got with their commo man and we fixed a battery compartment lid that was causing trouble. Then he told me that he needed more batteries because the ones we gave him didn't work. I showed him how to 'condition' the batteries by starting with the lowest power setting and increasing the power one step after another until they were working at 18 watts. This amazed him and he asked me how I knew this information, being an old National Guardsman. I told him that we learned this the hard way since we had been using SATCOM since 1989.

The teams on the river were living the hard life. The daily temperature was 120 degrees and there wasn't much vegetation out there that didn't do its utmost to bury its sharp ends into your skin at every encounter. The commo was good, and they called in a few suspicious sightings but no drugs were caught.

The SF team was having a great time. They rotated team members with four guys manning the OP while the rest worked in the CP or took the night off. One night, the team zoomed in on a desert romance that was being consummated on the hood of a car parked in front of their position about 500 meters away which was well within the viewing range of the LORIS we had loaned them.

I heard them laughing and talking about it after they were debriefed. I couldn't resist telling them that the BP agents in Presidio had gotten word from one of their most reliable informants that a large shipment of cocaine had been brought across the river right in the area they were supposed to be watching. "Sometimes they make a diversion, kind of like we do sometimes in war," I told them. They sat there considering that information for a moment before going into the full-on denial mode so I left them alone. Later in the day their CPT came up to me and assured me that if I thought his guys had missed a shipment because they had glanced at the couple on the hood of the car while they watched their assigned area I was wrong. "Nobody knows they are there," he told me with the assuredness of the uninitiated.

I asked him, "Are you sure that no one knows your team is watching Presidio?" He glared at me, "Of course I know that no one knows," he said. "You know that BP agent who's been taking you around everywhere since you got here?" I asked. "He was best friends with Rick Thompson, the sheriff who was arrested with 3000 pounds of cocaine about a mile from where we are standing. Why do you think he had so much time to spend with you?"

He looked at me with a slightly less arrogant expression as if he sensed that something was about to be said that was going to ruin his day. "The reason why he had so much time to spend with you is because no one trusts him around here, so they give him all the make-work jobs. I hope you didn't tell him anything about this operation." He stood there processing what I had just told him, and I watched him wilt before my eyes. After a few moments he joined me on the playing field of life where strong men are humbled by their experiences and gain wisdom. "This is a real shitty place, isn't it" he said.

RCPOC

After a year at G-2, I was learning the traditional analyst role on a large staff. There were quite a few other sergeant E-5s and it didn't look like anyone could get promoted. I wasn′t concerned about it until MSG Tony E stopped me in the hall at CD and asked me how long I had been in the service. When I told him that I had close to twenty years, he told me that E5s with twenty years were being discharged. "You need to get promoted," he told me.

I thought about what he said after we parted and realized that I might be unemployed soon. I asked the senior NCOs in the G2 if I had any chance of getting promoted. They told me that there were a lot of E5s ahead of me and all the E6 slots were filled by people who weren't going anywhere. I went to Steve during the next drill, told him my situation and he said he would look around. The next day he sent me to talk to MSG Dave.

Sergeant Dave told me that he had an E6 commo slot that I could have if I wanted to transfer to his unit. I said yes and put in for a transfer. The new unit was the Rear Command Post of the 49th AD or RCPOC. It was a staff unit that would be the HQ for one of the Assistant Division Commanders in wartime. RCPOC was commanded by a Japanese American LTC from Hawaii who had been an Armored Battalion Commander in his previous assignment, and I could tell he was more of a Patton fan than a staff guy. He did take his job seriously and we had some good training exercises under his supervision.

January 1996: BNCOC Phase I

I already had the commo MOS and after a period of time they sent me to BNCOC Phase One at the Texas Academy at Camp Swift. This was the second NCO school in as many years. The class was taught by army reservists and I was not impressed with their professionalism. I made the high score on the test and first time goes on the performance evaluations, but I couldn't get an 'exceeds course standards' on my 1059. This was a continual source of irritation to me because it seemed that there were two reasons why I couldn't get the distinction.

One distinct possibility was that they didn't like me and the other was that they were lazy and didn't want to do the extra paperwork involved. When I returned to the RCPOC after my graduation from Phase One, the full timer told me that because my MOS didn't have a Phase II, I would go Phase II BNCOC for Intel.

Remote Cameras-1st Operation

Lt Brian, Pat, and I had been talking about remote video equipment for certain jobs that were too tight for the normal surveillance team techniques that we had been using up to this point. I had done a job in

Houston recently where we were laying under a camo net in a vacant lot in downtown Houston, watching a liquor store for DEA. The OP was 20 feet from the sidewalk and 20 feet from a bank of pay phones at the gas station next to us.

We were there for a week and I rotated into the site along with the rest of the guys on the team. We had to move in before dawn, and we couldn't leave until dark. When you were in the OP you couldn't move around much, and you had to be careful when you did. We were all taking 800 mg Motrin tablets for the pain of lying on the ground for 15 hours at a time and I made sure that everyone had a day of CP duty between days in the hole.

Between the potential for compromise and the manpower needs for this kind of job, a remote video system could be useful for certain situations. This led to hundreds of hours on the internet and phone calls to vendors in our search for a video system that we could afford and would work. We finally found one that would give us the ability to watch suspects and record their activity using a relatively inexpensive low light camera with a zoom lens that could be panned, tilted, and focused, and zoomed by remote control from a distant location.

We played with it a lot when we got it and we were looking forward to an opportunity to use it. The opportunity came when a DEA Task Force on the border asked for our assistance in doing surveillance on some drug dealers there. They already had surveillance on the dealer's house, but they couldn't get the faces and license plates of the visitors.

I was given the mission and went to the Task Force office to talk to the narcs. We went out to the house and drove by a few times. I determined that we would have to put the team into a field near the house. They would have a good angle to see LPs and faces but there was wasn't much vegetation for concealment. Because of the situation I told the team to be prepared for a traditional surveillance mission, but, initially, we would employ the remote system.

The team went in and installed the new camera during the night and things were going well until the sun came up. We monitored both the cop's remote camera and ours from the same room. Unfortunately, when our guys installed our camera, they put it in line with a fence line corner-post and the target area. We couldn't get the license plate numbers of vehicles that parked in front of the target. I told them to go back in the next night and move the camera. This annoyed them. One of them said that they would move the fence post before they moved the camera. I ordered them to move the camera and that night they did.

The next day I was watching the monitors when I noticed a workman coming out of the suspect's house. He walked toward our camera. I was able to pan the narc's camera and watched with growing horror as he crossed the street and began climbing over the fence.

He walked toward the bushes where our camera was hidden. Meanwhile he was getting larger and larger on the monitor screen. He was filling the frame with his face when he stopped and began to look intently at our camera. I picked up the phone and called the Task Force. I told them that our camera appeared to be compromised and asked them if they could send a couple of agents to help me recover it.

The guy continued to stare into the camera for a few minutes and then he turned around and ran back to the fence and climbed back over. I followed him on the narc's camera, back to the house. The crook came out the front door and stood there watching as the workman gestured wildly, pointing in the direction of the camera.

The Task Force agents showed up and I told them that I needed their help to secure the camera. It had taken a long time to get it and I didn't want to lose it on the first operation. They seemed a little nervous, so I offered to loan them our ballistic vests if they wanted them.

This made them a little more assured, so I showed them on the map what I wanted to do. We would go in the same way the team went in and I hoped I would be able to find it in the brush. I could have called the team guys who were sleeping in the motel but I was afraid that the crook would wreck the camera before they showed up, so we took off.

We drove to the drop off and I led the narcs up a dirt road toward the area I hoped to find the camera in. I knew that the house was on the last street in the neighborhood and that the telephone poles only went another 100 feet past his street so when I saw the last pole, I left the dirt road and went into the bush. I could hear a lot of hooting and hollering in the general direction I thought we should be going so I continued on. I could see the fence and the street beyond it. I had been watching the area for a long time through the camera so now I could see what I had been looking at and that helped me find the camera.

I could see the workers and the crooks now. They were all out in front of the house yelling at us. The narcs were all sweating pretty hard and I wondered if it was the exertion or the fear of getting shot at by these guys that was causing the water to flow. Probably the heat; they had all put on the vests I had offered to loan them. I wasn't wearing one because I had loaned mine to one of the narcs. I noticed that the camera was not very well concealed and that pissed me off. The guys had decided to do a substandard job because they were mad that I had made them go back and move the camera.

I started pulling wires from the battery box and handed the battery to one of the cops. I picked up the tripod with the camera on it and told the cops to head back the way we came. The crooks and workmen hooting reached a crescendo as we disappeared into the brush. I returned to the CP where one of the Task Force officers was fast forwarding through the tape from earlier in the day.

He found what he was looking for. A UPS truck was unloading tires and giving them to the crook's workers. "He's getting new tires, damn it," he said, "I got to go" and left. I found out later that they had put a tracker on the crook's car because he always lost them when they tried to follow him. The narc who just left had realized that when the crook had a local garage mount his tires on his vehicle, they would probably do it on a lift and he was pretty sure that the tracker would be discovered. They followed him into town but this time they didn't lose him and when he stopped at a convenience store one of the narcs rolled under his car grabbed the tracker and got back in the car before he came out; never a dull moment in the life of a narc in Texas.

Marfa: Loadout - Brian briefs C-130/UH-60 Air Movement Plan

Tommy Reynolds SATCOM Check

C-130

'Shoe' Juan Hinojosa and Tom Reynolds Exits C-130

SOD Suburban Exit C-130

UH-60 Night Insertion

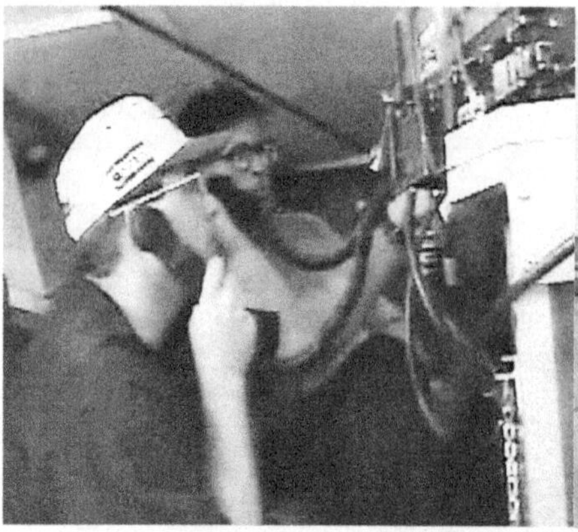

Pat C on SATCOM in CP

Extraction and Exfiltration

TROTEC Camera-Juan Hinojosa

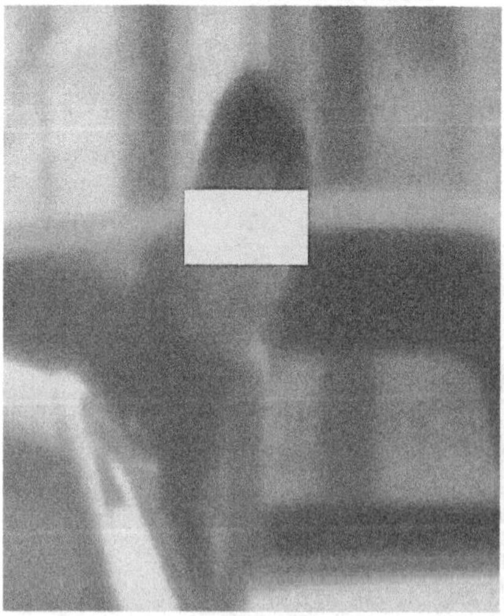
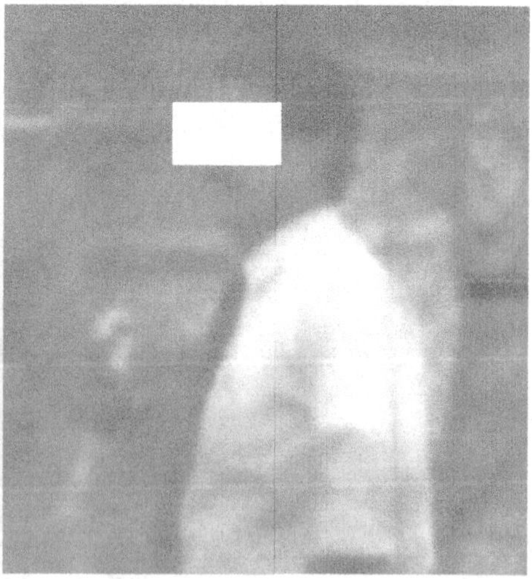

Suspects: The Border

For My eyes are upon all their ways; they are not concealed from My face, neither is their iniquity hidden from My eyes. Jeremiah 16:17

Chapter 17-March 1996

White Shark

It was Spring Break at South Padre Island and we were there for the party. Tonight we were going to get down on Boca Chica Beach. We bailed out of the Blazer just after the sun went down. There was no one on the beach near us. It was very cloudy and the only lights we could see were the lighthouse at the mouth of the Rio Grande and the hotels on South Padre Island glowing in the distance. We grabbed the LORIS and the batteries and moved them into the dunes. The Blazer drove off and the three of us were alone.

We struggled over the first line of dunes and found a small valley where we could set up the monitor. We placed the LORIS head on the low tripod on top of a dune, camouflaged it and ran the cables down to the monitor, hooked them up and powered up the Loris. The monitor brightened and Pat began adjusting the controls. Sergeant Dan was up on a dune next to the one with the LORIS. He was watching the beach with a pair of PVS-7 Night Vision Goggles.

I was walking around videotaping Pat and Dan when Pat said, “hey, come here and look at this.” I walked over to the monitor and saw a shark boat, low in the water, just outside the surf line, going north. I called the CP and said, "we have a shark boat approximately 100 meters in front of our position, heading north at this time.”

Pat was panning slowly and switching the near and far views and going from white hot to black hot. He was annoyed that I wouldn't let him talk on the radio while he operated the LORIS. He's like that. He was also the best operator I ever worked with and capable of doing everything on this mission at the same time. The CP called back and asked a series of questions that we tried to answer as we watched the boat. There were 5 men in the boat and there appeared to be a large box covered with a tarp or something in the front. We were fascinated as we watched them turn into the surf line north of our position and navigate their way through the three lines of surf and beach the boat. These guys were good.

Operation White Shark

Operation White Shark began for us when Al, the U.S. Customs Service, Resident Agent in Charge, Brownsville, TX, requested that JTF-6 support a USCS anti-smuggling operation on Boca Chica and South Padre Island beaches. Their liaison officer in our HQ decided that we could do a better job and routed the request to us.

Customs and the Cameron County Drug Enforcement Task Force had decided to work together to see if they could catch some of the shark boats that were landing on these beaches. The smugglers were unloading 1000 to 1500 pounds of marijuana and/or cocaine. The contraband was either buried in the dunes or loaded into vehicles for transport into Brownsville, its first stop on its journey into the U.S.

The Task Force had caught a boat, almost by accident, about a month before and they had convinced Customs that it would be a good idea to try to catch some more. They also brought in Emilio, a Border Patrol supervisor from the Santa Isabel station. Emilio was coming to the end of his career, but he wasn't ready to kick back. Every night he was driving a fast boat offshore or cruising the beach in a Blazer.

I went with him to Boca Chica Beach and he showed us around. He was from California and knew some of my old haunts in Newport Beach, so we spent the long drives reminiscing. He was working with Brad, the Customs guy who was assigned to the case. There was a senior Customs agent named Howard, a retired Lt Col from the Air Force, who would oversee our participation in the operation while Brad took care of the fieldwork.

Brad filled me in on the investigation of the colonel in our HQ who was suspected of being involved with drug traffickers. This investigation was the one causing so much turmoil when I started working at CD. Brad was running an investigation of the colonel's brother and was visited by some gentlemen from an 'other government agency' who told him to drop it for reasons of national security.

After my initial coordination visit, I returned with Lt Brian to Brownsville. We brought 6 operators and three LORIS sets. The initial plan was to set up one LORIS on the top of one of the very tall buildings on South Padre Island and see if we could spot the boats from there. Emilio had worked a shaky deal with the manager of the tallest building on South Padre where we set up the LORIS and we made a deal with the manager of a hotel on the island, to get 2 suites on the top floor and put our radio antenna on the roof.

We set a LORIS on the balcony of one of our suites and settled in to watch for shark boats. It was spring break on South Padre and every hotel including ours was full of wild drunken college students. There was loud music and people screaming until the early morning hours. One night some students were leaning out from their balcony and they looked up and saw part of the LORIS head on our balcony. "What's that, dude?" they hollered. Brian was on the balcony and said, "Dude, it's a beer cannon." "Far out, dude," one of them replied and they went back to their drinking and smoking dope.

The next night it was my turn to work the LORIS on the other building. I was watching the monitor after the sun went down when I saw a vehicle, it looked like a van, driving on Boca Chica beach. Because of the distance, the vehicle image was very small, and it was difficult to see much detail. There were other vehicles driving up and down the beach on Boca Chica at that time of night so I decided I would watch it for a short time before switching to something else. As I watched, the vehicle stopped, and four or five people got out and began running back and forth from the water's edge to the van.

Because the distance was so great and the figures were so small and I couldn't see anything else I just watched for a minute until they all got back in, turned around and drove off. I wasn't sure what I had seen but I decided not to call the Task Force guys who were hiding in the dunes. They were close to where I saw the activity and I figured that if anything suspicious was going on they would notice it. We shut down at midnight and went back to the hotel. I told Brian that we should go to Brownsville in the morning and talk to Al and Brad.

Boca Chica Beach Recon

I convinced Brian that we should see if they would let us move one LORIS onto Boca Chica beach because the distance was too great to really see with certainty any illegal behavior. He agreed and the next day we had a meeting with Al. He didn't know that we did things like hide in the brush. We assured him that this is what we were best at and he agreed that we could put a team on the beach. Brian and I drove out to Boca Chica to do a recon.

We drove south on the sand to the river mouth and then we drove north looking at various spots. We picked a few spots that looked OK and loaded the locations into the GPS. When we got back to Hwy 4, I suggested that we look at a few spots on the north end of the beach. We drove north for about a mile and Brian said, "This looks like a good spot." I looked at where he was pointing and saw that it would not be a good place because it had no high ground. I drove past the place and said, "I'll show you a better place."

We drove past a sedan with some people who were surf fishing and I turned left and drove toward the dunes which were higher at this point. I parked the Blazer close to the first line of low dunes and we got out. As we walked forward over the low-lying dunes, I saw what looked like boxes partially buried in the sand in front of me. At the same time, I noticed a man sitting to the left of the boxes.

I snapped that this was a load of dope being guarded and pulled my M-9 out of my fanny pack and faced the man. He reacted quickly when he saw the gun and took off running through the gully. I knew that I couldn't shoot him, and we weren't allowed to arrest people so we let him go. I suggested that we back away from the dope and park closer to the water. Brian agreed and as we did, we contacted the CP on South Padre and had them call Customs to let them know that we had a large load of dope and a suspect at large in the area.

I noticed that the people who were fishing had taken a great interest in us and the fact that I had a pistol out, even though I was dressed in my Hawaiian baggy swimming trunks. I then noticed a second group further north who also appeared to be fishing and were packing up quickly and preparing to leave. We got the license plate numbers and descriptions of the people and gave the information to the CP on the Customs repeater channel. Due to some problem with the radio programing, Customs could hear us, but we couldn't hear them. Several vehicles loaded with agents were hauling ass from Brownsville, but they couldn't talk to us. We waited and waited and finally one of the Task Force agents drove by and we flagged him down.

Within a few minutes, Al and about five agents showed up and we went back to the dunes to look at the load. The agents pulled their pickup trucks close to the dune line and we began to pull the boxes which turned out to be large bundles of marijuana from the sand. There was a total of 1300 pounds of weed at that site. Everybody was happy.

The Customs guys were kidding us about our luck, and we were telling them that this might be a worthwhile endeavor after all.

We went to the Customs office to plan for the next day's activities that would include a LORIS team being placed on Boca Chica beach. Brian called Austin and told the Detachment Sergeant to send down Pat with three more operators and another LORIS. Pat solved the communications problems when he arrived and after thinking it over, we decided to put one LORIS on the south side of Hwy 4.

Lt Brian went in with the first beach team the next night. I dropped them off a little south of Hwy 4 just after sunset. I went with them and took video of them setting up, using a set of night vision goggles which I held over the lens of my video camera. After they were set up, I got back in the Blazer and drove back to South Padre. I passed through the crowd of college students in the hotel lobby that were just getting revved up for the evening's activities and went up to our OP in the hotel. Nobody saw anything that night and at 1130 I got back in the Blazer and drove back to Boca Chica to pick them up. The CP contacted them and told them to prepare for exfil.

When I got to the end of the pavement, I called them on the radio and told them I was five minutes out. When they saw me coming down the beach, they asked me to flash my high beams and when I did that confirmed that it was me. I drove up to the drop-off point; they were ready to go and came out of the dunes carrying all the equipment. They loaded the Blazer and we drove off the beach.

As I drove them back to the Hotel, they told me that the wind had been strong that night and the sand was constantly in their eyes. Someone suggested that we buy some painters goggles at the store for the next night. When we got to the hotel at about one AM, the spring break college party was in full swing. We had rehearsed an explanation for this situation. When we unloaded our equipment and carried it through the lobby we were prepared to tell anyone that asked that we were working for the government; trying to protect the Ridley Turtles from extinction by spotting where they laid their eggs and protecting the nest. Our guys used this story while we were there, and the drunken college students loved it. We were ok and, would we like a beer?

Boca Chica Beach-O Dark 30

Lt Brian told me to make up a schedule and that was how I found myself with Pat and Dan a few nights later watching the shark boat land on Boca Chica beach. We watched through the LORIS as the smugglers took the bundles out of the boat and ran up the beach with them where they disappeared from our sight in the dunes. Dan look up the beach to check for vehicle lights and he didn't see any. We figured that they were burying it in the dunes like the load we had found a few days before.

Pat was counting bundles and giving a running commentary as he worked the LORIS. I was talking to Brian in the CP and trying to take video. My camera kept giving me a signal that I was out of tape. Here was a perfect opportunity to get some real good smuggling footage and the camera was acting up. I determined later that the RF energy from my radio was scrambling the brains of the camera.

Brian had alerted Customs and the Task Force. They moved closer to the beach on Hwy 4, hid their cars on side roads and waited. The boat crew continued to unload the shark boat and they were done in about 10 minutes. Pat and I watched as they pushed the boat into the shallow water and turned it around. Three of them got in the boat and two of them waded alone side and pushed the boat into the surf line.

When they got into waist deep water, they jumped into the boat and two of the smugglers began using oars to push the boat into deeper water. The man in the back of the boat lowered the propeller end of the outboard motor into the water and started the engine. When the prop started spinning the bow of the boat rose out of the water and the boat began to move swiftly through the three lines of surf, guided by the skillful hand of the coxswain.

The bow of the boat reared high in the air when they went over a swell with the group of men gathered in the stern. As the bow went up and down and the boat picked up speed it looked to me like the boat itself was waving bye-bye to us. I imagined that the smugglers were feeling pretty smug about the way they had delivered another load into the great United States and that they had beaten the Customs Service and the Border Patrol again. Not this time.

I told the CP that the boat had left. The cops wanted to know if we could see any load vehicles. I told them that the only vehicle that was on the beach on the south side was a big RV 500 meters south of us; it was dark and there was no movement around it. The task force guys had put a vehicle in a small overnight parking area at the end of the pavement and they advised us that no vehicles had gone onto the beach or left the beach from the time that we spotted the load until now.

Boca Chica Beach-Tree Finds The Cache

The cops asked if we could find where the dope was buried, and I told them we could. I decided that we would leave the LORIS aimed at the spot where the boat unloaded, and Dan would guide Pat and I to that spot so we could look for the cache. I told Dan that we would come back and get him when we found it. Pat carried a pair of night vision goggles and his pistol. I carried the radio and my bowling ball bag that had my camera and other gear. I had my pistol in my belly holster under my shirt. Dan had the other radio and after assuring him that we would be back, Pat led us through the dunes to the beach.

We began walking north up the beach near the dunes toward the boat landing site. I was watching the dunes to see if I could spot any smuggler lookouts before they spotted us. It was very dark, but my eyes were becoming adjusted. I noticed a small opening in a waist high dune to our left and I told Pat that I wanted to check it out. It took me behind the first dune which got higher and higher until it was shoulder high and concealed the area behind it from the beach.

There was a small open area back there and I found a large square hole in the ground. On the edge of the hole was a shovel with a broken handle and a small bundle tucked away on a kind of shelf that had been dug into the dune. I picked up the bundle; it was a white shirt and opened it. There was a book inside. Being an avid book collector, I put the book in my bowling ball bag and grabbed the shovel. I came back to the beach where Pat was waiting for me and told him what I had found. He said that this seemed to be a busy place.

We left the shovel at the entrance to this cache site and moved down the beach. I began to get paranoid about not being able to see and I was worried that there might be other smugglers in the dunes watching other loads or waiting for one. I pulled my pistol out and told Pat that we should move closer to the water so we could have some distance between the concealment of the dunes and ourselves. We continued our way slowly down the beach near the water's edge.

I called Dan every five minutes and asked him where we were in relation to the boat landing site. He told us that even with the LORIS on wide view, he couldn't see us. This meant that we were a lot farther than we thought from the boat when it landed. Those LORIS devices were good. I kept the CP and the cops informed as to our progress while we moved down the beach. Finally, Dan called and told us that he could see us on the edge of the screen.

Pat was using the night vision goggles to check for activity in front of us and to look for tracks in the sand. About thirty minutes after Dan told us he could see us, Pat found the boat crew's tracks. He led the way slowly up the beach toward the dunes. As the sand became drier the tracks became harder to see and then they disappeared. He said that it looked like they had brushed out their tracks as they were leaving.

Pat told me to stay in his tracks as we walked into the area that he suspected contained the cache. He walked behind a two-foot-high dune and stopped. He scanned the area with the goggles for two or three minutes. He walked forward a few feet and got down on his knees and began to push his fingers into the sand. He probed in front of him with his fingers and then move forward a foot or so and did it again. I realized he was doing what we would be doing in a mine field, trying to find the mines by probing. This went on for at least 15 minutes. The CP kept calling for updates and I was beginning to worry that we were in the wrong place when Pat exclaimed, "I found it".

The sand which covered the beach was packed and hard. The sand that had been used to cover the cache was soft. Pat probed down about a foot until he felt a package. I reported this information to the CP. Brian conferred with the cops and called me back. He asked if we could find a place where we could observe the cache until he could organize a team and come down to replace us. I told him that we could find a place to watch the cache but that we should stay to watch it. I said that any traffic on the beach might alert the smugglers that they had been detected. What I didn't tell him was that this was our load, we caught it and now we were going to skin it. They would have to get their own load. I informed him that I was going to send Pat back to help Dan pack up the LORIS and then they would rejoin me at the cache. I asked him to send the guy who dropped us off to the drop off point at dawn so that they could retrieve the LORIS.

It was about a mile and a half between the cache and the LORIS and I didn't think that if the smugglers saw our Blazer go by that they would associate it with the cache. Brian agreed. I told him to bring us some water, chow and some fresh radio batteries in a kit bag and throw it to us as they drove past. He agreed. It was now about 2:30AM. I called Dan on the radio and told him that Pat was coming his way and to guide him. I told Pat that when he got there, to pack up the LORIS and lead Dan back to the cache site. Then I told him to swing by the other cache site and pick up the shovel on his way back. He asked me why and I told him that we might need it.

Pat was always interested in the details and asked me why we needed a shovel, so I told him to just get the shovel. Pat left to take care of the things we agreed on and I took a seat on a low dune and waited. Pat took the other radio, so I had no communication with the CP. I looked around the area for hiding place for our OP. There was a high ridge line in the dunes about 100 meters from the beach and it overlooked the cache site. I thought I could find a spot up there.

Boca Chica Beach-A New OP

It took almost an hour for Pat to return with Dan. I saw them from a distance through my night vision monocular. Pat was carrying the shovel. I told Pat and Dan that I wanted them to stay at the cache site while I went to look for a hide. I took one of the radios and told Pat that I wanted him to turn on the red lens flashlight he had when I told him to "hit me." I would walk along the ridge line and try to find a hide site that allowed me to see his light.

I moved inland through the dunes to the south end of the sand ridge and walked up to the top. I told Pat to hit me and looked for the red light. I couldn't see it and told him so. I began to move north. There were some very low grasses on parts of the dunes but not good places to hide. I stopped every so often and told Pat to hit me again. About halfway across the ridge I started to see the light. When I got to the north end, I found a good spot. I verified that I could see the light and told him to bring Dan and the shovel to my location. Pat brushed out their tracks and I guided them to our new home, they were not impressed. I told Pat and Dan that it was LRS time and that we would have to dig a hide. They looked at me like I was crazy.

It was now about 4:00AM. Time flies when you're having fun. Pat and Dan where looking pretty tired, so I told them to take a rest. I took the shovel and began to dig. I dug a hole big enough for the three of us and our gear. The others eventually rallied and helped with the digging. The sky was beginning to lighten on the horizon and so one of us would watch the beach while the others dug the hole. I could see that the sand that we had thrown out of the hole really stood out against the rest of the sand behind us, so we pulled some grass and threw it over the spoil to conceal it. "Conceal it from who?" one of them said. I didn't know but I had a feeling, so I told them that we just had to do it. I'll admit that it didn't look like anyone would see where we were because we were at least one hundred meters from the beach and there was no reason why anybody would be walking around in our area.

The CP called and said that they were inbound and would be there in about 15 minutes. I told Pat to take a radio and go down to the edge of the dune line and find a place where he could grab the resupply bundle when the Blazer passed his location. Dan and I continued to improve our position, one of us digging and the other watching the beach. It was now very light out, the sun had not come up, but you could see very well, even in color. Pat had found a place to receive the resupply bundle and when the CP vehicle came down the beach he directed them to drive close to the dune line. When they were close to him, he waved at them. The vehicle never stopped or slowed down. One of the operators in the back seat threw the aviator kit bag right to him. He caught it and the CP team drove down the beach to get the LORIS.

Pat returned with the resupply bundle and we settled into our hole. I checked the bag and found some radio batteries, several small bottles of water, some pop tarts, and some grapefruit. By that time we were all pretty tired but I remember thinking that we needed to plan for these types of contingencies in the future. The CP vehicle came back north up the beach and as they passed our position, they told me that they had recovered the LORIS and were heading back to the hotel. They wished us luck and signed off.

Boca Chica Beach-A Day At The Beach

I told the guys to get ready for a long day. The sun was rising on the horizon when a vehicle drove down the beach from the north. It stopped about a hundred meters south of the cache and two people got out and began preparing to surf fish, a coincidence? Maybe, but I remembered the load we found that was being watched by people surf fishing on either side of the cache.

A few minutes later another vehicle drove down the beach from the north. It passed the cache, stopped, turned around, drove back north and parked about a hundred meters from the cache. Two people got out and prepared to surf fish. Both groups were looking up into the dunes and they kept looking around the area as they prepared their fishing rods. We were in our hole and had a small viewing place that was covered with grass so we could watch the activity without being observed.

I told the guys not to stare directly at the people surf fishing because they might sense that someone was looking at them. I had read about this type of psychic phenomenon and had experimented with it in Mexico, where I stared at the back of people's heads and 9 out of ten times they would turn around to see who was looking at them. Lee Justice had told me about similar experiences working on the river with the Border Patrol. The people on the beach would periodically scan the ridgeline we were on, but they seemed satisfied that no one was there and I wanted to keep it that way.

Everything was falling into place. I told Pat and Dan to put their heads down and sleep and that I would wake up one of them when I need to sleep. I was actually quite energized at this point. We had the load cache under direct surveillance, the crooks had their guards in place, and we had good communication with the cops and the CP. That morning was nice and cool but as the sun rose higher in the sky, it began to warm up. I was wearing a long sleeve shirt and so was Pat. Dan had a short sleeve shirt.

The guys continued to sleep while I watched. I knew about the value of a little sleep and I was determined to stay awake for as long as I could so that the guys could recharge their energy. There always seemed to be extra energy for the leader. That is one of the unintended benefits of leadership. I called the CP every hour and I called on the name of the Lord Jesus every few minutes and sometimes every few seconds. There was light traffic on the beach and the watchers played the role of surf fishers when a car went by. I don't think they had any bait on their hooks.

It continued to get warmer and the guys woke up around 11:00AM. I told Pat and Dan what was going on and put my head down until 1230. When I awoke it was quite warm and I was sweating. I looked in the kit bag for some water and there was none. The guys had drunk it and they had eaten the grapefruit. I had two bottles of Gatorade in my bag and I pulled one out and took a drink. Dan was complaining how hot it was and I wondered if he was prone to heat injury.

Pat was ok and said that a sedan had come down the beach earlier and had done donuts in the sand between the two watcher vehicles and then it had left. I looked through our observation port and saw that there were indeed some big donut tracks in the sand. We talked about the possibility that this was another technique of the smugglers or just some teenagers fooling around. The afternoon dragged on until about 2:00PM when we heard what sounded like motorcycles coming down the beach from the north.

We looked over the back of our hole and saw 2 four-wheelers turn off the beach and drive into the dunes behind our position. There was a path that we hadn't noticed the night before about 75 meters to the north of us. The path went from the beach to a higher ridge of dunes further to the west. The riders were young couples who looked like college students doing some spring break activity. They rode past our position and on to the top of the higher ridge to the west of us. They stopped on top and were sitting there looking down towards our hide-site. We all had to scrunch down in our hole and not move.

They stayed up there for about 10 minutes, looking around, talking, and laughing. I could just see them when I turned my head to the side. Whether they saw us our not was hard to tell but eventually they drove back down the trail onto the beach and went north.

It was really hot now and Dan was saying that he was feeling faint. He was talking like he was going down, and I had to consider the possibility that we would have to MEDEVAC him. I knew that if we did, we would probably blow the surveillance and I didn't want to do that. I began to tell him what a hooah guy he was. How he was a jumpmaster in an LRS Company and how he had been to a lot of places and had done a lot of things. I could see that his arms were getting red so I took of my shirt with the long sleeves and gave it to him.

I continued to encourage him, and he began to respond with some positive agreements. He was out of water and that seemed to be the big problem. Unless we got some liquid into him, we were going to have to have him extracted whether I liked it or not. I was pretty thirsty myself and was counting on my one and a half bottles of Gatorade to get me through to nightfall which would be in three hours.

I knew that after dark we could get a resupply of water and personnel if needed. I had no idea how long the crooks were going to wait before they dug up the load and moved it, but I was determined to stay to the end. I had been doing this mission for almost seven years and this was the most personal satisfaction I had had. I didn't like paperwork, I liked operating, and we were really operating. I knew I could count on Pat and Dan, but Dan was really going down now so I opened my bowling ball bag and gave him my full bottle of rip tide rush Gatorade. He protested for a moment, but I laughingly ordered him to drink it. Pat encouraged him also, so he drank about three quarters of the bottle slowly. The Gatorade did the trick. He began to feel better, his morale improved, and there was no more talk about going down.

Boca Chica Beach-The Smugglers Return

It was now about 4PM. A few vehicles which had been down south on the beach by the river mouth had been passing by the cache site on their way back to the Highway. The beach was deserted as far as we could see except for the two watcher vehicles. Then we noticed that the watchers were packing up their things like they were getting ready to leave. This was odd, maybe they weren't watchers after all. The two groups stowed all their stuff in their vehicles, got in and drove off. As they drove north up the beach, we observed them with binoculars and the spotting scope, trying to get their LPs.

As we watched them going north, we noticed a pickup truck with a camper coming south. The pickup parked on the beach in the middle of the donut tracks. Two guys got out and looked around. After a minute, they walked a short distance over to where the cache was. We could see them bending over the cache and then one of them stood up and ran back to the pickup. He got in the driver's seat and started the truck. He backed it up to where the other guy was and stopped. He climbed out of the truck and walked back to the other guy.

I had the video camera out and was shooting some of this. I gave the camera to Dan and he gave us a running commentary as he videoed the crooks. "You can't really see them, oh wait now you can, they're throwing the bundles into the camper, they're not even trying to hide them," etc. I was on the radio keeping the cops informed of the situation. Pat was pulling security to the north. We were rocking now. The crooks finished loading the camper. One of them jumped in the back and the other ran and jumped into the cab. The truck started and they drove off to the north.

Pat tracked them as they drove by and I relayed his information by radio to the cops. The truck turned north on Highway 4 and drove past the parking lot where one of the task force vehicles was waiting. The task force guys started following the vehicle and everything was going smoothly until it took off. They were right behind the load vehicle as it flew up the narrow two-lane road towards Brownsville at upwards of a hundred miles an hour.

The Task Force guys told us later that as the vehicle got close to town there were some spots where the river comes very close to the road. At one of these spots the load vehicle driver turned off the road and drove across a field at a high rate of speed. He hit a big bump and flew through the air. When he landed, they said that it was like his truck exploded. The truck came to a stop and the driver bailed out and began running for the river. The arrest report said that a passing car backfired causing the fleeing crook to believe that the cops were shooting at him and he surrendered.

Meanwhile we were still in our OP. The second guy that was in the pickup had been let out before they got to the highway and now, we saw him walking back down the beach toward the cache. Dan was still taking video and he raised up to shoot from the side of our hole instead of using the observation port and it's possible that the guy saw him. When you see the video, you can see him look in our direction and then turn around and walk back north on the beach. The cops arrested him later when they came down the beach after they got the load vehicle.

The TF contacted us on the radio. After we told them about the man on foot heading north, they asked us to go down to the cache and see if we could find any more dope. We packed up our gear and went down to the cache and waited. There was a Customs plane in the air circling around above us; the aircrew had spotted the man walking north and was directing the TF guys to his location. After a short wait the TF guys and some Customs agents drove up. We showed them the cache site.

They produced some shovels and started digging. Soon they were pulling large bundles out of the ground and loading them into the pickup trucks. Dan, Pat, and I were photographed sitting on top of a stack of bundles. It felt pretty good. I couldn't resist saying, "It smells good, let's see how it smokes." The cops and we were laughing and joking as we picked up the remainder of the cache.

The total amount of dope seized came to 1800 pounds of high-grade marijuana. Brian and the other operators showed up and we said goodbye to the cops and headed back to the Hotel. Brian had bought a couple of large pizzas and I took one and poured a half a can of spaghetti sauce on it, just the way I like it. The other guys didn't think much of my move and gave me some harsh words about my taste in pizza. Everyone was happy about the way things had turned out. We had learned a lot from this, and began sketching out some SOPs for the next night. I was pretty tired and went to bed.

Shark Boat and its Nemesis-The LORIS

White Shark: LORIS Ops Marijuana High Fives

Pat on the LORIS

"We have a Shark Boat…"

The Load Vehicle

Pat C

Dan S

Truck load of Weed: Dan S author Pat C

He who practices deceit will not dwell within my house
Psalm 101:7a

Chapter 18-April 1996

East Texas-Site Selection

I left South Padre two days later and returned to Austin for drill. During drill, Lt Brian called and told me that Rob and I needed to go out to East Texas on Monday and see Jerry about another job. Rob was a Ranger veteran of the parachute assault on Rio Hato during the invasion of Panama. Like many of us combat veterans he had a somewhat dismal view of life, but it was balanced by his sense of humor.

The following Monday we drove to East Texas and met with Jerry. Kenny and Hutch had moved on and Jerry introduced us to the case agent 'Bill' who told us that he needed video of a crack dealing operation that was taking place in a rural environment. He had aerial photos that he had taken using a private plane piloted by a friend of his. I usually took my own aerial photos of a target but his appeared to show everything we needed to see.

We went into the area for a look-see. The dealing was taking place at a crossroads in a heavily wooded area. There were major roads nearby but when you got into the woods it seemed like you were miles away from civilization. The dealers would hang around the intersection of two dirt roads and the customers would drive up and stop while the various dealers decided who would sell the crack.

It seemed that each dealer had his own clientele and although some of the clients knew all the dealers; they had some method of determining who would buy from who. The vegetation was very thick around the crossroads and it would be difficult to shoot video of these transactions. The dealers were weary of police activity and they had several lookouts that walked around looking and listening for the cops. One of them carried a shotgun. He became known as the 'chess-master', not because he was so smart, but because he appeared to be so dumb. He did have eyes and ears however, and he used them.

We went to the place that the cops had found for us to use as a command post. It was about five miles away from the target. There was a small hunting and fishing cabin on high ground, and it looked like we might make commo if we used our tallest portable antenna. We would have to erect it on the roof of the cabin and strap it to a tree. For the commo test we set up a jungle antenna in a tree and Rob, the team leader, prepared a jungle antenna for a commo test that would be conducted when he made a recon of the target area.

I drove Rob and one of the cops who knew the area, to a drop-off point about a mile away from the target. After I dropped them off, I went back to the CP and waited for Rob to call me. I heard static a few times but no clear signal. We had planned to try the radios at various points during the recon. We were talking on our own frequency, but we were using encryption which cut down on the range of the radios. Rob was supposed to try the radio without the jungle antenna as he moved around the target area. When he found a place suitable for a mission support site (MSS) he would throw the weighted end of a piece of parachute suspension line over the highest tree branch he could find and pull the jungle antenna as high as it would go. About two hours after I dropped Rob off, his voice came up on the radio, we had commo. No commo was a showstopper and I didn't want to have to look for another CP. He called for extraction, so I went back to the drop off point and picked them up. We went back to the CP and Rob briefed me on what he had seen on his recon.

"The place is a thicket," he said, "and it is very difficult to move without making noise." I could have told him that based on my previous experiences in East Texas. He went on to say that he found a good spot for a MSS but because of the commo situation, we would have to have someone in the MSS at all times to relay communications from the guys doing the surveillance to the CP.

The MSS was about three hundred meters from the target. The team would have to move through thick brush and try to find an OP where they could video tape the transactions. The cops also wanted us to see if we could spot where the dealers hid their dope. If they suspected that the police were coming, the dealers would hide their stashes near the target. The cops wanted to know where so that when they raided the place, they could get the dope and use it as evidence. Rob seemed slightly dismayed by the situation and I didn't blame him, but I told him that there was always a way to accomplish the mission and that we would have to figure it out.

We completed our target folder and agreed to return with a team in a week and left. On the drive back Rob reminded me about the armed man that patrolled the area and the difficulty of getting close enough to get the video the cops wanted. I agreed that it was going to be a hard one, but I told him that we would figure out how to solve the problems when we came to them.

We went back to Austin and I briefed Lt Brian on the operation. I requested a six man team and one person to help me in the CP. Rob was given five guys and Sergeant Al would be my assistant. The team prepared the majority of the equipment and Al and I assembled the CP equipment. I wrote the Operations Order and Memorandum of Understanding. We conducted commo checks with the team and tested all the camera equipment on Friday.

Everyone went home for the weekend and we met at the team room at 5:00AM on Monday morning. Because of so many mistakes in the past regarding equipment accountability, we conducted an inventory of everything we were taking, using the hand receipt. As we began loading the vehicles, we had a team member at each team vehicle, and me at the CP vehicle, checking off each item once again as we loaded it.

East Texas-The Operation

When we were loaded up, we drove toward east Texas and arrived at the Task Force at about three in the afternoon. I introduced the team to the cops, and they welcomed us. Jerry wanted to have the mission briefing at the CP, so we drove down there and went inside. The CP had boxes of soft drinks and snacks. Jerry told us to help ourselves. Sergeant 'Shoe' was with us and I whispered to him not to eat everything. Shoe had gone to Ranger School sometime before and while he was there, he had missed some meals which had made a big impression in his mind. He was much more than a chow-hound and he ate any food he could get his hands on including the food of those around him if they weren't looking.

The team guys unloaded their gear while Al set up the CP. I sat down with Jerry and Bill and went over the Operations Plan and the MOU. They were happy that everything we had agreed on during the last visit had been written down in the OPLAN. Jerry said it was good to go and signed the MOU for the Task Force and I signed for the National Guard. I called the guys over and we had the mission briefing.

With the signing of the MOU, the OPLAN became the operations order. Bill, the cop, gave the intelligence portion of the mission briefing. He told the guys that he wanted video of the dealers dealing their dope. He passed around a photograph of an individual and told us that the person in the photograph was 'Mr. Big', the supplier of the crack to all the dealers on the target. Bill wanted to be notified as soon as we spotted him. He told the guys to see if they could spot the hiding places that the dealers used to hide their stashes.

I gave the rest of the briefing. We decided to do the briefback after dinner and then we would infil the team at about 10:00PM. We used the rest of the afternoon for the team to conduct their planning while Al continued to set up the CP and I talked with Jerry and Bill.

We all went out to eat some East Texas BBQ and everyone in the restaurant knew we were with the cops because they all knew Jerry and some of them waved to him. After working with the cops for a number of years I had gotten used to the fact that Operations Security was considered differently by them. They figured that because they lived in the area and everyone knew they were cops, they didn't have to hide their presence, only their activities. Sometimes the guys would throw a fit about eating in public with cops, but I felt that rapport building was more important than certain considerations about OPSEC.

We finished the meal and by now some of the guys were yakking away with the cops like they had known them forever. This was a good thing and I was proud of the way some of them had learned about rapport building and its importance in the business we were in.

We went back to the CP, taking a route that would make it easy to see if we were being followed. The team gave us a briefback about an hour later. They answered all our questions regarding their contingency plans for problems such as compromise, loss of commo, medical emergencies, and take-downs. The team was good to go so we loaded them into the infil vehicle and Bill and I drove them to the release point and dropped them off. The team made commo from their hide site about an hour after we dropped them off.

We settled into our routine with Al and I rotating two hour shifts on the radio while our cop 'backups' slept away. We had a TV and VCR in the CP with a supply of movies. The cops would watch movies when they were up, and we would monitor the radio. I didn't allow the military people to watch movies or listen to music when they were on duty because I was concerned that they might miss a radio message from the team. I had seen this happen before, so I didn't mind the occasional argument with the guys who hadn't worked the CP with me in the past.

Rob called at about 5:30AM and advised us that he and another man would be leaving the MSS and moving toward the target to conduct a reconnaissance of the target area. He would look for places to set up OPs, where he could put two men and video equipment. It had to be close enough to get facial identification but far enough away so as not to be seen by the crooks. The recon would take several hours.

I woke up the cop who would respond to any calls from the team requesting a law enforcement officer. They would only do this if they had been walked up on by someone. If that happened, they would identify themselves as police and take the person into protective custody. We knew that the police could not get there in time deal with a situation like that, so we trained the guys to act like cops in those circumstances. They went to police schools and learned handcuffing techniques along with extensive training in the use of force, to include the use of deadly force. They all had to demonstrate a high level of proficiency with their M9 Pistol, 870 Shotgun, and M16A1 Rifle.

I believed that the better that an operator could shoot, the more likely he would use other techniques to control a confrontational situation. Cops routinely deal with situations where someone could be shot but they are not because the cop would talk them out of doing whatever they are doing and convince them to surrender rather than shooting them. This theory would be tested on this mission.

East Texas-"X-Ray! X-Ray! X-Ray!"

"Uniform Charlie this is Slayer (Rob's call sign), X-Ray, X-Ray, X-Ray." I yelled at the cop who was drinking a cup of coffee and told him to get ready to roll. I grabbed the microphone and called, "this is Uniform Charlie, what is your situation?" Tommy came back immediately, and said, "Slayer was spotted by two men that were walking on the road. One was carrying a shotgun. They were about 10 meters from his position. He pulled back into the woods. One of them threw a bottle at them as they pulled back."

I asked Tommy if Slayer was still in contact. A few seconds later he came back and said, "Slayer said negative, he is 50 meters from the road. He can hear them yelling things like, “we saw you cops and 5-0, 5-0. They are not pursuing him." Everyone was wide awake in the CP. The cop was driving out of the CP area and heading down the dirt road to the highway. I called him on the car to car frequency and advised him to hold up because the team was not in contact with anyone. He agreed that we should wait until Rob called us back with an update.

One of the hardest things about working as a mission commander in a CP is wanting to know what is going on the ground. It is a big temptation to want to get on the radio and ask for information updates every few minutes. I knew this because I had been out in the field under similar circumstances and when you are leading men in stressful situations, the last thing you need is distractions from the rear while you are trying to take care of the situation at hand.

A good leader knows these things and keeps the command element advised as soon as he has the situation in hand. Rob was a good leader and he called after a few minutes and recommended we stand down from the x-ray situation and wait to see what other actions the two guys on the road might take. Rob moved further into the woods and waited for about an hour. He called me again during the hourly Situation Report (SITREP) and told me that everything was quiet around the target.

It was about nine o'clock now and Rob could hear vehicles coming up to the crossroads and he could hear people talking. He had moved closer to the target, but he was deeper in the woods from the road. He could see a little of the activities, but he couldn't get any useful video. The other two-man element, had moved close to an abandoned building that was used by the dopers for a place to smoke crack.

A couple of the dealers had their client's drive into the parking area in front of the abandoned building. The guys were as close as they could get, and they were getting some video of transactions near them. They told us that the activity at the target was picking up, but they could only see a small part of it due to the vegetation and distance. The teams were relaying what information they had, license plates, etc., but they weren’t able to get close enough to get facial video, which was one of the things that Jerry and 'Bill' wanted.

They stayed at it until about 8:00PM. Then activity slowed down because it was too dark to see anything. Both elements moved back to the MSS. They consolidated the video tapes they had used and one of the guys walked up to the highway with them. He put them into a paper sack and crumpled it up so it looked like a piece of trash.

Al was with one of the cops and when the bag was on the ground they drove up. The guy on the ground told them when to stop. Al got out of the car and walked to the tree line like he was going to take a leak. When he finished, he walked back to the car and as he got in, he picked up the bag. They drove off and eventually returned to the CP. The guy on the ground went back to the MSS and we went back to monitoring the radio for the night.

When Al returned to the CP, he gave me the video tapes and I watched them to see how we were doing. The video that Rob had taken was useless and the video that the other team had taken wasn't much better. I called Rob and told him what I had seen. He wasn't happy but he said they would try to get closer tomorrow. The next day and the day after, Rob and his team tried different positions around the target. At one point, two of the guys were under the abandoned building that was the crack den. We picked up their video tapes that night and I showed them to Bill and Jerry.

East Texas-Working With Cops

The next morning Jerry took me outside the CP. He looked me in the eye and said, "These guys are cowards!" That shocked me. I listened to him rant and rave about how much it was costing to do this operation and how we weren't producing. "I specifically asked for you to come out here because I know that you will produce, so what about it?" I told him that I would get back with him that evening with my plan to fix the problem. Jerry's remark about cowards just wasn't true. The guys in the field were hampered by camera equipment limitations and they were not about to be drawn into a confrontation with the guy carrying the shotgun. I knew they could drop him in a heartbeat but that wasn't the issue. If we shot someone in East Texas, even though we might be justified by the standards of the Texas Penal Code, we would have a difficult time surviving the political storm that would ensue.

Our tactics took this into consideration. We were LRS and we knew that in wartime we would probably not be able to survive a compromise behind enemy lines, so we trained to be invisible. Rob was doing everything he could to accomplish the mission, but we were running out time with our law enforcement agency.

I went back to the CP and told Al that I would be busy for a while and that he would have to pull a double shift. I got the aerial photos and the map of the area. I sat down with a ruler and a magnifying glass. I looked at the target from every angle using every photograph. I used the map to help me figure the distance from one point to another and I used the ruler for figuring distance and line of sight. I spent about two hours with the map and photos before I finally came up with a plan of action that I thought would work.

East Texas-A Walk In The Dark

I decided I would go into the target and told Al that I was going to town for several hours and left. Twenty miles away was a larger town where I bought a few things that I thought I would need in the next few days. I could help the guys with solving the video problem, but I would have to be there and show them what I had in mind. I found the things I needed and returned to the CP after stopping at the Dairy Queen and picking up some burgers for Al and me.

While we were eating, I told Al my plan. I would go in that evening and he would become the Mission Commander in the CP. We worked out the details with Bill who agreed to have a cop assist Al with radio watch in the CP. The weather forecast said that a cold front was coming, and it was bringing thunderstorms with it. I didn't have any field gear with me, but I figured that the forecast was usually off anyway, besides it was springtime in East Texas and while it might get cold for a few hours, it would soon get hot and humid again.

Late that afternoon, Jerry showed up and I briefed him on my plan. He looked skeptical but he was happy that I was going in. I called Rob and told him to expect me about 9:00PM. He said he would me meet me just inside the wood line near the drop-off point. I checked my gear and went back to the photos and the map and continued to memorize as much detail as I could about the target area.

At 8:30PM I had a few last minute words with Al. Then I got into Bill's car and he drove me to the drop-off point. I bailed out and walked the short distance to the tree line. It was very dark out on the road but inside the tree line I couldn't see a thing. After a few minutes my eyes had gotten use to the dark, but I still couldn't see a thing. I heard some leaves being stepped on and Rob appeared out of the black, barely visible in the light from the road.

We shook hands and he told me to follow him. He was wearing a pair of PVS 7s and he moved quickly. Rob was a ranger and this kind of work was nothing to get excited about. I couldn't see a thing and moved by instinct, hoping I was following him. I had to cross a fallen tree and when I got on the other side, I had no sense of Rob's presence. After a short time, I heard him approaching me from my left front. "It's a good thing you stopped," he said. "There is a 15 foot drop off right in front of you." We took off again, this time I was right on Rob's heels and determined not to get separated from him again.

We arrived at the MSS about ten minutes later. The guys were just settling in when we got there. I didn't tell them what Jerry said about them but that the cops were getting antsy. Then I told them that I was going try some equipment of my own to see if we could get what the cops wanted. The next morning at about 4:30, Rob and I would go to the target area where I had determined that I could get some good video. Sergeant Juan had set up the guard shift and he put me on from 1:00 AM until 2:00 AM when I would wake up Joe.

East Texas-Sleeping On Guard

Everyone had just settled into their hooch's when the thunderstorm announced its presence with a bright flash of lightning and a deafening blast of thunder. Then came the rain, lots of cold rain. No Gortex, no poncho, and no shelter, I was screwed. I did have some large plastic garbage bags so I pulled two out and pulled them over my head. I poked some holes in them so I could get some air.

The ground was flooding so I couldn't lie down. The MSS area was covered with vegetation but eventually I found a large tree trunk near the team's garbage hole. I leaned up against the tree and listened to the storm around me. The rain was coming down hard, the wind was blowing, and there was a lot of lightning and thunder. I was soaked and I was really getting cold. The water on the ground was gathered into a small stream that ran along the ground, right where I was sitting.

I sat there shivering and tried to sleep. I know that I managed to nod off at least once because at 1:00 AM, one of the guys shook me and told me it was my turn on guard. I moved over to the place where the team had put up their jungle antenna and leaned up against the tree. I was soaking wet and really cold. I got another trash bag and pulled it over my head and over the other two, hoping that the added layer of plastic would warm me a little. I leaned back against the tree, made a commo check with Al, and nodded off to sleep.

I woke up with a start, 'what the blank was I doing?' I had fallen asleep on guard. This was as bad as it gets. Here I was, the Mission Commander, in the field with a team of guys who probably resented my presence in the first place, asleep on guard. I checked the time on my watch; it was 3:15 AM. I was mortified. I had slept for more than two hours, leaving everyone at risk with no one on guard. True, it was very doubtful that anyone would be out in this thicket at 2:00 in the morning during a thunderstorm with a heavy downpour but that was no consolation to me. I had let these guys down and would have to pay a price. There was still, however a mission to accomplish.

I got up and found Sergeant Jim, the man who had guard after Sergeant Joe, from 3:00 to 4:00. He was sharing a hooch with Joe who managed to wake up when I was waking up Jim. Joe became alarmed and wanted to know what was going on. "What happened," he said, "I'm supposed to be on guard after you and its 3:20?" I told him that I had fallen asleep on guard and that I was waking up Jim whose turn it was now. I told Joe that it was my fault and we would get into it when we got out of the field. "Roger that," he said, and lay back down. I walked with Jim over to the commo tree and made sure he was good to go before I returned to my tree by the garbage hole.

I asked Jim to wake me when he got off guard at 4:00 AM and he did. By then, the rain had stopped, and the clouds had moved on. The stars were out and I could see much better. I took off the plastic bags and put them away in my backpack. I oiled my pistol and magazines, then I checked my equipment and after eating a can of Vienna sausage; I was ready for the day. Rob came over to where I was sitting at 4:25 and asked me if I was ready to go. I told him I was and gave him a brief explanation of my failure on guard. I told him that I would discuss it with the team when we got out of the field and he agreed.

We took off with Rob in the lead as we moved to the point in the target area where I thought we would have the best chance of getting the video we needed. I took some video of Rob with my personal Hi 8 camcorder as we moved through the woods; it was light enough that I could just make him out in the view finder. We got to the dirt road that led down to the target and there we watched both directions for a few minutes and then we crossed together to the other side.

We crossed a short open area on the other side of the road. At the tree line there was a barb wire fence which we quickly crossed. On the other side of the fence the vegetation was sparse. We moved forward to the point where I thought we could shoot video. I moved right up to the fence and laid down to see what I could see. Rob was to my right on one knee next to a bush. He was watching the road while I was crawling around looking for the best spot. I found a spot where I could see the crossroads. We were about 100 meters away. The spot I had chosen was good but there was no vegetation to conceal us from anyone walking by on the dirt road to our left.

I crawled over to where Rob was and told him we would have to cut some vegetation and lean it against the fence. He and I took turns moving further into the woods where we would cut small branches off some of the bushes with our garden clippers. We brought them back to our OP and leaned them against the fence. We were taking a chance that someone might notice the change in scenery, but we only used a few branches and I hoped that no one would. One thing about working around dopers was that most of them were pretty oblivious to their surroundings due to the drugs that were attacking their brain.

East Texas-The Rail

I pulled out my small personal Bogen tripod and the rail that had been cut to my specifications several years ago by the machinist who worked in the maintenance shop at Camp Mabry. By mounting various lenses in front of my camcorder, I could zoom my camcorder lens into the mounted lens and get unbelievable magnification. The magnification was so powerful that it took a good tripod and a lot of practice to get any kind of usable images. I don't know the math formula for figuring the magnification, but we probably had the equivalent of a 4000 mm lens on the camcorder using the Nikon spotting scope.

Rob was pulling security as I lay there adjusting my equipment and moving the tripod around. I set up the rail using one of the Sony camcorders we had recently procured. I attached a Nikon spotting scope and I mounted the rail on my Bogen tripod with the legs folded out so that it was about 10" inches off the ground. I turned on the camcorder and began focusing the spotting scope and the camcorder. I had to get as much light as possible into the camcorder, so I didn't use any of the spotting scope's zoom capability.

I couldn't see anyone at the crossroads, so I focused on a small sign that was nailed to a tree at the entrance to the parking lot in front of the abandoned building. The small sign said, 'Keep Out' and it almost filled the screen. I used it to do my focusing and lighting adjustments. The camcorder was a Sony run of the mill Hi8s and it didn't have a lot of manual controls over exposure levels. Periodically, the viewfinder would grow dark and I would have to push buttons and spin small wheels until light returned to the view finder and I could see again.

"Psst." It was Rob and he was pointing toward the road. I turned my head toward the road and lay flat. I saw a man in his 20s walking toward the crossroads. I didn't look directly at him but used my peripheral vision to keep track of him as he walked by. He was about 30 meters to our left and he never looked our way as he passed by.

When he had passed us, I got back on the camcorder and continued adjusting it. When he arrived at the crossroads, I aimed the camcorder at him and was shocked at how close he appeared. I zoomed back as far as I could on the camcorder and still his head down to the center of his chest filled the viewfinder. I started taping him because he might be one of the dealers.

Soon after I started taping the chess-master arrived at the crossroads and began talking to the first man. I got some good shots of the chess-master, his shotgun, and as the morning went on, everyone who came into the target area. I was able to get the license plates of the vehicles that came down our road to the crossroads. I also got the ones that went in and came out of the parking lot.

I would have to search around the general area of the vehicle until I could see the LP. Then I would have to refocus quite often because my depth of focus was very small. My view was often blocked by vegetation, which hung from the trees along the road, mainly vines which grew fairly large leaves. I tried to focus through them, but they were two close to the target and they began to annoy me as the day went on.

Rob took a turn on the camera. I showed him how it worked. He practiced with it and soon began to get some good shots. I thought that Jerry would be happy with the new tapes, but I wanted to get more.

One of the other places Jerry and Bill were interested in was a house that was about 50 meters from the crossroads. Some of the dealers lived there. Jerry wanted to see if their mom, who lived there, was involved. I told Rob that I wanted to check it out so that afternoon we packed up the rail and moved deeper into the woods where we hid the camera equipment and then we moved toward the house.

East Texas-The Dogs

I could start to see the top of the roof of the house when a dog started barking. He barked louder and faster as we moved closer. We stopped and waited. The dog kept on barking and I could hear a women's voice, "What's out there boy?" she said. "What is it boy?"

Rob and I decided to move back because we were pretty sure that the dog had alerted on us. When we got back to where we stashed the equipment, we sat and ate some food. I told Rob that I was going to call back to Austin and have them send the TRONTEC to us so we could use it to watch the house. Rob asked how we could install it without being compromised by the dog at the house. I told him that we would get the cops to create a disturbance at the house and while the dog was barking at the cops, we would put in the camera.

Rob was okay with that for the time being and so we set up the rail and continued to shoot video of the target area until it was too dark for the camcorder. We packed up the equipment and pulled back in the woods and waited until it was dark enough to cross the road. While we waited, we whispered back and forth concerning the mission. I told him the cops would be happy with what we got today and that tomorrow I wanted to trim some of the vegetation that was hanging from the trees near the target. He asked me when and I told him that I would try to do it tomorrow morning before the chess-master showed up.

He said that he would give me Specialist Randy tomorrow because he needed to go out with one of the other groups. I was disappointed because Randy was a new guy and I hadn't worked with him before, but I didn't say anything. When it got dark, we crossed the fence and moved slowly up to the road. Rob observed the road with his PVS 7s.

After a few moments, he nudged me and we both walked carefully across the road and entered the treeline.

Rob moved at a quick pace and we were soon back at the MSS. The other guys were back, and Rob debriefed them as we sat in a small circle near the commo tree. The guys that were near the abandoned building said that they got some good video of some dope transactions that took place near where they were hiding. Rob told them about the rail and that he thought that we got some outstanding video. When the team guys finished talking, I told them about getting the TRONTEC sent up here so we could watch the house near the crossroads. There was some discussion about how we would do that and then Rob told Randy that he would be going with me tomorrow.

Randy didn't look very happy. I told him I would brief him on what we were going to do after we ate. Juan gave the guard assignments and nobody mentioned my failure that morning. I called Al at the CP and told him to tell Bill and Jerry that we would show him some good video in a few days. I ate and then got with Randy and explained what we would do in the morning. He didn't like the idea of my getting so close to the crossroads. Randy was a new guy and he seemed a little squirrelly. He had been with Rob on the first day when the chess-master spotted them and threw a bottle at them. He would have to overwatch for me while I trimmed the vegetation. I told him that all he had to do was watch and let me know when someone was coming near me.

I would have to low crawl about 50 meters through low vegetation and muddy ground to get to the trees that needed trimming. I would be about 25 meters from the crossroads and there wasn't any place to hide. I would have my pistol, the saber radio, and a pair of garden clippers. I was confident that I could deal with the chess-master or anyone else without using my pistol, but you never knew.

East Texas-Landscape Maintenance

We got up before daylight the next morning. I called Al and alerted him to the morning mission. He would wake up the cop and have him standing by if something went wrong. I knew the cop wouldn't be able to get there in time to intervene in a confrontation, but it would be good to have him there to inform the crooks of their rights or to deal with any follow-on activity by the other crooks.

After checking our equipment, Randy and I took off with me in the lead. We crossed the road and fence without incident and move to our OP. I led Randy to a place deeper in the woods where I put down the equipment. We walked back to the fence. It was beginning to get light and I showed him the places I wanted to trim. We went over the radio procedures and when it seemed like Randy understood what was going on, I re-crossed the fence and began moving toward the target.

I went down on my hands and knees after a short distance and continued ahead at a crawl. I wanted to get to the trim area as soon as possible, because if the chess-master was around, I knew it would take a lot longer if he heard something and came over to investigate. I found the first tree that needed trimming. Things looked a lot different up close so I asked Randy if I was in the right place and he said I was.

For the next thirty minutes I trimmed vines and small brush from the trees that stood between our OP and the crossroads. It took a long time because I was trying to be silent. Suddenly my radio gave off a rapid series of clicks. I froze. Randy's voicc came on and said, "The chess-master is moving toward you on the road."

I slowly lay down with my face on the ground toward the road. I could hear footsteps, but I couldn't see anything. I turned the radio down but not off. The footsteps stopped. I was frozen but my heart was pounding hard. I breathed very slowly and willed myself to calm down. I waited and he waited. I had been in these circumstances before. I knew he hadn't seen me and now he was just waiting and listening. He was probably afraid of snakes and that is why he didn't want to get off the road and walk through the low-lying brush that I was in. If he did, I hoped that Randy would remember his training and not shoot him.

I don't know how long we waited there, the chess master, Randy, and I, but eventually, I heard footsteps moving away from me. I waited until I couldn't hear them anymore and then I turned up the radio. I pushed the transmit button twice and waited. Randy's voice came on, "all clear, you can come back now." I told him I wanted to finish trimming and for him to move me to the next tree. He came back on and said I should return now. I answered and told him to direct me to the next tree and not argue with me.

Randy directed me to a tree to my left and very slowly I crawled to it. I started trimming and got some work done when I heard footsteps again coming toward me on the road. The radio clicked again, and we went through the same thing again. This time the chess master said with a deep East Texas accent, "I hear you cops." I waited until he walked away again and finished trimming the tree I had been working on. By this time, I felt that I needed to get out of there. I hoped that I had done enough trimming but anymore would have to wait for another time.

I told Randy that I was coming out. He said that the coast was clear and to come on. I put my garden trimmers in my pocket, checked my pistol and magazines and began crawling toward the fence. I crawled under the fence and started crawling up the hill towards Randy's location and in a few minutes, I linked up with him.

Randy moved to retrieve our equipment. He appeared to be relieved that I was back and told me how close the chess master had been to me. "He had a shotgun too," Randy said. We called the guy at the MSS and told him that all was well and to call the CP and tell the cop to stand down. We moved our equipment back to the fence and set up the rail cam on the tripod and settled in for another day.

East Texas-Takin' Care Of Business

The camouflage was still good. I set up the rail cam on the tripod and checked out the target area. I used the ‘keep out’ sign as a reference and scoped out the surrounding area. I had a lot more area now since we trimmed the brush from the trees near the crossroads. Business was starting and I got some good facial shots of the chess-master and some of the other characters who hung around the crossroads.

One of the dealers was named 'Pooty' by the cops. He was a young black man with handsome features. He had a small goatee and a cornrow hairstyle. The most striking thing about Pooty was the way he wore his pants. They hung down to the middle of his butt and his under pants were sticking out behind him. It looked really ridiculous, but I guess it was a style. Pooty was one of the busiest dealers at the crossroads and sometimes I could hear him shucking and jiving with his customers and the other dealers. He must have been a witty guy because as I was videotaping him talking, he would stop and then there would be a lot of loud laughter from the crossroads. Later, when we looked at the videos with the cops, one of them, a black female, when she saw Pooty standing around with his pants half down, would say, “Ah yeah Pooty, you just wait. We're gonna spank your ass.”

With the trees trimmed I could see that we would be able to video the people and vehicle LPs of most of the activity at the crossroads. I taught Randy how to use the rail and I felt that I could go back to the CP now and the boys would be able to continue the mission without me. That night I sat down with Rob and we went over the current situation. He would have two men work the OP that I had established, and they would use the rail cam with one of the team cameras.

I gathered up all the tapes that we had shot during the past three days. I talked with all the other guys, one at a time, to see how they were doing, and they all seemed ok. Al arranged for an exfil at 10:00 PM. Rob walked me out to the tree line on the edge of the highway.

A car drove up the highway about a half a mile away. I told him to blink his light and he did. When the car got very close, I told him, "slow, slow, stop," right in front of where I was waiting. The highway was deserted so I walked up to the vehicle and got in. Al was driving. I told him about how things had gone and asked him if everything was ok at the CP. He said everything was fine. I told him that I needed to get some sleep and then we would get into the mission details that were coming up. We drove to the CP and I went to sleep.

East Texas-Working With Cops

The next day, Jerry, Bill, and some of the other narcs showed up at the CP. I had found some good shots on the tapes and played them for them. One of the narcs said," Wow, this is really good video." They recognized the dealers and some of the customers. Jerry was happy and said things, like "I knew you could do it" and so forth. I told him that most of the shots were taken by the guys, but he would have none of it.

Then Jerry said that he was bringing in a couple of guys from Louisiana to make some buys at the crossroads. He said that these guys had guillie suits and that they wanted to crawl up to the crossroads. I told them about the problems we had with the dogs and the chess-master.

When they arrived, they said they wanted to go in and look around after dark. Jerry looked at me and said, "I want you to take them in tonight and show them around." I agreed to do so even though this group of strangers had never worked with us and they were not part of the OPLAN and MOU.

In this business you have to be flexible. The people back in Austin trusted me to make decisions on the ground. If you know more about what you're doing than your superiors, you want them to give you the freedom to make quick decisions in the field and not be second guessed by them afterwards. If not, then you are a yes man.

I told Jerry that I needed to brief the new guys on SODs part in this Operation and explain to them what we could and could not do. I would also tell them about our use of force policy. This was important because cops were always considering us as one of them. That was good as far as rapport building went but not good as far as getting involved in arrests, searches and seizures.

I sat down with the leader of the Louisiana Narcs. I decided to call him 'Crazy,' not to his face but here, which seemed reasonable after talking to him for a few minutes. He agreed to my briefing his people that afternoon. Jerry agreed to be there also. We had the briefing and then I sat down with Crazy and one of his guys and we went over the infil and exfil procedures. Crazy wanted to wear his guillie suit and crawl up to the crossroads. When I explained to him the situation regarding the terrain and vegetation, and our experiences with the chessmaster, he agreed to wait until he had looked the place over.

We went in that night, shortly after dark. It was cloudy and after we got inside the tree line, Crazy grabbed my arm and said, "I can't see a thing." "Neither can I," I told him. Rob met us and I asked him to wait for a few minutes until our eyes adjusted to the dark. Rob was wearing his PVS 7s and could see pretty well so after a few minutes we took off into the thicket and followed Rob to the MSS. Crazy was right behind him and I brought up the rear.

I kept an eye out for the deep drop off but Rob took us well around it so I didn't see it. Crazy stayed right behind him and I stayed right behind Crazy. It was so dark; you had to walk right behind the person in front of you. It seemed like you were moving by instinct rather than sight. I had spent a lot of time in the dark over the years and I always felt comfortable in it. I won't say I never got lost and I can't say I never walked off the edge of a deep depression or hole, but those incidents were rare. Moving in the dark has something to do with self-confidence and I guess I had some of that from all my after dark experiences.

Rob took Crazy up to the crossroads. Even after dark there was plenty of activity. Crazy saw how close a person would have to get to use a camera at the crossroads and agreed that it would greatly increase the risk of compromise and possible armed confrontation with the dealers. He came back and we exfiled around 2:00AM.

The next day, in the early afternoon, Jerry and Bill showed up at the CP. Crazy and I had a meeting with them, and it was decided that Crazy and his crew would start making undercover buys the next day. Their cars were rigged with small video cameras which would allow them to videotape the person selling them the drugs. Jerry wanted me to go back in and make sure that we videoed these transactions from the OP with the rail cam. Crazy gave me a description and LP for each of his vehicles. He would tell Al when he or one of his crew was going in and Al would call the MSS and the MSS would call me and let me know that an undercover buy was going down.

I went back in that night and the next morning Shoe and I were set up at the rail cam OP. Crazy and his crew started coming in every couple of hours and made buys. They were very good at what they did. They created a kind of competition among the dealers at the crossroads. Since they were new faces, they told the dealers that they had heard about this place from friends of theirs in Louisiana. They said that they would be coming to buy a lot more crack now that they found a steady source. This made the dealers very excited and you could hear them whooping and laughing every time one of Crazy's crew made a buy. Because I told Crazy the best places to park when they made their buys, we got every one of them on tape.

That night, after dark, I exfiled again and Jerry and Billy were waiting for me at the CP. They wanted to see the video. Crazy showed them the vehicle video and it was great. Although it was overexposed, you could clearly see the dealer's faces and the audio allowed us to hear every word that was spoken during the transaction.

I showed them my video next and everyone was happy to see the dealer's faces and in some shots, you could see the individual rocks in the plastic wrap as they were handed to the buyers. Jerry and Bill were very happy and told us that they had enough to take the place down. Jerry said that they would raid the place in the next few days.

He wanted our guys to keep the place under observation while the bust team went in. We had to pull everyone out that weekend for drill, so Jerry said that he would do the bust on Tuesday of the next week. He asked us not to tell anyone because he was worried about leaks.

East Texas-An AAR Clears the Air

We pulled the guys out that night and went to a motel about 50 miles away from the area we had been operating in. The motel was on the Interstate and no one was interested in a few more people checking in. We staggered the check-ins so that there wouldn't be 8 smelly men in the lobby at one time. Once we checked in, the guy's got cleaned up, had dinner, then we all met in my room for a quick AAR.

We went over the mission from planning to exfiltration. We identified several areas that need more attention and could be remedied with more rehearsals. At the end I asked if anyone had anything else to say. No one did so I said that I did. "The first night when I came into the MSS with you, I fell asleep on guard."

The tension in the room shot up and the guys began muttering things like, "It happens all the time", and, "Don't worry about it." I told them that I was very disappointed with myself for this failure and I apologized to them all. I told them that I would be informing the Detachment Sergeant about my failure when we got back to Austin. After that, I told them that they were welcome to complain to him and that if any of them felt uncomfortable with working with me in the future, that they should tell him that. The guys repeated their assurances that they weren't bothered by what happened and we ended the meeting. My stomach was in knots, but I was glad I had gotten my failure off my chest in front of everyone involved.

We drove back to Austin, cleaned our weapons and turned them in. I gave a report to Lt Brian. I talked to Sergeant Tommy and told him what happened with regards to my falling asleep on guard and my talking to the men about it in the motel. He told me not to let it happen again and I didn't hear about it again until sometime later when I used it as an example in a class on how to clear up an ethical dilemma.

East Texas-The Bust-Out

We went back to East Texas on Monday after drill and the guys went in that night. We had met with Billy beforehand and he told us that the bust would be sometime the next day. Jerry showed up at the CP the next morning and told us that the bust team would be at the CP the next day at 9:00AM for a briefing and that the bust would go down at 11:00AM, a peak time at the crossroads. Al dropped me off and I joined the others waiting for morning, the last day of this operation.

That night a dog came into the camp. He was big and hungry. I fed him some Vienna sausage and he seemed to be happy. He stayed next to me for most of the night and at four when I got up, I fed him some more Vienna sausage and after he ate, he took off.

By 7:00AM we were all in place. The day went on and the activity at the crossroads picked up until about 10:50 AM, when Pooty came up the road telling everyone to take off because 5-0 was coming. We heard cars starting and we saw cars driving off. A few minutes later; there were police and task force vehicles coming in from three directions. Some of the dealers and patrons took off on foot in the direction of the abandoned building and got away.

The bust only caught two people and neither of them had any dope. We exfiled a ways down the road and went back to the CP. Jerry and Bill showed up about a half hour later. Jerry was pissed. He said that one of the sheriff's deputies had called Pooty after they left the CP to start the raid. He couldn't do it before because Jerry told the cops not to make any phone calls until after the bust. That was why it was a such a short time before the bust that Pooty got the call.

East Texas: The Rail Camera at the Crossroads

Ranger Rob leads the way

Rob on Rail Cam

Suspect

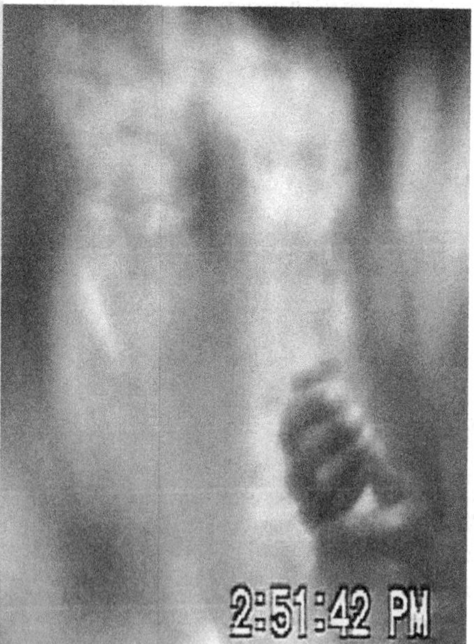

Handful of Crack

And do not be drunk with wine, in which is dissoluteness, but be filled in spirit. Ephesians 5:18

Chapter 19-September 1996

Trouble on the Island

We were still rotating teams to South Padre Island for White Shark. I was the mission commander in September when I got a phone call from the island on Sunday while I was at drill. There had been a car wreck on Saturday night and one of the Blazers was totaled. There were no injuries reported. The two operators that were in the vehicle had left it upside down in the field where it came to rest and got a ride back to the hotel where they went to sleep. They didn't report the wreck until the next morning.

It was my responsibility clean things up and determine what happened. I drove down Monday morning and talked to the other team leader about what had being going on. He said he didn't know so I decided to take care of the wrecked vehicle and clear up the rest when we returned to Austin.

From there I went to a wrecking yard in a nearby town where the DPS Lt told me the Blazer had been towed. I went to the office and one of the employees took me around the back where our shiny new Blazer was parked with the front of its roof crushed and not so shiny now. I paid the towing and storage fee and told them I would be back to get it. The operation was over in a few days, so I concentrated on making sure that we had all our things. On the last day, I said goodbye to the cops and we went back to Austin. I went to Tommy and told him that I needed his help to determine what happened We decided to interview everyone who had been on the operation. The first one we talked to told us the whole story. The rest either didn't know what happened or were evasive with their answers.

The next day Tommy called a meeting in the team room to go over a few things. Rumors and accusations were being whispered in the hallways and on the porch and he wanted to clear the air.

One of the guys asked to show a video that he thought would inspire us to be more 'hooah.' It was a History Channel program showing how the Marines selected men for the Force Recon units. It was a lot of rah-

rah with a good dose of PT. I sat there with the other guys watching this, but I was not interested. I felt that the integrity of the entire detachment had been compromised and I was furious.

A few days before, during the October drill, 29 years after I was first promoted to sergeant, I was promoted to Staff Sergeant E-6. My career was really taking off but I wasn't enjoying my elevation to next level of the NCO Corps.

Tommy stood up and thanked him and then he asked me if I had anything to say. Actually, I was ready to shout but instead I began to speak. "I just got promoted last weekend after 29 years as an E-5 which should have been a happy time," I said staring from face to face into each man's eyes as I spoke, "but this week has been one of the worst weeks of my life. I feel like a total failure as a leader because as soon as I left town for drill, two sergeants wrecked the Blazer and almost killed themselves. You want to watch a video?" I asked.

I popped a cassette into the VCR. It had a clip of the wrecked vehicle I had made when I went to the wrecking yard.

"After they wrecked the vehicle they went back to the hotel where some of you saw them but decided to cover up for them. Meanwhile some of the other guys went to a party off the island which was absolutely against the instructions I gave all of you before I went to drill. On top of everything else all but one of you lied about what you did or what you knew about what happened last weekend. I think Sergeant__ should resign today and find some other line of work and I think the rest of you need to realize that I don't trust any of you right now. Trust, like respect is earned and I have lost both for the rest of you!"

I was trembling I was so mad, and I told Tommy I was done. The place was really quiet and after a minute I walked into my office and closed the door. The place was not its usual noisy rumpus room for the rest of the afternoon. Tommy came to see me afterword and told me that the guys really respected me for saying what I said but I was not consoled. I felt we had to do something to restore the ethical climate in the detachment, but I didn't know what.

The Ethical Dilemma

I went back to the office and began thumbing through the Army Leadership Manual. The whole Army Ethic was right there. The commander of the unit was the ethical standard bearer. He needed to create an ethical climate that made it difficult for anyone to give into their weaknesses. I started reading and typing. I worked on the document for the whole weekend and brought it to the command group Monday morning. I assured them that every word was out of the Army Leadership manual and we needed to tell the detachment that this is where we were coming from so that we could all have the same frame of reference when we were confronted with an ethical dilemma. Tommy told me to give the classes.

I went ahead and gave the classes. A few months later I was returning from an operation and didn't feel like counting every bullet in the arms room again for the millionth time and told the young PFC with me that the count was good. He said, "Remember that class you gave us about integrity; is this what you were talking about?" On one hand it was like a spear through my self-righteous heart but on the other hand I was elated that this young man who was 30 years younger and who I out ranked by three grades had the moral courage to call me on my own integrity violation. I thanked him for being a man and counted all the bullets and never considered lying about the count again. The ethical indoctrination program worked.

I did learn something about crime and punishment. When a soldier breaks the rules, he needs to be punished. This upholds the standards for the rest of the unit. If the person is given a punishment that fits the crime, such as extra duty, or reduction in rank for more serious offenses, and is able to admit to himself and those involved that he did what he was accused of, then the way was clear for him to be restored to the fellowship of the unit.

In the old army, we used to say that a sergeant should sew his stripes on with zippers because it was so easy to get busted. When a sergeant did get busted and showed up for work the next day ready to soldier at the same level of proficiency that he had had when he was a sergeant, he was quickly restored to his rank in a few months.

In the modern Army, getting busted was a career-ender as were most disciplinary censures. A letter of reprimand could stop you in your tracks as far as promotions and assignments went so the climate of fear pervaded the military on both sides. The leaders were afraid to punish and the troops were afraid to be punished so everyone was covering one another, and people were giving into their weaknesses so easily that when it came time to revise the leadership manual in 1999 they removed the term 'ethics' and replaced it with 'values.'

FY-96 Roll Up

A tally of our stats for FY-96 showed that we did 29 operations resulting in 7.7 million dollars in seizures. The seizure totals were misleading on many levels. If a pound of cocaine was seized in a multi-agency investigation, each agency claimed a pound of cocaine. Many of the other cases we supported were still on-going and would take so long that no one would remember our participation when they finally nailed the bad guys. We had some of our biggest successes and our worst failures during this period but there were no funerals, so it was a good year.

SOD Reunion

For if we believe that Jesus died and rose, so also those who have fallen asleep through Jesus, God will bring with Him. 1 Thessalonians 4:14

Chapter 20 - January 1997

Tommy Reynolds Leaves

Sergeant Tommy Reynolds was leaving. He had been hired by the FBI. He had been a stabilizing factor for the detachment during the storms on SPI last fall. Tommy had come here from G Company and put in a solid performance as an operator, team leader and mission commander, and finally as the detachment sergeant. A veteran of the 82nd Airborne Division, he knew his stuff and he was an in-your-face leader. When he first came to the detachment, he used his training from the Recondo School of the 82nd to build jungle antennas that were able to be connected to our Saber Radios, greatly extending their range. He was always experimenting with our photo/video equipment and made improvements in their employment.

Tommy liked training and he put together an intense Recondo school for the cops. He also developed Rappelling and Rappel Master Courses that were quite popular with SWAT teams around the state. He was an all-American and proud of his Korean heritage. When we were looking for a detachment symbol for our t-shirts, he suggested a saber-tooth tiger, a popular symbol in Korea. His leaving was going to leave a big gap in the leadership of the detachment.

Acting Detachment Sergeant

Brian was now the operations officer in G Company. A sergeant who had been with us five years before was still in the company and expressed to Brian an interest in coming back to SOD. He had quit in the middle of a training exercise and I hadn't seen or heard from him since. Clayton, who had left at the same time, had since returned and was now our admin sergeant. He had gone to Ranger school and was tabbed but decided that he didn't want to spend another night in the woods for a while and became our clerk. He had remained in contact with this sergeant. Brian and the sergeant had been spending some quality time together and before we knew what was happening, he was back in SOD, part of our largest OTC class that graduated 10 men. The buzz among the team guys was that he would replace Tommy as soon as he finished OTC.

"The word on the street is that you are going to make him the next detachment sergeant," I said to Brian. Before he could say anything, I continued with the suggestion that he make me the acting detachment sergeant for a few months while he gained some experience in the field. He thought about it for a minute, and then he agreed. The next day he published a memo to that effect. We had had yet another season of turmoil after the SPI fiasco and Brian was looking to tighten the screws on the teams. He had put the word out, along with some other directives that everyone was to turn in their pagers whenever they were off-duty and this was causing a small storm in the team room.

"The Men Have Grievances"

I called a detachment meeting in the team room. The men were wondering what I wanted. When I told them that I wanted to hear their complaints, the damn broke. For the next 30 minutes my right hand was busy as I wrote down their demands, suggestions, complaints and denunciations. Of course, they wanted to keep their pagers, and among all the petty complaints was one issue that they all agreed on, one that had been gnawing at us for a while and that was —they wanted permanent teams. I told the men to standby and went back to Brian's office with my notes. After closing the door, I quoted one of my favorite lines from the movie, 'Glory' when the sergeant major tells the colonel that "the men have grievances." We went over the list and he agreed to rescind some of his directives, including the one regarding the pagers. I went back to the team room and had another meeting where I gave them the revised directives and sent them home.

Permanent Teams

Since we had become so high tech, permanent teams seemed like a good idea. We only had one expert with all the equipment and that was Pat. We didn't have more geniuses like him, so the best thing was to establish fields of expertise and train the men to become experts.

I sketched out a team plan based on the SF concept and our own unique mission. I figured we would have a team leader and assistant team leader, a photo/video guy and an assistant, a commo guy and an assistant, a night vision/thermal guy and a medic. They would learn their jobs and cross train the other members of the team. Eventually we would have the technical skills for operations that were rapidly outweighing the need for our tactical skills.

I talked to the supply guy and he gave me a digital list of our equipment. I went through it and divided all the equipment including film camera systems, video cameras, radios, medical bags, and everything else we had enough of to divide up for two or more teams. I printed out the list and wrote up a basic description of a team and the duties of each member. When I had finished, I went and had a meeting with Brian. He had made note of the grievances and told me he was considering forming permanent teams. I had my team description and equipment list with me, and I handed it to him. He gave a quick glance at both documents. "I figured we better do this, so I wrote up something to use as a starting point." He was staring at me, so I said, "What do you think, sir?" He looked back at the team description, then named two of the operators as team leaders and told me to get it done.

Gulf Shield-A Visit from the Puddle Pirates

The Coast Guard decided to call for a surge in the Gulf. We had struck some good blows to the transportation organizations that were moving drugs onto the beaches of south Texas and had disrupted their operations during the 1996 harvest/smuggling season. The Coast Guard felt that our success was making them look bad because they had the mission of preventing the drugs from landing. Since we were catching so much after it landed, they were afraid that they would look bad in the eyes of the suits and uniforms in D.C. They moved in and changed the tactics. Claiming supremacy due to the Federal Mandate that they were the lead agency in offshore deterrence and interdiction, they invited all the White Shark players to join them in Operation Gulf Shield. There would be a big press conference announcing the 'surge.'

A Coast Guard Captain named Al notified Buckwheat that he would be coming to Austin for a meeting. Buckwheat asked for SOD to attend so LT JT, another sergeant, and I attended. JT was the XO and because, Brian wasn't here; he commanded SOD. We filed into the conference room and sat at the end of the long table. Buckwheat, Gladys and the Captain came in and we all stood at attention until Buckwheat told us to sit down. He introduced the Captain who was in a sharp looking khaki uniform with all his ribbons.

The Captain began by telling us what a great job the Coast Guard was doing to stop drug trafficking on the coast and how they were now considering bringing their expertise to Falcon Lake at the west end of the Rio Grande Valley, more than a hundred miles from the coast. Due to some past friction with the CG I was skeptical but now I was really getting mad. We had done some successful operations on the lake and now they wanted to screw those up too.

He went on and on about how their expert seamanship and their skill in the air were what was going to put a stop to drug trafficking in south Texas. I was having none of it.

I had already dealt with the chief of the Port Isabel Coast Guard Station, a weasel named Jackie who tried to settle a disagreement with me by trying to pull rank and I was sure that this guy was cut from the same piece of cloth. The Captain finished his Falcon Lake fantasy and proceeded to the nut of what he had come here for; to demand our support for his great idea, Operation Gulf Shield on South Padre Island.

He wanted us to deploy six LORIS teams from the Mansfield cut to the Brazos River Channel for the next six months. I sat there fuming while he gave us the details of command and control: the Coast Guard would command and control everything, as for logistics; we would pay for everything. He finally stopped and asked Buckwheat if he could have his assurance that we would support him and his operation.

I couldn't have been more surprised when Buckwheat turned to me and said, "Sergeant Murry, what do think about Captain Al's plan?" JT felt that he should speak for SOD and he said, "Sir..," but Buckwheat stopped him by saying, "I want to hear what Sergeant Murry has to say." Since I had not discussed anything in advance with Buckwheat or anyone else about this meeting I was pretty much on my own.

"You have some interesting ideas, sir," I said to the captain. "As you know, last year during White Shark, a cooperative effort between RAC Brownsville, the Border Patrol, the Port Isabel Coast Guard Station, the Cameron County Task Force, DPS, and the Guard, the operation seized more than 10,000 pounds of marijuana, arrested 10 people, and seized some cars and a boat. The agreed upon strategy was for the Coast Guard to maintain a presence north of the end of town on SPI while the Customs and Guard teams used the LORIS systems on Boca Chica Beach and the Task Force and DPS did interdiction."

"Now you want us to participate in an operation that has completely ignored the successful approach that was used during White Shark and adopt one that has alienated all the other major players because there is nothing in it for them. As for the overt operations you have on Boca Chica, there has already been at least one instance when a load got by your patrols and that's because they are using night vision goggles that give them only a few hundred meters visibility. When the shark boats come in they take only about 3 minutes to unload into a vehicle while your patrol is at the other end of the beach."

I didn't mention the fact that the Coasties on beach patrol had told me that they were hiding in the dunes instead of driving back and forth on the beach. They hoped to spot a load, like we had numerous times before, and take it down. I told them that they were pissing up a rope because we owned the beach after dark with two LORIS teams and they could see only a few hundred meters. They admitted that it was a long shot, but they were tired of driving and they hoped to get lucky.

I felt a nudge against my right hand and looked over out of the corner of my eye while facing Captain Al at the other end of the room. There was a yellow post-it being pushed toward me by JT. On the post-it were the words, 'I'm not comfortable with this!" Buckwheat noticed this and said, "Go on, Sergeant Murry." Oh well, in for a penny…

"If we deploy our six LORIS systems for the next six months, we will be unable to support agencies we have already committed to. Since the shark boats are still landing drugs on Boca Chica, I think we would be better employed there because two LORIS teams can cover the whole beach and we'll catch every load that comes in."

Captain Al was not looking too happy so I decided I better bring up the legal thing, "Sir," I said to Buckwheat, "I think the lawyers were wondering if the Coast Guard was actually considered a law enforcement agency as far as our legal requirements go." Buckwheat decided to end the meeting on that note, and we went back to the team room. JT was concerned about my speaking so negatively to the Coast Guard captain but I assured him that if Buckwheat was ok with me saying whatever I wanted then everything was ok. The next week we got a request from CORE to work with them on Gulf Shield.

CORE was the Covert Operations Response Element, a TNCP special missions unit. We had worked with them many times so this would be good for us and satisfy the Coast Guard that we were supporting Gulf Shield. We would bring two LORIS teams and work mobile OPs north of town on SPI. JTF-6 managed to get in on it and was putting an SF team in the sand on the north end of the island. We would run the CP in the same hotel as last year and we would put a LORIS on the balcony again to keep an eye on the beach north of the motel.

Charlie Foxtrot Pays A Visit

We moved our stuff down to SPI and I set up the CP in the hotel. I noticed some other people that looked military and after we put the teams out, I was standing on the balcony and heard a two-way radio below me. I leaned over the rail and saw a LORIS on the balcony below me. I brought this to the attention of one of the Customs guys and he found out the Coasties had gotten their hands on some LORIS systems and were watching the same area that we were.

This was typical of Jackie the weasel. He put a team in the same hotel watching the same beach without telling us that they were there. I didn't know if he didn't trust us or if he was trying to spot a shark boat before we did to get the credit, or what he was doing, but I called down to the guys there and invited them to come up and talk.

They were CG guys on TDY from the East Coast and had no knowledge of the animosity between the chief of the Port Isabel Station and the rest of the cops that were on White Shark. They came up and we sat around and talked about the operation. I decided to invite them to combine our OPs and CP so we could work together for a common cause and they agreed. Their LT said he would talk to the weasel and get back with us tomorrow. The next day they knocked on the door and said they could, so they moved their equipment in and we worked together. The weasel was happy because he could keep an eye on us and look for an opening to shove a screwdriver in but we were running a tight operation and I didn't think he would get the opportunity.

The next problem came when one of the SF guys came up and told me that one of our OPs was parked on a dune above a large crowd of college students who were cavorting around a huge bonfire on the beach. "Your guys couldn't be more visible if they painted themselves with chem-light juice," he said. I hated it when we looked bad in front of anyone but especially in front of the SF.

I thanked him and went downstairs, got into one of the SUVs and took off down the beach. About three miles from the hotel I found them, and the SF guy was not exaggerating. Mike, one of our new guys that just didn't get it some of the time, had parked his SUV on a large dune about 40 feet above a large crowd of young people. He had the LORIS head on the roof of the SUV and in the firelight that was reflected by the large lens of the LORIS he looked like some high-tech sentry guarding the crowd of drunken kids from an invasion.

I had told him to hide in plain sight; meaning to run his LORIS out of the back window of the SUV while he and the others cooked dinner on a fire, simulated drinking beer, surf fishing, or doing something like everyone else on the beach, not standing out like a sore thumb. I called him on the radio and told him to get the LORIS off the roof and to move to a less conspicuous location. He grumbled a little as he packed up and I left. The SF guy was going back up the beach to drop off something for his guys so when I thanked him for giving me a heads-up about our OP I asked him to keep an eye out and let me know if he saw any more problems. This surprised him and he became much friendlier after that.

One night we were sitting in the CP/OP when a Coastie came up on the radio announcing that they had spotted a shark boat. We put the LORIS on wide angle and turned on the tape recorder. A few moments later the shark boat came into view. It was traveling north at a high rate of speed. A minute later a Coast Guard small craft appeared going north in hot pursuit. Then there were two more Coast Guard boats and finally a larger cutter that came in from the Gulf, all moving at top speed. The shark boat got smaller and smaller as it headed north and when the vessels in pursuit had passed us, we lost sight of the shark boat in the cluster of boats on the screen. We sat there watching as the boats slowed and then turned south and began chasing the lead Coast Guard boat. It got larger on our screen as it raced towards us with a small flotilla right behind it. The radio traffic was almost continuous as the Coasties moved closer to their target.

Then, from the back of the pack, a boat, moving much faster than the rest, raced through the pack, and passing the lead boat that everyone was chasing, raced south towards Mexican waters. It was the shark boat. A Coast Guard helicopter flew in and hovered right above the shark boat with its high intensity searchlight illuminating the boat and the water ahead of it. The closest boat behind the shark boat was driven by our old friend Emilio of the BP. The shark boat and the helicopter remained together until they reached the mouth of the river and the helicopter broke off. Emilio fired a parting burst from his MP-5 as he broke off, and the shark boat continued south where they could take a break, refuel, and wait for another night to try it again.

One on One with the Commandant

The Commandant of the Coast Guard came to the Port Isabel Station and the Cameron County Narcotics TF Commander briefed him on the situation. He mentioned our past successes with White Shark, and he singled out SOD as the keystone of those successes. After the briefing, the Commandant invited me to sit with him on a couch so he could ask me some questions. When he asked me how I thought things were going, instead of complaining, I tried a different tact and reminded him of the great success the Coast Guard had had in the Caribbean as part of the South Florida Task Force in the 1980s. "I just finished reading 'A Sea of Grass: The Maritime Drug War 1970-1990.' It detailed how the Coast Guard pretty much stopped the drug smuggling in the Caribbean by seizing drugs and arresting violators. I believe that is the best deterrent." He stared at me, smiled and the meeting was over.

Minnesota-Bil Enney Pulls The Plug

On May 7th I ran into Major Akers in the hallway after I finished PT. He pulled me into his office and said, "Bil is dead. He blew his head off with a shotgun up in North Dakota yesterday." I was saddened by the news and I went down to the team room and told Brian what I had heard and before long Gladys and Buckwheat announced the bad news at the morning staff meeting. Bil had been gone for several years and some people didn't know him or what he had done for the program.

Buckwheat was in touch with Bil's daughter Jeanette and his son Billy and together they planned for a memorial service for Bil at the Camp chapel. Buckwheat told me that Jeanette wanted me to give the

Invocation. SOD would form an Honor Guard that would fire a 21-gun salute. We had a man in the detachment who had served in the Old Guard and he worked with the guys and got them squared away.

I prayed with my wife and son that morning and when it was time to speak I got up and told the full chapel that Bil had prayed with us to receive the Lord, and then, when my remarks were finished, I invited friends and family to come forward to say something good about Bil. I warned everybody that if anything bad was said about him that a group of paratroopers and the Special Forces Association were there, and it wouldn't be a good idea.

New Detachment Sergeant

After a couple of months Brian told me that he had decided to put the new sergeant in as detachment sergeant. When he told me, I was devastated for a moment because I felt that we were making some real progress; raising the professionalism and the morale of the detachment. I didn't see any use in arguing so I acquiesced and continued to put my energy into helping the team leaders develop task, conditions, and standards for the employment of our various photo/video and communications systems. We wanted to have these standardized so that our next OTC staff could use them for training new operators. Some of the work had already been done, like JT's excellent photography class and range/lens/film/fstop/shutter speed explanation. He had spent days with his team doing practice shots of people and vehicle LPs and documenting all of it so when we had an OTC, the graduates had a solid understanding of photography which enhanced their understanding of video and their use in low light long distance situations.

Redford, TX-An American Tragedy

On 20 May a U.S.M.C. corporal shot and killed an 18-year-old goat herder in Redford, TX. The furor that ensued put an end to our long relationship with the Border Patrol. Beside the fact that an innocent young man was killed for no good reason, the basic problem that I have with this shooting was the lack of understanding by JTF-6 of the difference between the terms, 'Rules of Engagement' and "Use of Force.' This I blame on the inability of the military leaders to ask for and consider input from lower ranking people like myself who were on the ground and had conducted these operations for years.

The facts are that the Marines left their designated LP/OP and crossed a paved road so that they could get closer to the action. If they would have stayed where they were supposed to be this wouldn't have happened. After they left their LP/OP they spotted a man herding some goats and carrying a rifle. This was 18-year-old Ezekiel Hernandez, a local high school student with aspirations to join the Marine Corps. Hernandez fired his rifle several times into the brush and the Marines determined that he was firing at them. They maneuvered closer to him and when the team leader saw Hernandez aim his rifle in the possible direction of another member of his team who he couldn't see, he fired his M-16 rifle at Hernandez, the bullet striking him under the armpit and fatally wounding him, he died a short while later.

I walked the ground where this occurred, accompanied by a BP agent from that area that was intimately familiar with the facts of the case. The Marines were in a dry wash that was at least three feet deep which gave them total defilade from the direct fire of Hernandez. They didn't call out to him and ID themselves as Border Patrol, Police, or even Marine Corps and call on Hernandez to put the gun down. This was because they had not been trained to do so before they were sent on a counterdrug surveillance operation in the United States.

The blood of this young American is all over the hands of the many arrogant officers that came and went through JTF-6 and never inquired from the National Guard about their training and lessons learned. We had been operating on the border and all over Texas for eight years and we had had a number of armed confrontations with criminals, all of whom were taken into custody with no shots fired.

I had discussed the need for information sharing with several of the JTF-6 Liaison officers who camped in our HQ for years at a time, but even if they thought my suggestions were useful, the people in El Paso weren't interested. This episode of American military incompetence was quickly swept under the rug after the Grand Jury sent back a no-bill for murder. The government gave more than a million dollars to the Hernandez family in a settlement for a wrongful death suit. The BP quickly put out a policy prohibiting the use of armed Military Forces in support of Border Patrol activities which more or less put an end to our operations on the border out west for a while.

September 1997: FY-97 Roll Up

I did a quick tally of our stats for FY-97. We did 21 operations and were credited with 19 million dollars in seizures. The body count for seizures meant something to the people in D.C. The 21 operations without compromise or injury meant something to me.

December 1997: Austin-The Academy

After spending a year in RCPOC, Sergeant Tony E saw me one day at CD and asked me if I had considered being an instructor at the Academy. Tony seemed to be appointed by God to guide me in my Army career and when I showed interest, he told me to make a packet with my NCOERs and school certificates and he got me an appointment with the CSM of the academy who was also the Commandant of the school for the Non-Commissioned Officer Education System (NCOES).

The Texas Army National Guard Academy was currently running the Phase One Basic NCO and Advanced NCO courses, one weekend a month for six months. In addition, they were running Phase Two courses for the cavalry, infantry and armor MOSs during the summer at AT. CSM Larry W had recently taken over the position of Commandant and was trying to establish order out of the chaos that the Academy had lapsed into after the retirement of CSM Boehm of the airborne mafia who had run the Academy like a drill team for many years.

The CSM who replaced Boehm had lasted for a couple of years and the place had gone to hell under his less than inspiring leadership. The new CSM came from the artillery and was determined to square the place away, but his opportunities were limited to the weekend drills.

I kept my appointment with the CSM. He asked me why I wanted to be an instructor and I told him that I was getting a little older and that I wanted to pass on what I had learned to a younger generation of sergeants before I got out. He took my APFT card and remarked that it looked like I liked PT. He said that I shouldn't expect the others in the academy to be that enthusiastic about PT. I told him that it was my understanding that as long as a person stayed fit and could pass their PT test; it wasn't my place to bother them. He seemed ok with that and said that he would give me a chance to be an instructor.

I told him I was interested in teaching BNCOC Phase I and he said that all the instructors were to be instructor qualified in their MOS as well as NCOES. He directed me to the admin section to start my paperwork to transfer to the academy. I knew several of the sergeants at the academy and told them of my interest in being an instructor in BNCOC. They told me that the Army Reserve provided the instructors and the Texas National Guard Academy administered the course. I might have a chance later on, they said, but right now it looked like I would be an MOS qualification instructor for 11M (Mechanized Infantry).

In order to be qualified as an instructor I would have to take the Total Army Instructor Training Course (TAITC), the Small Group Instructor Training Course (SGITC), MOSQ for 11M, the Combat Lifesaver course, and I would have to be certified by boards of instructors for the various courses I would be teaching. I had committed to become an instructor but then in October the STPA results came out and I was selected for an E-7 position. I declined it because of my commitment to the Academy and trusted that my promotions would come from God's arrangement.

And you shall teach them the statutes and the laws, and make known to them the way in which they should walk and the work that they should do. Exodus 18:20

Chapter 21 January 1998

The Academy

For my first weekend drill with the Academy, I had to report to Camp Swift. I walked into the HQ building to find out what I was supposed to be doing and they didn't know me. I stepped outside and met an instructor who asked me if I was looking for the TAITC course. This was one of the courses I needed so I said yes and the instructor invited me to join the class. At lunch time I ran into one of the admin people who asked me where I had been, and I told him that I was in the TAITC class. He said "cool, I didn't know we had one; " so much for command and control at the academy.

The TAITC taught you how to teach performance oriented or hands-on training. It only lasted two weekends and at the end of the course each student had to make a presentation. There were a lot of requirements for teaching military classes according to the standard, but it seemed like the administrative details were emphasized more than the actual content and the way it was presented. With most military subjects, the knowledge of military history was critical because it gave the instructor the way to illustrate the point in a way that kept the student's interest. Most instructors I observed had very little knowledge of military history and their classes were pretty boring.

After I passed TAITC, I was qualified to teach MOS qualification courses if I had the MOS. The weekend that the TAITC course was completed, an announcement was made that a Small Group Instructor Training Course would start the next weekend. I signed up for it because you couldn't teach BNCOC Phase One without it.

Small Group Instruction was a somewhat new instruction technique that the army had adopted. I read the manual that they gave us at the beginning of the course. It was written by a PhD in education for other PhDs in education. It had not been translated into Army Speak. Later, I became quite interested in the small group instruction concept and attempted to translate some of the concepts into language that the troops could understand; it was an uphill fight. The main difference between small group instruction and the rest of the army training methods was that the instructor was no longer the only one talking in the class. The instructor was there to facilitate the conduct of the lesson that was being taught by a student discussion leader. The instructor was also there to be the subject matter expert and to be able to give the doctrinal solution to any questions. The instructor, being unfamiliar with the material, struggled through the class and we all managed to pass.

The next event was combat lifesaver and I passed that, having had it before and having had extensive medical training in SOD. The next NCOES class was starting in November and I asked the Branch Chief of BNCOC if there was any chance for me to be an instructor. He told me that the Reserve was providing all the instructors and I might be able to be an assistant instructor if I wanted to. I said I did, and he told me that he would keep me in mind.

East Texas-TV Shootout in the Motel

We were doing a lot of gigs with CORE. This time we were back working with them and the same East Texas Task Force I had worked with for six years. We had several different locations we were watching. One was some apartment houses that had a lot of undeveloped land behind them. I decided to try out my new warm weather guillie suit that my wife had sewn for me, so I went in with one of the teams for a night in the woods. I brought along a rubber snake that I purchased at Wal-Mart just in case there would be a need.

The team crept into a place where we could watch the target area. We were laying on the ground watching and there was nothing going on, so I decided to have a little fun and give the snake to the guy next to me. I tapped him on the shoulder, and he rolled over toward me to see what I wanted, "Take this," I whispered and handed him the snake. He took it and when he saw and felt what he thought was a large coral snake he jerked his arm in a manner that his elbow almost broke my jaw. When the other guys found out what happened there was a lot of amusement had from my snake trick; but all I got out of it was a sore jaw, a fitting consequence for my dumb-ass prank.

A week later I was still working the same operation but now we were sitting in CORE's high-speed surveillance van watching some more crack dealers down the street. I was pulling 12 hour shifts with Specialist J, one of the many guys who strode in and out of SOD that year. He came from G Company on someone's recommendation, but we never were able to pin down just exactly who. He was in a zone of his own and the guys working with him were having a hard time getting him to follow SOPs and simple instructions. At one point a team leader dummy-corded his pistol to his belt for failure to keep track of it.

He and I would take turns on the camera videotaping vehicles, license plates and the faces of the people that were going in and out of the crack dealer's yard. Whoever was on the camera would call out a vehicle description, an LP, or people description and the other guy would enter the information into the log. There was a big picture camera that we used to watch the whole scene and my video camera on the rail with a spotting scope to get the details.

When J was calling out various bits of information, I glanced up at the TV on the wall of the van just to see what was going on. I began to notice that as the day went on, J's descriptions were becoming less and less like what I was seeing of the same scene he was describing. I mentioned a few of his discrepancies which he denied so the atmosphere in the van became less and less conducive to teamwork and more and more confrontational. I was really glad when that day was over, and we got back to the motel so I could get away from J for twelve hours.

I was sitting in my motel room after dinner. I had just gotten a non-linear video editing machine and I was editing my surfing videos that I had taken in Hawaii several months before. All of a sudden there was a loud pounding on my door. I jumped up with my pistol in my hand, took a defensive position next to the door and asked who it was in a loud voice. "It's J; I need to talk to you!"

I opened the door and let him in. "I was in my room cleaning my pistol and it went off." "Is any one hurt?" I asked him. "I don't know," he said. "Where did the bullet go?" "It went through the TV." I felt a smile come to my lips, but I repressed it; that takes care of J. I told him to follow me and we went to his room. I went to the door of the room next to his and knocked and asked at the same time, "Are you alright in there." "Leave me alone," came a voice from the other side of the door; "I'm not opening the door." "Ok," I said, "No problem, everything is ok," and ducked into J's room and closed the door.

There was a hole about 2 inches in diameter in the middle of the TV screen. I repressed the smile again as I looked behind the TV. There was no damage to the back of the TV or to the wall; that was good. The telephone rang and I answered it. It was the front desk and they wanted to know if we had heard a gunshot. "No, I haven't heard anything," I told them. "Thank you," said the desk clerk and hung up.

I had J tell me his story and he stuck to the fairy tale that he was cleaning his pistol and it went off. He didn't have any idea how it could have happened because he had followed all the safety procedures. It looked to me like he had been doing dry-fire drills using people on the TV as his targets, but his weapon wasn't dry.

I took his weapon and directed him to pay for the TV with his credit card when he checked out in the morning. I went back to my room and called Brian at home. I gave him the report and told him that it appeared that we were OK. I told him that I wanted to send J back to Austin in the morning and I was going to keep his weapon. Brian agreed and after I got off the phone, he called LTC Gladys.

The next morning, I realized that the small hole in the TV would lend itself to the possibility that there had been a gunshot in the room the night before. I took my tripod down to his room and walked over to the TV and smashed it into the TV screen. The hole became a little larger. I did this several times until the hole was about ten inches in diameter. It looked much less like a bullet hole now.

When J checked out, he told the desk clerk that he had knocked a hole in the TV with a piece of heavy equipment that he had brought from his car to the room. He told the clerk that he wanted to pay for the TV and the clerk wrote him up a bill and didn't ask him any questions. Either he had not been told about the reported gunshot the night before at shift change or he chose not to make any further inquiries.

Then J drove back to Austin and into the lore of SOD. We were all glad to see him go but his final act as an operator was done smoothly when he checked out and paid for the TV without arousing any suspicion. I checked out of the motel that morning and that evening I checked into another one on the other side of town. When I turned J's pistol and ammunition into the arms room, I included the empty cartridge case. His plastic ammo box with 49 rounds and an empty cartridge case stayed in our ammo safe for several years after that, throwing off our even count.

OTC

We had a new OTC coming up and the new detachment sergeant decided to get involved with running it. One of the new guys was Sergeant Stephen M, who I had first seen over at personnel when I was trying to get all my documents added to my personnel file before the annual review that determined the order of merit list for promotions. He was talking to a clerk about his records and I overheard him say 'Task Force Ranger.' I knew that was what they called the unit that had gotten into the Battle of Mogadishu in Somalia. I looked at his uniform and saw Special Forces and Ranger tabs along with a CIB, jump wings and HALO wings.

There had been some interesting cases of inflationary military history recently; a major had recently showed up for an airborne operation with a uniform full of badges and tabs and busted his ass on his first parachute jump. I admired his desire to be a paratrooper and if he had learned how to land, he might have gotten away with his fantasy. I wasn't sure about Stephen that day but when I got to know him in SOD, I found him to be every bit of what his uniform said and much more.

Stephen M was what I considered to be a classic professional soldier. He went to Ranger school in the 70s but missed Vietnam by being assigned to Korea as the Vietnam War was coming to an end. He got out and became a fireman in Austin. In the early 80s he went back in the Army to be a Green Beret. He qualified as a SF commo sergeant and was stationed at Fort Bragg. He heard about an interesting unit and walked over to the Delta Force compound. There he spoke with the commo supervisor and told him he wanted to serve there. Next day he had orders transferring him to the most high-speed unit in the Army where he worked in a section that supported the operators with commo and other electronic equipment used in that unit.

He was HALO qualified and he had been involved in a number of classified operations which to him meant they were classified. He did tell me a little about his time in Somalia and I, reading between the lines, determined that the loss of the six Delta operators and twelve Rangers in the battle that was portrayed in the book and the movie, Blackhawk Down, had disturbed him. The whole thing had bothered him so much that he got out and opened a very successful doughnut shop that he put together from scratch. He had decided to go back in the service to get a retirement and that's how he found himself at SOD one day, appearing before our hiring board.

We had had a number of Special Forces men come through the detachment in the past. Some of them made great contributions and some of them led cleverly organized rebellions against the chain of command that left smoking craters in the ranks and long periods of strife while we reorganized and consolidated. I asked him some pretty hard questions at that board, and he answered them all so calmly and clearly that I wished I hadn't asked them.

He was hired and during OTC he had shown himself to be extremely capable. He was one of the smartest and most experienced people we had ever had and the fact that he had served with Delta Force gave him a great deal of credibility. He was very familiar with remote surveillance systems and their employment. He never talked down to us and applied himself to learning all our tactics, techniques, and procedures. He had the respect of OTC students and the instructors.

I was running OTC, but the detachment sergeant and another guy were involved with some of the practical exercises. They were emphasizing shooting solutions to all the compromise scenarios we were using. I had to intervene, which made them mad at me for questioning their judgment in front of the new guys. The other guy let me know that he didn't appreciate my interference with his scenarios by pushing a ball point pen against my Adam's apple to make his point.

I punched him in the face and knocked him down. He sprang back up and I was hollering at him to back off. Some guys came in and separated us. I was pretty distraught. Brian chewed me out for making so much noise and everyone stayed away from me for a while. I apologized to the man I hit and burst into tears. He tried to apologize by saying that he could see that I was stressing out and we left it at that.

Austin-Crazy Stuff

We had always had a five-day training schedule for OTC while the rest of CD including SOD was on a 10 hour a day, four day work week when we were in town. The detachment sergeant decided to put us all on a five-day work week and things began to slide downhill. Brian didn't say anything, so it was obvious that he had convinced him that we needed to be flogged until our morale improved. PT went into overdrive with long runs on Town Lake and the OTC guys were being pushed hard.

One day, the CD IT officer accused someone of looking at porn on one of the computers in our area. No one would admit it, so we were all locked down for the day and then the detachment sergeant ordered us to go home and get what we needed to stay in the detachment area until someone confessed. Then we started being ordered to do push-ups in cadence every time we got a page. Things were getting really crazy and they culminated an hour after the OTC graduation ceremony when Steven M went into Brian's office and resigned from SOD. I was there and Brian really looked shocked. Steven told him that this was the stupidest outfit he had ever seen, and that since he had only so much time in his life, he didn't want to waste another minute of it here. Brian asked him to reconsider before Stephen left the office.

After Stephen M left, Brian called the detachment sergeant in and told me to be there while he reamed him a new one. "I don't want this place run like a recon platoon in the 82nd," he said. Now it was his turn to be dumbstruck. He sat there while Brian went off and I got the impression that he had been sure that Brian was backing him all the way and now this!

The detachment sergeant was already in the dumps. He told me that another operator had gone through his phone while he was taking a shower to get his girlfriend's telephone number. When he told me about it so I could join his hate club, I reminded him about love and war and that finished any casual conversations between me and him.

There was some talk going around that he wanted to get rid of Stephen M because he was afraid that Stephen would take his job. Stephen may have considered that as a possibility before OTC but now he just wanted to get away from our little circus. I was embarrassed by the whole thing and tried to persuade Stephen to stay. He did stay for one operation and showed the guys how to use a step ladder for going over a barb-wire fence rapidly during infil.

Tradecraft

In my conversations with him, Stephen told me about participating in tradecraft exercises while in Delta. They were run by the CIA to prepare their officers for operating in places like Moscow. At my request, he gave us a class on the skills of a spy and helped me set up a tradecraft lane that we would use for a detachment training exercise. I didn't think that Brian would approve this unless we incorporated a video camera exercise with it. It took a lot of planning before we scripted the exercise. My son and I validated the entire lane, riding the buses, and walking from point to point. The script went like this.

The operators, who were chosen to walk the lane, were told to stay home the day it started. The others were organized into two-man video teams that would tape the operator walking the lane. Tom Stephens paged the first operator on the lane and directed him to a bank of pay phones at a medical center near the camp. The operator was instructed to pick up the phone five minutes before the call came, pretending he was talking while holding down the lever. When the call came he was directed to go to a nearby bus stop and catch a certain bus to a shopping mall. The video team was then directed to go to the bus stop and video the operator waiting for the bus.

We incorporated a dead-drop that was hidden in a restroom in the mall. It was under the bowl of a certain toilet and it had a message giving the lane walker the next task. He had to go to a certain address a mile away and sketch a house before catching another bus downtown to another pay phone to wait for further instructions. The video team was directed to each of the operator's stops. They had to find a place where they could hide and video him.

The phone call directed the operator to cross the river and go to the Gazebo on the south bank where he was directed to conduct a personal meet. The video team had to remain on the north side and shoot the meet from a long distance. This required them to use the 600mm lens on an LX-1 camera. This was a difficult shot and the best one was made by Lt Smith, our new XO.

The operator had a code phrase and a clothing description of his contact. On one of the meets we used Colonel Buckwheat to be the contact. After exchanging bona fides the contact surreptitiously passed the information needed to complete the next task. The operator had to cross the river again and walk fifteen blocks to the hospital where I used to work. He had to go to the top floor of the parking garage and find one of our vehicles. There was a key hidden underneath and inside was a video camera, long lenses, and a tripod. He had to assemble the camera and video a meeting of two of our people in the park across the street. After he was finished he had to put the equipment away and walk over to the Texas Capitol where he would catch a bus back to the camp.

We ran everyone in the detachment through both walking the lane and taking video. At the end we had an AAR. The things that came out were interesting. Most of the men were freaked-out riding on the city bus, a new experience. A few of them stressed out on the dead-drop event. Craig did well but when he finished his last task and was on his way to the capitol, he had to march in a gay pride parade so that he wouldn't miss his bus. The video taking was appreciated by most of the guys and added to their decision-making skills.

Stephen M also gave us a class on meditation, a skill he had been trained on by the unit psychiatrist. We had the guys sit in a darkened room while he tried to explain about breathing and mantras. My mantra was to call on the name of the Lord Jesus, who had brought me this far. The other guys were drinking too much Rip Fuel to calm down enough for the more subtle psychological techniques of rest, so that exercise was wasted on most of them. Try as I might, I couldn't convince him to stay. He did come by and observe the tradecraft lane when we ran it and then he was gone. The detachment sergeant left us shortly after and since I was out of town, I didn't get to say goodbye.

Camp Shelby, MS-MOSQ: 11M-The Bradley Infantry Fighting Vehicle

In May I was sent to Camp Shelby, Mississippi for MOSQ in 11M, mechanized infantryman. I thought I knew all about the subject, having served in a mech infantry unit in Germany in 1965 but times had changed and so had the infantry. The squad was now organized around the Bradley Infantry Fighting Vehicle, first introduced in the early 1980s, as part of a new series of weapons systems known as the 'big five;' which included the M-1 Abrams Tank, the Blackhawk helicopter, the Apache helicopter, and the Patriot Missile.

The Bradley had gone through a long series of modifications during its development. Finally, when some brilliant officer in the Pentagon was hoping to add a flight capability to everything else, the congress said enough and ordered it tested before it was put into full production.

The story of those tests was told by Colonel James Burton in his book, "Pentagon Wars," and portrayed in a movie by the same name.

He was a disciple of Colonel John Boyd, a Pentagon Reformer now famous for the OODA Loop which would help us to win all our wars if we could only understand it.

The Bradley was a small tank armed with a 25mm automatic cannon, a TOW missile launcher, and a machine gun. It had night vision and thermal capabilities and could move pretty fast because it had to keep up with the Abrams tank that could really move. It had a three-man crew with a vehicle commander, a gunner, and a driver.

In the back it could carry a six-man infantry element. There was a ramp that raised and lowered and along the sides, some benches to sit on. Once inside with ramp up it was like being in a horrible nightmare. It was dark, the engine was loud, and you were crammed in with a bunch of other guys all wearing the same equipment.

The drive system for the tracks would grind loudly and the Bradley would lurch this way and that throwing you into each other. After a meal of MREs, the ride stimulated our digestive systems, causing the troops to release the most noxious of gases. In combat the troops in a Bradley are surrounded by fuel and ammunition, both of which can really get excited when disturbed by too much heat. The Russian designed RPG-7 was the perfect foil for the Bradley, so the early models were equipped with small firing ports for some modified M-16's that the guys in the back could use for spraying and praying as they rode across the battlefields of the future.

Camp Shelby, MS-A Real Soldier

I drove to Mississippi with Stephen M, he was still at the Academy as an instructor. When Stephen joined the Guard, he was assigned to the academy as an instructor and since he was an infantry instructor, he too had to attend 11M MOSQ. The course was designed for privates just coming into the Army. The course in Mississippi had some privates and a few more sergeants, who like us, were reclassifying our MOS.

We stopped in New Orleans so he could show me some artwork by a painter that specialized in expressing lighting in his paintings. Stephen was originally from southeast Texas and he liked the Cajun food and atmosphere.

Most of the training had to do with the Bradley and its armaments, the 25mm Chain Gun being the beast in the room. The instructors decided that we should be able to pass the Bradley Gunnery Skills Test in order to pass the course. This included wrestling that heavy thing out of the turret, correctly disassembling and reassembling it, wrestling it back into the turret and hooking up the ammunition feed chutes to the gun, all within a certain time. I was totally befuddled by all the mechanical requirements as were most of the rest of the class.

Stephen figured everything out quickly and became an assistant instructor who helped us dummies until we were able to pass the hands-on performance evaluation. I hated it at the time because I was 51 years old and didn't want to know another thing about infantry, but I felt a little better about myself as an NCO after I passed.

While we were there, we got a message from the academy back in Texas. At the end of this course, we were ordered to go from Mississippi straight to Fort Hood, Texas where we would be instructors teaching the same course that we were in. I started paying much closer attention and asked for reviews of anything I had tuned out.

On the last day of the actual course we were assigned to a work detail moving wall lockers and other heavy furniture from one barracks to another. Stephen and I went to the senior NCO and asked if we could leave a day early because we had to be at Fort Hood in two days. He became very angry when we tried to play the 'professional courtesy' card as fellow instructors and refused to release us.

We returned his denunciations with some angry words of our own and after another hour of working, one of our instructors gave us our 1059 Course Completion certificates and told us to hit the road. We drove back to Texas and spent the next two weeks as assistant instructors teaching soldiers the secrets of the modern mechanized infantry.

Ft Hood-11M MOSQ Instructor

We were billeted with all the other instructors and it was like being back in the Stone Age. The academy administrators had not gotten the figures right and there were more instructors than there was room in the barracks. They found extra bunks and crammed them in until it was more like a troop ship than a barracks. Each night the drunks would come back from the club arguing and bumping into things. One instructor would get up in the middle of the night and, while sleep walking, urinate onto a wall locker or into someone's boots.

This caused a lot of laughter among those at the other end of the barracks and some serious complaining from those who slept near him. I witnessed him doing this one night when I woke up and saw him walking toward my bunk. I jumped up and stopped him with my hand against his chest. He just backed up and continued down the center aisle until he found a more peaceful spot and I heard him void his bladder like the proverbial racehorse in someone else's area.

The 11M instructors were a seasoned bunch with a lot of time in mech infantry. The senior instructor kept saying, "stone cold" all the time. He was a Hispanic sergeant from the Valley, and I couldn't tell if he was speaking a Spanish phrase I had forgotten or some Bradley terminology that I had missed in Mississippi. I asked the instructors a lot of questions and questioned their judgment regarding safety.

There was night driving and the students were required to drive a Bradley around a two-mile course using a night vision device. There were a few terrain features that made me feel like we were going to have some injuries or deaths, and I asked the senior instructor to change the route. "Stone cold," was his reply and he and the other instructors let me know that the mech infantry was not a place for pussies.

My worst moment was when I had to drive a Bradley onto a lowboy trailer for transport back to the vehicle storage area. You have to carefully drive up two ramps that you can't see, being guided by someone you don't know. I prayed pretty hard that day and managed to do it without hurting anyone or destroying government property.

The two weeks went by pretty quickly. Brian was doing AT with G Company at Fort Hood that year, so I tracked him down at their COB. He told me that I should get ready for deployment the day after we got back from AT. For the first time since I came on the Counterdrug program I complained and told him that I had spent the last month away from home and that I would like a day off.

Changing Seasons

Beside the burnout, our relationships were changing. When we started the program, SFC Tom Payne insisted on meritocracy as the deciding factor in leadership. NCOs ran the unit for the first year and after the train wreck we had a captain who was content to leave most of the decision making to the NCOs. During all this time, I occupied the operations position on the staff. I learned a lot about these operations because I was immersed in them. Eventually I was known as a subject matter expert even though I wouldn't have made that claim myself. I never felt like I knew what to do but when it came time to do something, I usually had an idea and went with it. I guess I did something right because I was kept in that position for the last nine years.

Captain Brian had come up in that system and had risen from operator to unit commander in five years. He was maturing as an officer and he wanted to run the unit the way he saw fit. I wasn't the best man for that. If I didn't agree with him, I let him know and this was a source of increasing irritation to both of us. I couldn't stop arguing with some of his ideas and he was starting to ignore my recommendations. Most of my time was taken by preparing lesson plans for OTC, making sure that the mission commanders did their paperwork, and running an operation every now and then.

Camp Ashland, Nebraska-ANCOC

Three months after AT, I was back in an Army school. This time it was the Advanced Noncommissioned Officers Course, Phase One, which was given at Camp Ashland, Nebraska. The academy sent me there so that if I passed it and the Phase Two, I could be promoted to SFC E-7. That was pretty exciting for me because I had been promoted to SSG the year before after 29 years so I had figured it would be at least 15 years before I got promoted again.

I reported to Camp Ashland, on the banks of the Platte River and was immediately impressed with the professional operation of the school. The first thing that I liked was my room assignment. I had a room to myself with cable TV and my own bathroom. That was a change and the rest of the administrative procedures were just as impressive. The classrooms were equipped with all the modern audio/visual equipment and all our manuals were ready to be issued.

The instructor, was a SFC who had spent years in the Air Force. He wasn't very hooah by SOD standards, but he guided us through the course. The subjects were many and far away from what I thought was important for a sergeant to know.

There was more emphasis on Risk Management which was becoming the most important part of any operations plan. Developed by aviators out at Fort Rucker, it is a complicated system of identifying the hazards, assessing risks, developing controls, making decisions based on this process and making sure the controls are implemented to prevent death and injury to the troops. What's wrong with that?

Nothing is wrong with that in itself but when you really get into it you find that everything you think you need to do could be considered by someone to be too dangerous; too dangerous for you or for their career? I began to appreciate the system, but it was complicated and required us to be able to articulate the measures we would take to prevent the hazards we had identified from hurting our troops.

Once I watched James F, one of our LRSD team leaders, brief the Division Commander on a mission that included a nighttime movement across a lake in some rubber boats. At the end he gave his risk assessment and like a true paratrooper let the general know it was a low risk operation. All the general could see was a rubber boat full of troops, sinking in the middle of the lake in the middle of the night and him, waking up in the morning to the gentile shaking of his aide who was telling him that six LRSD men had drowned last night. He jumped up and said, "Sergeant, I think your assessment is too low." The team leader took him back through the process showing all the measures he had taken that would have ensured the survival of the crew of the Titanic if they had been on his team. The general came to the realization that we either crossed the lake that night to complete the mission or we didn't. He manned-up and told the sergeant, "Your right, go ahead."

With the careerists everything is high risk and their main concern about any operation they are associated with is the risk to their career. This is the second primary principal of war, the search for scapegoats, or making sure you have a chair when the music stops. With the youngsters everything is just another opportunity for a rush of adrenaline. For the professional every operation is an opportunity to use everything that they know to accomplish the mission, take care of the troops, and themselves. Risk management means taking all of this into consideration and using your knowledge of history to assess the risks and your common sense to mitigate them and your leadership skills and willpower to make sure that the requirements for safe operations are met.

I got into a little pissing contest with one of the younger sergeants who was wearing a Ranger tab. He was full of himself and wanted us to sing some inappropriate cadence when it was his turn to march us between classes. This led to some lively discussions about ethics which helped me to have a deeper understanding of the role they played in the climate of any unit. Changing the culture was impossible for an NCO but anyone in a small unit could have an effect on the climate.

I had the high score on the final exam, but I didn't get an 'exceeds the course standards' on my 1059. I asked the Master Sergeant who ran the course why and he said he would look into it. Of course, I never heard from him again.

One of the former SOD guys was in the Nebraska National Guard LRSD and was based at Camp Ashland. I went over to their HQ before I left the camp but he was out in the field, so I missed him. He came to the unit a few years before and had gotten into a brawl down on 6th Street in downtown Austin. He started the fight but managed to escape being arrested while some of the others were charged with assault with serious bodily injury, a felony. I felt bad for the guys, but I had often warned them about getting drunk on 6th Street from my experiences as a police officer. Once when I was ranting about their needing to grow up, one of them stopped me by proclaiming, "We have to grow old, Sergeant Murry, we don't have to grow up!" After they were arrested, I heard about their travails from time to time as they spent thousands of dollars they didn't have on lawyers as their case dragged through the court system. Getting drunk and having a barroom brawl was an expensive bit of fun but the whole experience did help them grow up.

Change of Seasons Continues

We were coming to the end of an era. Captain Brian had gone into high gear in his quest for remote surveillance systems. We started out with our Trontec which was a quantum leap in technology solutions for our surveillance operations. By installing a remote system that could pan tilt and zoom by remote control while being monitored from a location miles away meant we could watch a target with two guys for a week at a time and remain out of the danger zone around the target area for days at a time depending on how much battery we used.

The systems that Brian was looking at now were costing $100,000 but they were nice. Based on micro-wave transmission, these systems could be concealed in a number of ways and would provide as good a coverage as what a team of 4 guys could do and the ones watching the monitors could do so from a nice hotel 20 miles away.

Captain Brian had me teach some classes to the guys on the duties of a mission commander and after he returned from a staff officer course, he insisted that we begin using the Military Decision-Making Process (MDMP) as part of our mission planning. This is a logically thought out system of steps that should allow you to come to an almost scientific conclusion on the best way to accomplish your mission.

I was too old-school to really get excited about the prospects of choosing three courses of action and putting them through a vetting process using weighted criteria to determine which one was the best but, because he involved me in all of this, being I was the operations guy, I was forced to learn it. I could see some value to the younger guys, and it would certainly protect us from criticism by an investigation into our planning if something went wrong in a big way.

I still liked flying by the seat of my pants but as we went more for the remote surveillance technology the planning process would be more closely supervised by the CO. We built some dry erase boards with an MDMP matrix and room to write in information about the target and its limitations as it became available. The teams would have one in each team room and it would serve to keep everyone on the team informed on the progress of the plan.

September 1998: FY-98

In Fy-98 we did 30 operations and unknown seizures.

October 1998: Central Texas-Old School

We got a request to do an operation in a nearby town and Captain Brian gave it to me. I did a map recon and realized that Tom Stephens had some land near there. He showed me on the map where his property was, and it looked like I could walk across country to where we would set up an OP. We drove out there and Tom waited at the vehicle with a radio while I took a long walk through someone's property. It was pretty much unused land with clumps of mesquite and open areas. It had been used for cattle at one time, but I didn't see any fresh sign. I came upon a house that was in a wooded area. After waiting and watching to see if anyone was around, I walked up to the open front door and looked in. The ceiling was falling in and the inside was trashed. Someone left there 20 years ago and hadn't returned.

I continued on my way and after another 30 minutes, I entered into a thicket with large clumps of prickly pear cactus and brush. I broke through the last line of brush and there 100 yards in front of me was the front of a large house, our target. I ducked back in the brush and watched the house for 10 minutes while I let Tom know that I had arrived at the target. There was no activity that I could observe so I took some video and headed back. This was my ninth year of doing this kind of work and I had a feeling that I wouldn't be doing it much longer.

Tom and I returned to Austin and I showed Brian what I had come up with. I wanted to take a team in and do a traditional mission. Brian told me that with a new system we had just received, we could microwave our video to a town nearby and do the whole thing by remote. I didn't think we could get line of sight to the town so I pushed for my plan and he relented though he came to me several times suggesting that we try the new systems but let me go ahead with my plan.

The team was chosen, and we did our planning and rehearsals. Sergeant Dennis, the team leader, was going to set up the OP and Pat was going to climb a tree and install the Trontec up high for a look over the wall that surrounded everything but the front of the house. We ran the CP from a motel in the nearby town and use one of the jungle antennas in the field and a vehicle antenna on the roof of the motel for the CP.

I led the team in and when we crossed the first barb wire fence my pants split open at the crotch and so I was almost naked from the waist down as I led them through the woods to the OP. When we got there, we dumped the rucks and sent out our security guys on both flanks of the OP where they could watch the road and the target.

It was a good thing we did because right after we started working on the OP, we got a series of clicks in our earpieces that caused everyone to freeze as per SOP. We waited patiently until one of the security guys came up on the air and told us that a couple had been walking their dog down the road in front of us and now, they were quite a ways down the road so all was clear. We were way out in the country, but you never knew, and our security plan saved us from an early compromise.

Dennis had looked at all the possible locations that they could get a good shot of the front door and the only place that would give us what the cops wanted was right in the middle of a large patch of prickly pear cactus. Dennis was a real professional and one of the only ones in SOD who wanted to stay in the field after returning from Ranger school with his tab. He was the team leader and he led the way into that patch, carefully cutting a path into the center where he would place his video camera on a tripod. The team took turns developing the OP and pulling cactus spines out of their bodies until Dennis was satisfied that they could do what we came for.

Meanwhile, I was helping Tree who had climbed the largest tree in the area and was installing the Trontec on a limb that gave it a good shot into the walled compound. I passed him the components from our rucksacks, and he hauled them up using a bag tied on to a length of 550 cord. He finally told me to turn on the camera and check using the monitor which was hardwired into the camera for this operation.

Everything was working so he climbed down, and we ran the controls and the monitor over to the OP that Dennis and his guys were putting the finishing touches on. I made sure that Dennis didn't need anything else and he released us to go to the CP. Pat and I went through the brush a ways and called for a pick-up. One of the CP guys drove up and we signaled him with an IR source and he slowed down long enough for us to jump in and then drove down the road.

The CP was set up when we got there, and we settled into another long wait. We were doing this for the DEA task force which was why we were able to hide on private property. The 'Open Fields' doctrine in the U.S. law defines certain aspects of the 4th Amendment search and seizure requirements. The Feds were able to gather information and even evidence if there was no legal expectation of privacy. Whatever you see across an open field is not protected by the 4th Amendment, even if the cops are trespassing. A shaky status for sure but we had been doing this for so long that as long as the Feds kept us under their umbrella, we were willing to do the deed.

We kept that place under tight surveillance for a week waiting for a drug shipment to come in or the place to get some visitors. I rotated the guys out every few days to take a shower, eat a pizza, and get a good night's sleep. Sonny went in with a resupply once and got lost.

I told him that he should have rubbed his tab a little harder before he attempted such a difficult navigation which didn't do anything for our relationship. I almost told him that I too had gotten disoriented momentarily on my own resupply run a few days before, but I didn't because Sonny would have considered that a sign of weakness and made my job more difficult than it already was.

On the last day I told the DEA guy that we were pulling the guys and he was disappointed with the results. No drugs, no suspects, no joy. "Maybe we'll get them next time," was all he said.

Academy-BNCOC Phase 1 Instructor

BNCOC started on the first weekend of November that year and I showed up for drill hoping to be chosen as an assistant instructor or AI. I was with the Infantry section that Saturday morning when Larry Rayburn, the BNCOC Branch Chief came up to me and asked me if I still wanted to AI a BNCOC Course. When I told him I did, he said to follow him. He took me to where they were in-processing the new students and introduced me to a SSG from the Reserves and told me that he would be the primary instructor and I would be his assistant.

This guy looked pretty stressed out so I asked him what I could do to help him. He told me that he couldn't find his lesson plans for the class and that the first lesson was in an hour. I began to check around the various classrooms and found that most of the instructors were in the same boat.

This was when I began to get educated on the real situation at the Texas Army National Guard Academy. It was the most screwed up operation I had ever seen in the military and after having been in a number of regular army and National Guard units this was saying something.

I had attended PLDC in Mississippi and it was a professional operation; this was not. Two years before when I attended BNCOC Phase 1 at the Texas Academy, it was run by USAR instructors and administered by the Texas National Guard. It wasn't as well run as the Mississippi Academy, but it was adequate. Things had deteriorated and now it was several orders of magnitude worse than anything I had ever seen. Just being associated with this operation was the most gut-wrenching experience I had had since Vietnam.

Finally, a sergeant drove up to the row of classrooms where I was standing and brought out a number of binders with the lesson plans. I got one set for my instructor and brought it to him. He wasn't looking very happy and I asked him about his experience as a NCOES

Instructor. He told me that this was his first time. He had just come off active duty at Ft Hood and had joined the Reserves. He was appalled at the disorganization in his unit and in the National Guard Academy.

As I listened to the instructor complain, I began to see that the Reserves thought that all the admin requirements were the responsibility of the Guard and the Guard thought that the Reserves were responsible for all the instructor material which the reserve considered part of the admin requirement. Eventually I realized that the root problem was twofold: one, all the leadership was lazy and two, the leadership had not established SOPs for the conduct of BNCOC/ANCOC.

I saw a lot of senior NCOs blaming each other for the train wreck we were in. I did what I could to help my instructor get through the day. Fortunately, the Branch Chief had prepared the student materials which consisted of file boxes of thirty Field Manuals each that were issued to the students on the first day. They would have to take these boxes home with them at the end of each weekend and bring them back for the next training weekend.

At the end of the day my instructor took off and I went home for dinner. I told my wife about how messed up everything was but, after saying, “Well, what did you expect?” She could only nod her head because she had heard it all before. "No, really, this is really bad," I told her as she went to bed.

I got up the next morning at 4:30 so that I could be there for PT and when I arrived, I found the class I was assigned to but my senior instructor wasn't there. I led the group through the PT training and marched them to breakfast. I saw my Branch Chief in the mess hall and told him that my instructor hadn't showed up yet. He told me to stay with the students and that he would have the Reserve NCOIC call my instructor to find out why he wasn't there. I was in the classroom during the first hour of class when the Branch Chief came in and told the class that the original instructor was no longer with the Reserves and that I was now the primary instructor.

For he is a servant of God to you for good. But if you do what is evil, fear; he does not bear the sword in vain; for he is a servant of God, an avenger for wrath to him who practices evil. Romans 13:4

Chapter 22-January 1999

Austin-BNCOC Instructor

Being the primary instructor was not what I thought it would be. The students were not impressed with the way the academy was being run and it was difficult to get their minds off the pathetic situation and onto the task at hand which was to learn the material. Fortunately, the first class was physical fitness and so I introduced them to the current SOD view of fitness which was, 'you can't do too much PT.'

I had them down on the floor with me doing push-ups, in cadence, every time it seemed like it was necessary. It wasn't for punishment; it was to motivate them. Someone in the class would give a good answer and I would jump up and direct the class to get down and do push-ups with me while I counted the cadence and they counted the repetitions. Some of the class didn't like doing this but most of them did. There were several paratroopers, some motivated infantrymen, and several female NCOs who supported my enthusiasm and the rest of the class went along with them. Soon, we were a group with an identity and a goal which was for everyone to pass the course.

When I told them that we would have study hall every Saturday night, there were a few groans and complaints; and I told them that I would be there with them. Some of them were hoping to party on 6th St in downtown Austin. I told them that they could, after study hall.

Besides teaching many of the lessons, I had to evaluate each student performance on tasks which included; squad level drill and ceremony (marching), conducting an in-ranks inspection, leading a PT session, teaching a class, conducting an after action review, conducting height and weight measurements, and performing maintenance on a piece of equipment. They had one evaluation and a retest, if needed, for each task. I filled out an evaluation form for each person for each task. I also had to counsel each student several times during the course. This was a time-consuming process and I had no system.

The administrative requirements for instructors kept them busy and the experienced ones developed systems to cope with the workload. I was overwhelmed by the administrative requirements and when I was inspected by the Branch Chief I found myself constantly behind the eight ball. I shouldn't have because I had attended the same course but then I was only concerned with learning the material. My encounters with my instructors were exactly what I now was expected to do but the shoe was on the other foot and I had to learn as I went. I did have the thought that there should be a training session for new instructors to help them organize for the administrative requirements.

In the middle of the course, the admin sergeant told the Branch Chiefs that 'someone' had screwed up the training schedule. Louisiana called and told him that we had too many weekends and that we had to cut one. Instead of graduating in May, we would have to add two days of material to the last three weekends and graduate in April.

On top of all the other chaos that we were just beginning to sort through, this almost flipped me out. I asked Kenny, the admin sergeant, if he would come to a meeting of the instructor cadre of the school and explain what happened. It must have been too painful, so he, the CSM, the Branch Chiefs, and the senior small group leaders, stayed away and let this slide downhill onto the heads of the instructors.

Through it all I stuck with keeping the class motivated. I used military history to illustrate various teaching points in the lessons. I was amazed at how little the soldiers knew about American military history. Most knew little or nothing about Operation Desert Storm, or Panama, or Grenada, not to mention Vietnam, Korea, and the World Wars. I encouraged them to read books like 'Into the Storm', by General Fred Franks and Tom Clancy. I told them about the controversy between General Franks and General Schwarzkopf. They were fascinated with some of the examples I gave them and asked many questions.

To prepare them for the final, I made up practice test questions for study hall and gave them key word drills. I gave them a key word like 'frequency,' a factor of fitness in the PT class, and they would have to use their memory and their notes and then look it up in the correct manual. The final examination was an open book test and I wanted them to practice using their notes, the index and table of contents of their manuals and anything else they could use to find the answer quickly.

I taught them my strategy for test taking, which was to answer all the questions I was sure of all the way to the end of the test. Then, going back to the beginning, I would answer all the questions that I knew I could find the answer for quickly; the questions were all multiple choice and some of the choices were very close to another choice. When I went to the end the second time I would go back and do it again, this time answering the harder questions but never spending much time on questions I couldn't remember much about. By the time I went through this process 3 or 4 times I had answered all but 2 or 3 questions and I had time to do the research for their answers.

We gave the students a PT test at the start of the course and at that time, if you failed, you had one re-test. No one failed from my class, but a large number failed from the other classes and after the re-test they closed one class and consolidated the students in the other classes. I got an older NCO from El Paso who I became close with. A Vietnam veteran, and a believer, he was in the Counterdrug program.

He had a difficult time reading and because of him I began to pay close attention to the other students when they read something out loud in the classroom. I identified several weak readers and talked to them privately. I encouraged them to study with someone else that was in the class from their unit. They all managed to get study partners. Later I found that weak reading skills were the number one cause of failure in the NCOES courses. It was mainly older black and Hispanic soldiers of my generation who had the problem. They had quit school to go to work when they were young and never learned to read at an adult level. I told the weak readers that they had to study the material diligently between the class weekends using the outlines that were given to them.

Each weekend I would show up for drill, meet my class at formation and take them to the helipad for PT. I would evaluate their performances and encourage them to push their own PT programs because, I told them, their troops would have more respect for them if they were physically fit. I told them that the PT test was one of the last dividing lines between soldiering in the reserve components and being a slug. The troops know if someone is cheating on their PT test and even if they go along with it because they are afraid to speak up, they know who the real soldiers are. I stayed with them doing PT until 0650, the time to stop on the training schedule.

Soon after the training schedule fiasco, the CSM and Branch Chief called a meeting of all the instructors. They told us that we would be inspected soon and that we all had to be on the same slide in the same lesson at the same time. This inspection would be by the Louisiana National Guard Academy, our higher headquarters. They would inspect us this year in preparation for the accreditation inspection by the Sergeants Major Academy next year.

There is always a conflict between the desire to train to standard and the need to train to time schedules. The inspection pushed this conflict to the center of my brain and made me angry. On the day of the inspection, for some obtuse reason, I had the class out marching, doing performance evaluations, longer than the schedule allowed.

We got back to the classroom about 10 minutes after the class was supposed to start and the inspectors showed up about the same time. They were all excited about my class being way behind. They took me outside to spend more time telling me what a poor instructor I was and how if I didn't straighten out, I would be relieved of my duties. I was given a negative counseling statement and my Branch Chief told me that I had better watch my step because they were watching me.

I knew the inspectors and my feeling was that they were a bunch of sorry senior NCOs who had allowed this place to lapse into a military disaster. Instead of coming around to find out what was wrong, they stayed above the problems and dumped their observations on the instructors who were forced to execute the mission. Apparently, the inspection didn't go well because I noticed an ambulance in front of the Academy Headquarters when I marched the students to the mess hall for lunch. I looked into the office and saw one of the senior NCOs being strapped onto a gurney, in the midst of a heart attack that began when the Louisiana guys gave him the results of the inspection.

At the end of the course we gave the final exam. All but one student in my class passed the first time; the older NCO from El Paso failed by several points. The high scorer and several others volunteered to coach him that night and the next morning he took the retest and passed. My branch chief came up to me later and told me that my class had the highest average score of all the BNCOC classes plus a 100% pass rate. That report made me feel that I could be an instructor and it caused my superiors to see me in a different light. The other classes didn't do so well with 40 students failing the first test and 25 students failing the second test and the course.

Austin-STPA

In the middle of the BNCOC class, I was ordered to attend ANCOC phase two. This time I would go to 11M, mech infantry. This was my third trip to Camp Shelby and while I wasn't looking forward to another school, I knew that if I wanted to get promoted, I had to pass this course. After I was promoted to E-6, after 29 years as an E-5, I started to pay attention to the Texas Army National Guard promotion system, STPA.

Select, train, promote, and assign was how it was supposed to work. Each year a records board was held. Last year, I was ordered to work on one of these boards. Three NCOs were given a stack of personnel files. Each file was scrutinized by each of these NCOs and points were assigned based on the contents of the file. Points were given for awards, time in service, time in grade, military education courses attended, civilian college hours, letters from a General Officer, PT tests, marksmanship scores, etc. The points were averaged and were the basis for that soldier's standing on an MOS specific 'order of merit' list. The list was published once a year on-line.

When there was a vacancy in a unit, instead of promoting someone there, now, the person at the top of the order of merit list would be offered the position. If that person accepted the vacant position, then they were reassigned, sent to a NCOES course, and after they graduated, they were promoted.

Since I had been on active duty with CD, I had accumulated a lot of documents that would mean points on my STPA standings. I had to go to the personnel office at State Headquarters several times with my documents in order to get everything entered into my records correctly. When I was in the personnel office, I saw row after row of file cabinets full of records for every person in the Texas Army National Guard. Since I lived in Austin and worked at Camp Mabry it was relatively easy for me to square away my records. The clerks that worked in personnel were doing the best they could by their own estimation, but the rest of the Guard was sure that they weren't doing enough. As I went through this process, I realized that the Army had changed dramatically since I first joined its ranks.

Promotion then was a mysterious process that for me was meritorious in nature. If you endeavored to be the best at whatever task you and your comrades were given and you accepted responsibility for some task when it was offered by your superiors, you would be promoted. Now, an NCO had to work on his career more than he had to impress his superiors with his worth as a member of the unit.

The whole system was designed to counter the good old boy system that favored those who ingratiated themselves with their superiors, rather than those who were the best soldiers and best able to bear the responsibility for leadership. Now you could be a low performing soldier but because you went to college, attended military schools and did some of the other things that got you points, you could be promoted faster than a younger, better leader with fewer points.

Since most of the Guard was in units far from Austin, the soldiers had to rely on their full-timers, the unit clerk and the readiness NCO, to make sure that all their documents were added to their personnel file before the annual boards were held. I never had a high regard for most of the full-timers that I met in the Guard. The one exception was SSG Janie at the RTI. She was the most efficient and motivated full-timer I ever saw. She made sure that everyone's files were up to date.

Many soldiers are waiting for well-earned promotions that won't come, thanks to the laziness, incompetence, and sabotage of the full-time mafia that runs the Guard. SSG Janie was repaid for her efforts by being harassed by a certain LTC until she left the RTI and went to another unit where I'm sure she set the standard for honesty, competency, and efficiency unlike the LTC who did none of those things.

Camp Shelby-ANCOC II

ANCOC Phase II for 11M Infantry was designed to train mechanized infantry squad leaders to be platoon sergeants. The mech infantry platoon consists of four Bradley Infantry Fighting Vehicles and thirty-six infantrymen. The platoon sergeant is responsible for the NCO business of the platoon and he is a vehicle commander of a Bradley in combat. Much of NCO business is taught in the phase one course so phase two would be more Bradley business.

The course was difficult for me because once again I was learning about mech infantry, but I did not have 11M BNCOC. I had gone to phase two BNCOC for intelligence analysts and missed a lot of Bradley technical information. I decided to try for the top 20 percent of the class for personal reasons and that helped motivate me through a lot of classes. We went to the field and drove around the Camp Shelby training areas. I learned a little more about the Bradley Fighting Vehicle and that would soon be useful.

I had some interesting talks with my fellow students. There was a sergeant at the course that I knew from LRS days. He had moved up north and let me know that Waco had convinced him that Texas had fallen into the hands of the communists. Another sergeant from Alabama told me that he felt like the rest of the country looked down on the people of Alabama and thought that they were backwards. I told him that the civil rights era had given me that impression but when I told him I was a Christian he seemed a little softer.

A Georgia sergeant told me about his adventures at the National Training Center (NTC) with the 48th Infantry Brigade of the Georgia National Guard. They were a round-out brigade of the U.S. Army's 24th Infantry Division. If a war came, they were supposed to deploy with the 24th and fight. When Iraq invaded Kuwait in August of 1990, the force that eventually invaded Iraq in February of 1991 included the 24th ID but not the 48th Brigade.

They were sent to the NTC for two months and after the four-day ground war to liberate Kuwait was over they went home. This was the end of the 'total army' façade that had been promulgated by General Abrams who determined that we would never go to war again without the reserves like we did in Vietnam. The concept was enhanced during the Reagan era buildup of the armed forces, but none of the National Guard maneuver units called to active duty ever saw combat and that was a slap in the face to the enhanced brigades like the 48th.

On the final I had the highest score in the class but once again I failed to get a distinction on my 1059. I was told I would be teaching 11M BNCOC that summer, so the course was a good preparation.

The Border

When I returned to Texas, I was sent down to the border to participate in an operation on Falcon Lake. We spent a lot of time in that area working with CORE, the TNCP drug commandos. There was a lot of large load smuggling on the lake and it became a familiar area to us. We incorporated RAID and the C-26 into the mix and were starting to get pretty good at coordination between all the elements. Several large loads of marijuana were seized and the trafficking on the lake continued like nothing had happened.

Bosnia

The 49th Armored Division was chosen to command the peacekeeping force in Bosnia. MG Halverson, a Vietnam vet, was the commander. He had been the G-2 for a time when I was assigned there and was a good officer. Steve, an SFC now and the head of the counterintelligence agents at G-2, asked me if I would go as an Intel Analyst. Since this was a peacekeeping mission, I didn't feel to abandon my commitment to the Academy, so I declined to participate.

This deployment generated a lot of excitement. It was the first time in recent history that a National Guard unit would command a regular Army unit, in this case, the 3rd Armored Cavalry Regiment. I heard from Steve several times asking me to reconsider but by then I was in the thick of things at the Academy and told him that I was committed to fixings things there or else.

When they returned, I heard the story of the cat the Intel people kept in a small room near their work area; when they started stressing, they would go and 'pet the kitty' until they felt better. They said that the soldiers of the 3rd ACR announced that they had discovered the significance of the number 49, as in the 49th Armored Division. It was the average age of a specialist (E-4) in the National Guard.

Austin-BNCOC II Instructor

After I returned to Austin from the border, I was ordered to AT at Fort Hood. I was assigned to be an instructor in a BNCOC Phase 2 course for 11M. There were two classes of 16 students per class. One class was from Louisiana and the other was from Texas.

The Louisiana soldiers came from an enhanced brigade and were quite proficient with Bradley operations and tactics. Their unit had been mobilized for Desert Storm and was now one of 15 National Guard 'enhanced' brigades which meant that they received the newest equipment and most intense training. They spent the war at Fort Hood, where they had gotten some bad publicity when some of them became weary of our Mexican cuisine and returned home for a little gumbo which made them AWOL.

The Texas soldiers were all from the Rio Grande Valley and considered themselves 'dismounts' and professed to know little or nothing about the workings of the Bradley. Many of them had also been mobilized during the Gulf War and sent to Fort Hood to be part of a round-out brigade of the Mississippi National Guard. They had not had a pleasant time at Fort Hood either. One of them told me about the fate of their Re-Mobe binders, a labor-intensive compilation of unit plans for mobilization. The unit commander gave them to the regular army officer overseeing their training who gave them to a sergeant who threw them into a dumpster outside the orderly room.

I received my instructor materials and was partnered with SSG Craig, a fellow SOD operator and experienced Bradley infantryman who had been reclassified from being an anti-aircraft missile crewman. We read through the course schedule and material and realized that there was a big problem up front. One of the first tasks was a series of hands-on skill tests of the Bradley weapon systems. These were skills that were a prerequisite for attending this course.

This test was a detail that the south Texas units had failed to mention to the sergeants that came to the course. When we briefed the classes on the requirements of the course, we got a lot of shrugs and a few words in Spanish that everyone in Texas understands but never repeats in family gatherings except when drunk.

In our first instructor meeting this issue came up and it started a lively debate which developed three courses of action: fail them and send them home, pencil whip their documents and give them a go, or try and train them to meet the standards in whatever spare time we could find in a pretty tight schedule. I advocated the third choice and through force of will managed to assume responsibility for the extra training even though my class of men from Louisiana was not the problem. Craig supported me as we started the course and we ignored the bad vibes from the other instructors.

On top of everything else we were being accredited by the Sergeants Major Academy. A MSG was the Branch Chief for this course and seemed to be distracted. He had become quite fond of drinking beer every night at a Karaoke bar and took me with him once where I had to endure several hours of entertainment without the anesthetic effect of alcohol since I had stopped drinking 10 years before.

He decided that we had too many machine guns and turned several of them into supply just before the accreditation inspectors began to inventory the on-hand equipment required for the course. We failed that inspection and the SGM who was over all the courses went ballistic. After ripping off a piece of his forth point of contact, he sent him back to get the machine guns, causing a minor disturbance in the back of our classroom while we were unfurling the mysteries of Bradley tactics to men who knew more about the subject than we did.

The barracks situation was as bad this year as it was last year. The same sleepwalking drunk was still peeing into people's boots or on their wall lockers and the same instructors were crammed together with a foot of space between bunks. Craig lived in Killeen, which is next to Fort Hood, so we stayed at his house every night. We got up at 4 each morning so that we could wake up our class for PT. Physical fitness means something to infantrymen, and Craig and I were in pretty good shape thanks to the maniacs at SOD who considered physical fitness as the be all and end all of special ops.

We led the first session and then the students would lead the exercises every morning. Craig and I joined the students and did the same exercises as they did. This established our credibility with the students, especially the ones from Texas. There had always been complaints from the soldiers from the valley that their leaders never set the example and so they responded very positively to our participation. Some of our fellow instructors chose to 'observe' training rather than participate and since they were teaching the Texas soldiers, they created a dissonance that continued to the end of the course.

One of the blocks of instruction involved the adjustment of indirect fire using binoculars with an incremented reticle, a compass, and the 'WORM' formula, which is used for determining range when adjusting mortar or artillery rounds onto a target. This requires simple math which caused me to totally panic even though I had done the similar exercises a month earlier at my ANCOC class. Craig was not very well versed in this procedure either, so we decided to work through all the problems in the lesson plan the night before we taught the class. It took us several hours but when we went to sleep that night, we slept the sleep of the prepared.

The course required us to go to the field with the Bradleys and when we got there, we did the skill testing. I took the guys from the valley and we went over the skills they were supposed to know but didn't. We went over each task and gave them some time to practice. When they were ready, they would come to me or Craig and we would test them. If they didn't pass, we sent them back to review their procedures then they could come back and try again.

This didn't conform to Army training doctrine that prescribed one test and one retest before total failure and the requirement to take the course again. I felt that as long as they performed the task to standard they were trained, and this compensated for the negligence of the Guard leaders who had failed to prepare them for this course.

The problem with performance-oriented training like this is that these tasks have to be repeated many times under varying conditions before the soldier is truly trained. This is why the National Guard soldiers require a longer pre-deployment train-up. They did it once years ago but hands-on skills are perishable.

After we finished skill testing, we started a series of small scenarios requiring the students to maneuver their Bradley squad across the training area and attack an enemy position. The Louisiana guys did a great job, masking their movements with terrain and vegetation, exposing themselves for brief moments while firing then reversing and moving to another location and firing again. They did a credible job of attacking an enemy position according to doctrine and since no one was shooting real ammunition at them they all received a go.

The Texas soldiers went next and when they left the assembly area they got on line and charged across an open area for about a kilometer and drove right up to the objective, proud of their Texas heritage and ignorant of conditions on the modern battlefield where even a terrorist with an RPG would have had them for lunch. We used to do the same thing when I was in Germany in the 1960s, using enthusiasm as a replacement for competence.

It is the knowledgeable leader's duty to correct mistakes like this and several of the instructors called them on it during the AAR which we had right after they finished. They assured us that this was the way they did things in the valley. We told them that they were a long way from home and after going over their tactics we made them do it again. Eventually they got the idea, fire and maneuver, and did a reasonable job of following doctrine even though in their hearts they were still dismounted 'grunts' who didn't know or care about the Bradley that they rode into battle.

At the end of the course we had the final exam. Everyone in our class passed except one who missed by one question. He claimed that one of the questions had an element that we hadn't covered. After I checked all of the instructor lesson plans, I went to the SGM and made a request for him to disregard that question. He was skeptical about my claim and went through the same lesson plans until he had to admit that the question covered material that wasn't taught in the course and that allowed my student to pass. Needless to say, the students from Louisiana were impressed that we Texans had done everything we could to make them successful in their course and they all thanked Craig and I after graduation.

A few months later at an RTI assembly I was given an Army Achievement Medal for my work as an instructor in that course. I was disappointed that Craig had not received one, but he had transferred to another unit and the SGM who recommended me for the award had a limited capacity for the paperwork involved in those things.

A Trip to NICI

Captain Brian called me into his office and told me that I was going to the next NGB class on our mission, Ground Observation and Reconnaissance. It was presented by the National Interagency Civil-Military Institute. Designated Mission 5A, ground observation and reconnaissance was the term we were forced to use instead of surveillance because of political sensitivities. Surveillance was defined by the JP 3-0, the DOD dictionary as, "The systematic observation of aerospace, surface, or subsurface areas, places, persons, or things, by visual, aural, electronic, photographic, or other means." The lawyers didn't like 'systematic' so we didn't do surveillance we did 'observation.'

The sensitive ones wanted everyone to believe that the guardsmen conducting this mission were being sent to some general area to watch for whatever might come into view. If they observed something suspicious that could be construed as illegal drug trafficking, they were to call the law enforcement agency that had requested their support and report what they saw.

That was all well and good as far as it went but we had evolved in our support capabilities and the LEA's needed us to systematically watch a specific point on the earth for particular people, vehicles, and activities that when reported would give them reasonably trustworthy information that they could use to make traffic stops, and search vehicles and persons for drug contraband without a warrant.

The course they were sending me to was usually conducted near San Luis Obispo in central California, the land of my people. While I wasn't excited about listening to people tell me how to do this mission, I was excited about going to California and visiting friends who live nearby. I could go to the beach to take some surfing video for my surfing movie called, 'Pacific Coast Highway' which remains 'in-progress' at this time. Brian dashed those hopes when he told me that I would be going to Fort Indiantown Gap (FIG), Pennsylvania.

My friend Stephen who had recently left SOD after telling Brian that he didn't want to waste a second more of his life there, resulting in the departure of our detachment sergeant which brought a huge sigh of relief from the rest of us, was now working for a big defense contractor in Maryland. I asked Brian if I could visit him before I came back and get some information on surveillance equipment. Brian was good with that so a week later my wife dropped me off at the airport and I flew to the Baltimore/Washington International.

Fort Indiantown Gap, PA

Driving a rental car through Maryland and southern Pennsylvania was a treat. It was fall and the leaves of the trees had turned. I hadn't seen such a dramatic change in seasons in years and I stopped and took pictures of some of the more spectacular explosions of color that lined the roads. When I arrived at FIG there were signs at the front gate directing students to the area where the course was taught.

The Northeast Counterdrug Training Center (NCTC) was one of five regional schools that were funded through the NGB Counterdrug budget and were designed to train both military and law enforcement personnel in subjects related to the war on drugs. The 5A course I was attending was required by NGB for any state that was performing the

5A mission. Brian and some others had attended it in the past and now, 10 years after I started this work in Texas, I was finally going to learn the correct way to do it. Only a fool is convinced that he knows everything so for the sake of the program I determined to keep my mind open in hopes of learning things that could better our operations.

The facility was brand new. I checked in and was given instructions on where and when the first class would be held the next morning. The room they provided was comfortable and after getting my uniform ready for the next day I went out and drove around the post. I visited the Indiantown Gap National Cemetery, a peaceful place adjacent to FIG and took some more pictures of the trees.

The next morning, I reported to class in my Class B uniform. There was a seat with my name on it and a porcelain coffee cub embossed with the NICI letters and symbols that were on the letterhead of instructions they sent me before I left Texas. The students were arriving, some captains and lieutenants, but most of them were Air and Army Guard sergeants from the states who conducted the 5A mission.

The course was introduced by a LTC who ran NCTC. He was followed by a captain from California's NICI, the first of the five regional Counterdrug training centers. He and his staff which consisted of a 1SG from Kansas and a SF Master Sergeant from Florida, began to lay out what tactics, techniques, and procedures we should be using for the 5A mission. There was nothing new or earth shattering here.

Most of the information was either obsolete or irrelevant to what we were doing in Texas. Most of our operations were utilizing remote controlled video systems and they were not being used by the other states at that time. The course didn't get interesting until they brought in a LTC JAG officer to tell us about the rules for being armed.

The 'ARMED' Forces

The LTC was a lawyer from the East Coast and he let us know that he and the NGB hierarchy didn't much care for arming the troops. They felt, he said, that the potential for harming the program with a bad shooting incident outweighed the need of the troops to defend themselves if they were set upon by drug traffickers. I was wide awake now. The language that this guy was using led me to believe that I was looking at one of the authors of the new NGB regulation, a draft of which had recently made its way through the Texas CD HQ where we in SOD were asked to review it and give our recommendations to Buckwheat.

As I listened, my suspicions were confirmed that the new rules they were trying to put on us in the new regulation were for the purpose of dis-arming the troops performing the 5A mission. We had some discussion after the first hour of the lawyer's class and I found out that the states west of the Mississippi River were armed and the ones to the east were not. One of the members of an East Coast 5A team told us about a surveillance operation where he was discovered and had to run for his life through the woods while the bad guys fired shotguns at him.

During the next hour the lawyer described a scenario on the border between Mexico and the U.S. where a team was hiding in a hole and watching a crossing. It was dark and some smugglers carrying large bundles walked up on the team and one fell on top of one of the team members. The question the lawyer posed was what should that team member do? He answered himself and told us that the team should drop anything that would impede them and run away.

I had enough of this and raised my hand. He acknowledged me and I said, "Sir, you say we are on an international border, in the dark, and what appears to be a drug smuggler in the commission of a felony falls on top me and I'm to drop everything and run away?" The LTC acted like I was making a repeat of his statement to support it, but I moved quickly to disabuse him of that notion. "Sir, we teach our guys to take control of the situation by handcuffing the suspect and calling the cops we are supporting. We're not going to turn our back on a suspect in the same hole with us and attempt to run away!"

He didn't like that and told us that he was a lawyer and he knew more than we did about the legal ramifications of bad shootings. Before I could reply, another sergeant, a police officer in civilian life from one of the East Coast states stood up and said he agreed with me. That started a lively discussion and the lawyer had lost whatever agreement he might have had with the East Coast people, who were now saying that they should be armed like the guys out west.

I was pretty mad now. We had to spend a lot of valuable time researching all the doctrine and regulations in order to present a substantial objection to the proposed changes in the NGB regulation regarding the arming of troops for CD work. This lawyer was one of the ones behind the gun grabbing intent of the regulation and I wasn't going to let him have a free ride after what he and his buddies had put us through. The overlying concern of these NGB commandos was that the cowboys from out west would shoot some innocent person and congress would cut the funds for CD and they would be out of a job.

The LTC lawyer finally shut down the discussion by giving us a long break for lunch. I joined some others and went to the nearby hamburger place where we traded war stories and complained about the idiots who were trying to micro-manage us from the Beltway.

When we returned, we had classes from the 1SG and the SF MSG. They were mainly illustrating the need for detailed planning and contingency planning for these operations. The 1SG made several references to the way Texas did things. He had been with Brian during a weeklong skull session a few years ago when some of the states came together to provide the people who put this course together with some real lessons learned from the field. He seemed a little embarrassed by the lawyer's presentation, but he was assigned to the course as an instructor, so he had to steer a wide birth around the unpopular opinions of the legal people.

I considered the whole weapons thing that night. When I was in Vietnam, we were issued our weapons when we got there, and we turned them in when we went home. In between we always had them within arm's reach. The ammunition stayed with the weapons along with the hand grenades, claymores, and pyrotechnics. My buddy Alan, who was there several years after me, told me that halfway through his tour with the 173rd Airborne Brigade, his unit tried to take their weapons away from them in the base camp due to all the turmoil in the ranks at that juncture in the war. He said they refused to be disarmed and were allowed to keep their weapons, but the writing was on the wall.

In 1983, when the Marine Barracks in Beirut, Lebanon was blown up, the Marine sentries at the entrance were still trying to put magazines in their rifles as the suicide bomber drove his explosives laden truck past them and into the lobby of the barracks where he detonated the explosives killing 241 service members. The sentries were following rules of engagement that had been dictated by a General in Europe.

During Operation Desert Storm I heard about sentries in Europe not being allowed to put magazines in their rifles. That wasn't as bad as the stories that came back from Iraq where soldiers were not allowed to load their weapons, even after they crossed the border into Iraq.

The culture of the Armed Forces was changing, and it was looking more and more like the dis-armed forces. We had faced this at the beginning of our participation in the war on drugs when our Adjutant General faced down the Chief of Staff of the Army, General Vuono, and told him that if Texas was going to put men on the border they would be armed. Even then, the NGB people tried to saddle us with the doctrine for domestic disturbances and have our troops carry their weapons unloaded and if trouble came, they would be issued the magazines by the team leader; we ignored that. I was worried that if a war came in the future, the troops would have a tough time transitioning from this mind set to war.

We had a mini-FTX on the last day and after that we got our certificates. I had had several run-ins with the LTC lawyer and told him that he needed to consider what the title, U.S. Army that we wore on our uniforms actually meant. "We're the ARMED FORCES," sir! I said with as much emphasis as I could safely voice. "If not us, then who will bear arms for the defense of the country?"

I cringed inwardly at my 'patriotic' rant, but I despised these lawyers who were tying the military in knots not found in the Ranger Handbook. He glared at me with a baleful eye the last time I upbraided him in the hallway during a break and after that he managed to keep enough distance between us that I wouldn't be able to talk to him without witnesses to my disrespectful if not insubordinate attacks on his legal stance regarding the arming of troops.

We took a group picture on some bleachers and I decided to keep my 'secret squirrel' attitude apparent to the others in the class by bending over slightly and hiding my head behind the guy's head in front of me. I gave my email address to several of the sergeants who asked for it and said goodbye. I'm afraid I made a poor impression on the captain from California and the two E-8s that were teaching the class. I let them know that I was pretty sure that Texas had the best 5A operation in Counterdrug and I was annoyed that this 5A course was teaching TTPs that were not up to our standards. Intellectually I still think I was right, but emotions still play a role in military life.

These guys took offense that a student in the class thought he knew more about the subject then they did and that's a difficult situation for a saint to overcome, much less an infantry captain and a 1SG and a SF MSG. I wished I could have made myself a more pleasant individual so I could enjoy the camaraderie with my fellow soldiers like I had before I went to Vietnam; but I found myself becoming more and more bitter about what I perceived to be fluff replacing substance and I couldn't ignore it.

Washington, D.C.-The Technology Exchange

I checked out, dumped my things in my rent car and drove to Washington, D.C. The trees along the road were even more spectacular in their fall colors than they had been a week earlier, and I had to concentrate while driving so I didn't have a wreck while admiring them and shooting video.

I was going to meet Stephen for a 'technology exchange.' He was working for a large defense industry company as a communications equipment trainer. He taught Special Ops commo people from SF, Navy SEALS, and Air Force Special Tactics how to use the new radios and other high-speed devices being made by his company. He traveled around the world continually, training troops in the use of equipment that was so complicated that the services couldn't retain enough experts to do their own training.

He was staying at a big hotel in Crystal City. They called his room from the front desk and he told me to come up. We sat in his room with a view of the Potomac and caught up. I gave him a rundown of the 5A course. He laughed and said that I would be complaining about the Army while the undertakers were trying to stuff me in my coffin.

We took the Metrorail from a station near the hotel and went downtown. He wanted to see some things at the Smithsonian, and we took a look at some other places on the mall. Eventually we got tired of walking and took the Metrorail back to the hotel and ate. He told me about some of the trips he had taken recently and what he was planning to do next. He was in love with a wonderful woman in Texas, but they just couldn't agree on certain conditions he had so they could get married. I encouraged him to give in to whatever it was she wanted because it was better than being alone. I could see that he was miserable, but he was a stubborn guy.

I had to go back the next day and he had to go back to his office in southern Maryland. He gave me a couple of brochures from a company owned by a friend of his from his days in Delta. This was our 'technology exchange.' Brian was totally involved with a manufacturer of microwave video equipment on the West Coast so I didn't think these brochures would go anywhere but to the stack in Pat's office.

I said farewell to Stephen, one of the best soldiers I had ever met, and drove toward Baltimore. I turned in my rental car that afternoon and continued to read 'Blackhawk Down,' by Mark Bowden. Stephen had been with Task Force Ranger in Mogadishu during the events described in the book and I was looking to see if he was mentioned but Stephen was a true 'Quiet Professional' and while his work was mentioned he managed to stay out of the limelight.

When I told him that I was reading it and offered to let him read it he declined and expressed disdain for the Delta operators who had obviously talked to the author. I told him that all those 'secret' units needed some publicity when it came time for the budget and for officers to get promoted but he was adamant and said he wouldn't read it.

He had a particular dislike of a certain General officer who had been in the unit when Stephen was there. During Desert Storm, while working in the Delta TOC he overheard the CO telling a senior officer, who had gone into Iraq against the CO's orders, that he could submit his name for a CIB (2nd Award) but that if he did the CO would court-martial him for disobeying an order. When this officer left the unit he gave an autographed picture of himself to each squadron's lounge for their picture collection and Stephen delighted in telling me how he watched another senior Delta officer collect every one of those pictures the next day and deposit them in the dumpster.

Promotion to SFC E-7

After AT I began to sniff around about my impending promotion to SFC E-7. I had attended all the schools, and even though the Academy was exempt from the STPA order of merit list, my name was right at the top for my MOS. After some unsuccessful attempts to elicit information from the unit clerk, I began to badger my Branch Chief who was now MSG Tony, the same man who told me I should transfer here. He seemed to enjoy my sniffing around the issue and after I broke cover for a frontal assault on the question, he laughed and told me to go to the CSM. I was really perturbed.

First of all, I never talked to the CSM unless he spoke to me and secondly, in my old Army you would never ask about your own promotion. MSG Tony kept prodding me to go to the CSM so finally I decided that this must be the way the Army did promotions now. I went to the CSM about the matter and he told me it was none of my business and to get out of his office. I was really confused now because we had an ANCOC class starting in November and Tony told me that he wanted me ready for the podium. I stopped talking about the supposed promotion after the CSM shamed me and threw myself into getting the student materials ready for the next course.

When the November drill came along, I went to formation with everyone else and was sitting in the soft seats of the auditorium nodding off when I heard my name called. I looked up, startled, thinking that I had been spotted sleeping by whoever was talking. One of the guys next to me was amused by the look of alarm on my face and told me, "Get up there, they're calling your name." I jumped up and made my way to the front and stood next to one of the other sergeants in the unit. "Attention to orders," said the adjutant. "The following individuals are promoted to Sergeant First Class." When he said my name, I felt a wave of shame and relief; shame that I had become a bit of a careerist in the last few months and relief that I actually got promoted. I vowed at that moment that I would never behave like that again.

I went to work for MSG Tony who was now Branch Chief for both ANCOC and BNCOC. He needed a senior small group leader to help him support and ride herd on the small group leaders for both courses so he told me that I would be a backup instructor but my main mission was to help him with the admin side of NCOES. This was the start of my education for the work I would be doing until the war.

The Academy-The Ethical Dilemma

The problem of who was responsible for preparing the student materials had not been resolved. The full-time admin sergeant and MSG Tony were at polite logger heads about this issue. The A and BNCOC instructors were almost all from the Army Reserve unit. This convoluted arrangement was the result of some past reorganization which resulted in the National Guard providing the facilities and administration and the Reserves providing instructors.

It sounded good but the reality was that the Guard was held responsible for the success or failure of the academic outcome of the classes, but they had no authority over the instructors, mainly their NCOERs. If you can't rate a soldier all you can hope for is the soldier's professionalism in a case like this and that was the problem.

The USAR contingent was run by a MSG. He was a slick, slightly overweight NCO 'king.' He ran his unit by doing 'favors' for the troops under him in return for their total allegiance and loyalty. These 'favors' included promotions and schools and if they didn't kiss the backside of this guy they weren't going anywhere in their careers. I began to have contact with him when I delivered messages regarding instructor issues from MSG Tony. He let me know the first time I talked to him that he outranked me. That told me what I needed to know whenever I had to evaluate anything he said to me regarding his responsibilities for the success of our common mission. He was a snake, and I kept my eyes open whenever I was around him.

The instructors were a mixed bunch; some good and some pathetic. They were caught in the middle between the demands of the Guard's administration of the Academy and the USAR administration of their careers. As I began to sense the friction of war, in this case, soldiers rubbing each other the wrong way, I eventually came to the realization that the whole Academy was stuck in an enormous ethical dilemma.

The leadership manual we were using at this time was FM 22-100 (July 1990). On page 30 are the Ethical Responsibilities of an Army leader. One of those responsibilities is to avoid creating ethical dilemmas for your subordinates. Because the National Guard was in command of the Academy, the majority of the ethical dilemma rested at its door. The USAR had their share of this pie but much of what was wrong could only be corrected by the Guard. I saw this as a duty and a challenge and decided that since no one seemed to be that interested in doing anything to fix the problem, I would do whatever it took to improve anything that I was involved in.

The first event that took place when the students arrived was enrollment in the course. This was a major hassle because each student was required to have a number of documents before they could be enrolled. One of the biggest problem documents was the physical examination. Getting a physical was almost impossible unless you came to Camp Mabry and then it was just very difficult. The Guard only gave physicals every four years and the requirement for these courses was a physical within the last year.

All of the documents had to be placed in the student's file that was kept at the Academy and would be inspected by the accreditation team from USASMA every two years. Incomplete student files could cause the Academy to lose its accreditation and that would be unacceptable to the Academy CO so the CSM did everything he could to make sure that every student file was complete and correct.

The biggest problem here was that he had to use the USAR instructors as a part of the in-processing because of the large number of students who showed up on the first day of the course. They were supposed to attend a briefing on the procedures and standards that were to be followed when in-processing began on Saturday morning, but the USAR MSG informed us that they wouldn't be available until 30 minutes before we started. The admin people would try to brief them as best they could in that short amount of time but that's where things started downhill. Many times, the students would be missing a certain document, or it would be out of date. Some of the reserve instructors had a pretty cavalier attitude toward these details and let them slide.

When the students arrived, they were checked for height and weight. This was always a drag because many of the students were older guardsmen who were right on the line as far as their weight in relation to their height and their age. Horror stories are told about the extremes that soldiers went through before they did height and weight for a school like this. Hanging upside down from a contraption that held you by your feet for a few hours every day was one and starving yourself for a few days was another way of getting past the gatekeeper. Another way was to stand on your tip toes when they were measuring your height which I caught one plump SSG doing that first morning.

"What the hell are you doing, sergeant?" I yelled at him in the latrine where we were doing the measurements. Just like the criminals I used to arrest he immediately lowered his heels and said, "Whaaat?" I told him to get measured again. He did and now he was overweight. This meant that he had to be taped. The regulation prescribed a complicated formula for determining body fat content by a series of tape measurements of the neck and abdomen. The CSM had a couple of sergeants handling the taping and I didn't hear of anyone failing that, so I guessed that he didn't want to be the bad guy for what was a Guard/Nation-wide problem, too many fat boys.

After in-processing came the PT test or APFT. The student was required to do a certain number of push-ups and sit-ups in two minutes and run two miles in a certain amount of time. The standards were determined by the soldier's age and gender. The biggest problem with the PT test was the push-ups. The description of the correct push-up was read to the ones being tested and it was demonstrated but when the timekeeper called "go," any number of permutations of the correct push-up could be seen. Many times the one being tested started out good but as they got tired their form changed but it changed so slowly that the grader would be hard pressed to determine when the push-up no longer met the standard and most would count every one that looked like a complete push-up.

The sit-ups didn't present as much of a problem and the integrity of the two mile run was pretty easy to maintain except where the track had a loop into the center and it was possible to run across a short distance and avoid the loop thereby gaining some time. Anyone who failed the PT test was given a retest on the fifth weekend of the course which gave them four months to get in better shape.

After the PT test the students were issued their box of manuals and turned over to their instructors who started their training. During the first weekend I spent much of the time running from classroom to classroom getting lists of missing manuals from the student boxes. I would take these lists to a warehouse where we stored all our manuals. As soon as I filled an order, I would deliver them and go to the next class and get another list. Some of the manuals were not available and I had to make copies on our copy machine if we only needed a few or I would have to give a request to the full-timer to take to the print shop at the Camp where they could do large print jobs.

Attention to detail is a term that is bandied about in the Army, but the Academy was where I learned the practical application of that slogan. If we had 120 students, that meant that we needed 120 boxes of thirty manuals each. Each of those manuals had to be up to date with all applicable changes. Some of those manuals were ten years old and had six changes. Getting all that straight was just part of the Gordian knot I was trying to untie during that course. Since we were shorthanded, I was doing these things for both BNCOC and ANCOC and by the end of the weekend I was glad to get back to the war on drugs.

South Texas: Pat C finds the Keys

Obey the ones leading you and submit to them, for they watch over your souls as those who will render an account, that they may do this with joy and not groaning; for this would be unprofitable to you. Hebrews 13:17

Chapter 23-1999

SOD-Command and Control

At the same time I was beginning to feel like I could have a positive effect at the Academy, I was beginning to feel like I was losing my effectiveness as the Operations Supervisor in SOD. Captain Brian and I were arguing about everything. He said, "Why do you always argue with me?" I replied, "Why do you always make me argue with you?"

Since we had created permanent teams, I was not going to the field as much as I used to. Brian had me give a series of classes to the detachment on the duties of the mission commander (MC) and we had developed a checklist that could be used by anyone that the CO deemed ready for this responsibility.

The MC duties, which were critical to the success of the mission, both in rapport building and planning, and in the execution of the mission, were now being carried out by almost anyone in the detachment. We had achieved the highest goal of command and control in modern warfare which is unity of command and decentralization of execution. I had worked myself out of a job.

The Unit had been started by NCO's schooled by Bil, a Special Forces operator who had retired as a Master Sergeant. He appointed the officers and told them that their job was to interface with the senior officers in the Task Force and to make sure that the NCOs had what they needed to plan, conduct, and command the missions.

The first officer performed this role and made improvements while having to deal with the effects of the train wreck on the operators. The next officer was younger and he was a tanker. He was able to see some things that could be improved and as a result of his leadership, they did.

The Captain was promoted and left in 1996. Now, Brian and JT were spreading their wings as officers. I think that an officer should be an NCO before he is commissioned. In practice this can only work if the new officers are guided by older officers in their chain of command.

Brian and JT were in a unique situation. The CD command trusted us to run our operations with minimal oversight on their part.

G-Company, where Brian and JT had been assigned, was commanded by a captain who had to run as fast as he could to keep up with this dynamic duo, so, as a result, Brian and JT received very little critical feedback when they were 2nd Lieutenants and needed it most.

When Tommy was the Detachment Sergeant, it wasn't like this and he and Brian had a healthy dialog in which Brian respected Tommy's abilities and stayed out of his lane. Tommy was long gone now and working as an FBI agent in California. The detachment sergeants since then found it difficult to stand up to these young officers, by now captains and full of their 'officerness.'

SOD-Officers

I got the impression from the modern rangers I had met since I came back in the Guard that the Ranger Regiment is a world of total uniformity. Every ranger carries his equipment at the same place on his body or in his rucksack. This is useful in combat when you might have to resupply your ammunition needs from the dead body of one of your comrades while in the dark.

In order to achieve this kind of uniformity in SOD they sat down and designed a load bearing vest system with a number of features that allowed the wearer to configure it for either a shotgun or rifle. It had some pouches and was functional for our mission but when they proudly issued them, the guys decided that they hated the vest, mainly because they had not been consulted about the design which was different in appearance from most vests being worn in the special ops world. The guys had been working with the freewheeling narcs too long where the rule of thumb was, 'let those who ride decide.' When Brian found out about how the men felt about their design he got mad.

The next thing they tried to develop was the computer program that would do everything. It would track everything in SOD and connect it all together. The only thing like it that I know of was the machine built by a team of engineers in the Soviet Union in the 1930s. Before Stalin had them all shot, they designed and built a machine that was supposed to run the Soviet Union. The story was told brilliantly by Adam Curtis in a documentary called "The Engineers Plot."

I digress. Brian and JT took over the backroom of the SOD area. They hung large white dry erase boards on all the walls and went to work. For several weeks, after the staff meetings, they would retire to the backroom and work on their masterpiece. I looked in there a few times out of curiosity, making sure that they were out of the area and not wanting to get involved. There were diagrams that went all the way around the room with lines going up and down and sideways out of blocks and dotted lines running behind the solid lines and blocks.

I could see that they were trying to track personnel, equipment, operational information, vehicles, leave, and many other aspects of our life. They began to run out of gas after a couple of weeks and their irritation with the rest of our activities was becoming a source of contention within the detachment. After while everyone learned not to inquire about their progress as their grim faces told us all we needed to know. One day Brian took part in an operation that took him out of town for a couple of weeks and when he returned, he never went to the back room for the big project. The boards were left in case they decided to pick up where they left off, but they had had enough.

1999: SOD-High Tech Rednecks

The unit was going more and more High Tech. In 1994, we received six complete Thermal Image sensors which allowed us to detect and monitor a human being in total darkness at ranges up to 5 miles. These devices greatly enhanced our capabilities on the border. We also began to use more high-tech devices that belonged to the law enforcement agencies we were supporting. Sometimes they would show us some equipment that they had but that no one in their office knew how to use. They had systems that used video cameras that could be concealed near the target and they could transmit the video images using microwave to a distant location where it could be monitored without fear of discovery. The operator could pan, tilt, zoom in and out, focus, record, and make video prints from his location.

We began to do research on these types of systems and eventually we purchased one. As we employed this system, we began to see that some of our operators, who were very good at packing a rucksack and hiding in the brush for 10 days at a time, calling in spot reports and conducting resupply operations, were not technically oriented. They were airborne rangers and proud of it. High-tech was a different matter. We had to spend a lot of time in training to raise the level of proficiency to a satisfactory level with each new piece of equipment.

Our Motorola radios could be programed to any frequency in their range but most of the operators couldn't grasp the proper way to do these tasks and eventually we had to have all the radios programmed by Pat, our tech guy, before we deployed. If there was the need for a change in the field, we were on hold until Pat could get to our location and make the changes. This is an illustration of the climate in the unit. Rangers like to push themselves physically and coddle themselves intellectually. Most of the tabbed rangers and those who aspired to be tabbed didn't like to read. The high-tech that they were most interested in was the newest video game system.

The problem was that it was hard to work with high-tech systems when you have been awake for twenty hours. The best way to prepare for a mission where you are going to have to install a high-tech video system at 3 in the morning, one that will transmit the video signal to a distant location by micro-wave, is to go into a reverse cycle mode three days before the install time so that when you are trying to set up a complex system, you are at the top of your work cycle not at the bottom of your sleep cycle. The ranger mentality insisted that you should overcome everything that tried to prevent you from accomplishing your mission by the force of your will. Brian was the leader of the unit and he knew that the men preferred to get the weekend off prior to the mission; even though he knew that they would drive all day and perform the complex tasks required after being up for 20 hours. The guys wanted their weekend off before the mission because they had been sent on back-to-back missions enough times to believe that they wouldn't get any down time after the mission.

SOD-Sleep Loss

I was studying the information that the Army had on this situation. It was in a manual titled, "Soldier Performance in Continuous Operations." The first edition of this manual provided a wealth of information based on tests done by the U.S. and U.K. armies. The fact was that you become less efficient when you don't get enough sleep.

We were having continual problems with broken equipment. The guys would make up all kinds of excuses but in the end, it was obvious that they failed to follow proper procedures when setting up these systems. A common problem was attaching the wrong power cable to the wrong terminal. There would be a flash of bright light, a cloud of blue smoke, and a useless piece of high dollar equipment that would have to be repaired, costing a lot of time and money. I tried to explain to the guys that if they were honest, they would admit that many of their mistakes that were made during the crucial installation phase of the operation were made because they were exhausted by the long drive and no sleep. Many of them agreed with my analysis but they insisted that they were tired of being burned on their days off after the mission was over so they would rather take the chance of screwing up the operation than to miss some downtime with their families.

I wrote a long paper on the subject of 'Time Management' and gave it to Brian and he came back a few minutes later with the words, "you don't have time for this," scrawled across the top of the page. It was funny, his choice of words, but I wasn't laughing at the time.

It wasn't until after I retired and re-looked at the whole sleep deficit problem that this was a major contributing factor to what had happened to my unit in Vietnam. CRS, or 'can't remember shit' is how Colonel Hackworth put it in his own memoir, "About Face." The institutional memory from one war to the next doesn't exist.

SOD-Gun Policy

NGB had been adding more restrictions to their regulations. They wanted the law enforcement agencies to request that we be armed when supporting them. This was a blatant attempt to disarm the states that armed their CD personnel. The lawyers on the Potomac knew that most law enforcement agencies would see this as an attempt to put the liability for a guardsman's actions on them and they had plenty of lawyers too. We went to Buckwheat and he agreed to fight this move and he was eventually able to get it changed. We operated in an environment where everyone was armed to defend themselves and it was unreasonable to expect men to be able to make decisions under the extreme stress of an armed confrontation when they had to think about whether they were carrying their gun that day.

SOD-CD PAO

In August, I saw an announcement for a public affairs officer position on the CD staff. I applied for it but they gave the job to someone else. I imagine that this gave Brian the impression that I wanted out of SOD, the die being cast, so to speak.

SOD-My Last SOD Operation

In early December I was told to help Sonny, one of the SOD staff. The APD narcotics officers wanted us to help them do surveillance on a suspected drug dealer's apartment in one of the city's low-income housing projects. Pat, Sonny and I did a recon and decided that we would park a vehicle with a concealed remote-controlled camera in a parking space that had line-of-sight on the front door. I was chosen to drive in and park the car. We had to get there early enough to get that particular spot so I put on some old clothes covered with paint and made myself look like the rest of the workers who were doing some remodeling work in some of the other apartments. I had my favorite bowling ball bag with a shoulder strap around my neck. In it I had my Beretta, a Saber Radio, and several 'tall boys' for window dressing.

I entered the complex early on the morning we started the operation and got the parking place we needed. Sonny was nearby and I adjusted the camera according to his directions while he watched the picture in a small monitor attached to the receiver that was picking up my video signal. I was trying to look casual, sipping on a 'tall boy' that was in a small brown paper bag, as I adjusted the panning limits of the camera. Finally, he called me on the radio and said, "Good to go."

I had to walk through the HQ in order to get to my office where I could change my clothes. I was still carrying my 'tall boy' and when Brian saw me, he called to the other SOD guys in the area and told them to look at the best urban camouflage he had seen lately. That made me feel good enough to overcome my anger with Sonny, who, the next day told everyone he could that I had screwed up his operation by aiming the camera in the wrong direction. Brian asked him if he had told me we were "good to go" and while he did admit that he thought it was good, later he found out that it wasn't. "So, whose fault was that?" asked Brian. Sonny just walked away.

Remind them to be subject to rulers, to authorities, to be obedient, to be ready unto every good work Titus 3:1

Chapter 24-December 1999

Counterdrug Operations Section

In December of 1999, Captain JT approached me in the hallway and asked me if I would be interested in transferring to the J3, the Operations Section of the Counterdrug Task Force. Looking back, I would guess that Brian must have told him about the problems we were having and for some reason, JT decided to invite me to come to his section. JT was a hard man to get along with and he knew that I wasn't the most compliant person in the building. I had known him for 12 years, from the time that he was a rear rank private at LRSD, all full of himself like the rest of the young paratroopers. He was a good soldier though, graduating from OCS and successfully completing the Ranger Training course at Ft Benning.

On-Call Operations Sergeant

Captain JT told me that he wanted an experienced operations person to handle the on-call missions. These included the C-26, RAID, and SOD. My job would be to review the mission request and make sure that it met all the requirements for our support. It had to have a drug nexus, it had to be on an agency letterhead, and it had to be signed by a law enforcement agent. I agreed to transfer to the Task Force operations section after I came back from leave.

I decided to go to the land of my people, California, for the millennium change that was causing a lot of concern in certain sections of the Guard. Everything from the end of the world, the second coming, and anarchy in the streets, not to mention the fact that our computers might have a hard time computing the new date. Before I left, I was ordered to call in every few days so that they could recall me if all hell broke loose on New Year's Day.

That was the end of my Sneaky Pete operations in the war on drugs. I cleaned out my office in SOD. I had been there for ten and a half years. It had been a wild ride occasionally and I enjoyed working with the best soldiers in the Texas National Guard. Sonny did get his revenge by screwing up the engraving on the shotgun that was presented to me as a going away memento along with two different plaques that were also screwed up with misspelled names etc.

The CODEL

Right after I transferred to the CD Operations Section, we were told to prepare for a CODEL. After finding out that that was a congressional delegation, I also found out that I was assigned the task of writing an operations plan for how to support it. Nine congressional staff people and six military escorts were coming to Texas to look at how we did Counterdrug. One was from the office of the House Majority Leader and another was from Senator Strom Thurmond's staff.

We were to give them briefings, take them to the Capitol, take them down to the border, and feed them some Mexican food and BBQ. The entire visit from picking them up at the airport, taking them to places in Austin, flying them to Marfa and El Paso, and returning them to the airport was an intensive logistical effort which required detailed planning and coordination.

My years at SOD had prepared me for this and I finished a basic OPLAN with contingency plans for all possible problems to include search and rescue operations if their plane went down in Marfa. The staff did all the driving in the Austin area and the CD people in El Paso took care of the CODEL when they visited the El Paso Intelligence Center and looked at cargo inspection at the Port of Entry.

Buckwheat was a veteran of political activities after his one-year stint at NGB so he made sure that everything was done to make these people happy. I drove one of the vehicles and ate with the staffers several times. Most of them were young political science majors who were getting their first taste of politics by doing the staff work for the senator or congressman that hired them. They were supposed to absorb information about the Counterdrug program to help their boss know which way to vote on budget and policy issues in the congress.

We tried to show that we were doing great things to stop the tide of drugs and if we had more money we could do more. I talked to a few of them while we ate but I kept my more contentious views to myself. The whole thing was becoming more than I could stomach as far as what we should do tactically. I wished that the government would start at the top and make the different agencies work together, clear up the ridiculous body count for the budget shell game and a few other bullets on my shit list of what's wrong with the drug war. I was touched by their youthful enthusiasm and hope for a better world, but my own experience had led me to believe it was hopeless.

CDTF-Staff Warfare

After the CODEL, the Operations section stayed busy. Just before I got there, Captain JT had two of his SOD buddies who were now LTs, working in operations. Then a LTC was added as the Operations officer and Major Sue, a long time CD member who had been an Intel analyst in the field was made the Deputy Operations officer.

The LTC only last for a short time before he was shipped out to El Paso where he finished his time and retired. Major Sue remained as the Operations officer and she let JT handle most of the operations issues.

Captain JT wasn't lazy and in his spare time he figured a way for the Counterdrug Task Force to partake in the Texas Civil Codes Asset Forfeiture procedures. This meant that on the cases we supported, we would receive our share of the convicted drug trafficker's money and the real estate and property that was seized and auctioned

CMIS

Preparing our paperwork for inspection by the National Guard Bureau was one of the things we did along with keeping up with CMIS, the NGB Counterdrug Information Management System, a database that tracked every aspect of the CD program. NGB claimed that we wanted the program and our superiors claimed that they didn't want it.

The data standards were non-existent. One requirement was for each state to show how many CD missions they were completing each year. Texas assigned a mission number to each Port of Entry on the border. There were up to 30 soldiers and airmen inspecting cargo at those ports. The port got one mission number for the year. Another state had a person inspecting packages for the U.S. Postal Service. That state would assign a mission number for every package that person inspected. Texas got a mission number for 30 people working for a year and the other state got a mission number every time their person picked up a package on the conveyor belt.

When CMIS went on-line, the system was very slow. It would take five minutes to get the page you needed to put data into. This was on a fast day. There was a young sergeant named 'HeyZuez' who was the data entry guy and he was totally frustrated with the situation. He would take his work home with him because he had high speed internet access at home. Captain JT would run through the building trying to catch people who were watching TV on the internet at their desks because they were clogging the bandwidth. All of this was a drain on morale and a source of irritation between me and JT. I would go into his office and yell at him and he would yell at me and then he would go talk to the higher ups. This went on and on.

Candidate N, who was my supervisor, was going to OCS at the time and he was totally burned out. He couldn't bear to look at CMIS or the mission paperwork that he had been responsible for. He left me two file cabinets with a huge mess and acted like it was my job to fix it. Shortly after we got that problem solved, Captain JT developed a paperless system for operations and purchased an HP Digital Sender that could scan fifty pages at a time. As soon as we got a fax or some other paper document, we scanned it into a file system that was backed up daily and then we shredded the hard copy. The file cabinets were reduced to one and the remaining one was used to store two or three data DVDs which contained all our records.

When the NGB inspection team came to evaluate our program, they were upset that we didn't have any hard copies of our required paperwork. Captain JT took them on a journey through all the regulations that NGB had written for the administrative requirements of a CD program and challenged them to show him where anybody needed to have a 'hard' copy of anything, and they couldn't.

SGM Salt was the lead inspector and from what I heard, he was a guiding light in the CMIS boondoggle. I told him that the standards for data entry were non-existent and JT hammered the same point only a little more politely. This resulted in the SGM arranging to have a conference with representatives from some of the larger state programs, including ours, where they could 'work' these issues.

The Academy-BNCOC/ANCOC

Midway through the ANCOC/BNCOC courses, a new MSG, was assigned as the ANCOC Branch Chief and was now my boss. I had been running around like a man with his hair on fire, trying to keep up with MSG Tony's requirements. One of the biggest was the need to get the student materials completed. The boxes of manuals were incomplete and after getting through the student enrollment and PT test, I spent the next four weekends going from one classroom to the next and having the instructors stop training and conduct an inventory of the student boxes so that I could get a list of manuals that were needed so that each student could have a complete set as per regulations.

Each time I had compiled a list of missing manuals, my new boss would tell me to give it to him, he would take care of it he said, and then he would proceed to lose it. He was a nice guy with lots of airborne badges on his uniform and a hooah, hooah attitude when he spoke to the troops, but I had lost all respect for him months ago. Since he was my immediate supervisor, he was the biggest problem with ANCOC. This happened three times before I finally kept the list, found the manuals and issued them to the students one weekend before the final exam. Needless to say, the test scores were low and 30% of students failed the exam twice, and so they failed the course.

The last thing to be done at the end of the course was to make sure that the student records were straight. This was a major tooth pulling operation with some of the USAR instructors. Two of them failed to give end of course counseling and the forms were either not signed or not in the student folder. The Academy would have an accreditation inspection during the next class, so this was a big problem.

The two instructors started out by playing dumb and then, when I pressed the issue, they became defiant. I went to MSG Jeff about the insubordination of one of them and he told me, "He's like that with everyone." We ended up mailing the counseling forms to the students who signed them and sent them back. I stayed on the records until we got them all correct. Then I told my new boss to ask the CSM to remove the two instructors in question from the ANCOC podium and he did. They went to BNCOC where they continued to be sub-standard sergeants teaching sergeants. I wrote a memo to him recommending that we get rid of the USAR instructors and put Guard instructors on the podium where we could train them and write their NCOERs.

2000: The Academy-Training Future Leaders For War

We were given a heads up that there would be a completely new Course Management Program-Program of Instruction that would be required to be taught during the next class in November. I got my hands on the course material and noticed that they had taken out rifle marksmanship, added Joint Operations, and increased the hours of Sexual Harassment and Equal Opportunity Training. The people at TRADOC had a vision of things to come but they weren't giving any direct information about it so these were the tea leaves that I had to read and the message was pretty dismal.

Since I wasn't able to have any effect on the direction of the Army I decided to do what I could to improve the way we taught. One of the problems included the lack of an SOP for the instructors on how to teach the course at the RTI so I wrote one. In it I covered everything that I had encountered as a BNCOC instructor and as a senior small group leader helping MSG Tony and my boss.

One of the most time-consuming tasks for the instructor was student counseling. Each student was required to have at least 4 counseling forms in his student file at the end of the course. If there were any problems, there would be more. I decided that things would go quicker if we standardized all the counseling forms from Initial Counseling through End of Course Counseling. This was a massive effort but after teaching myself to create forms using MS Word, I was able to make them correct in both appearance and content.

I was on a roll and made them for BNCOC also because I was attempting to get NCOES standardized for the RTI. The new BNCOC Branch Chief was happy to take and use the counseling forms but he resisted many of my other initiatives to the point where we would have had a fight and he would have kicked my ass. I was a lot older than him and decided that it wasn't worth going to war about.

The next thing was to get time with the USAR instructors to teach them the SOP. Standing Operating Procedures are something near and dear to all staff people. They are required in many units and I had written several for both LRSD and SOD. The thing with SOPs however, is that they are not an SOP unless everyone in the unit knows the procedures and practices them.

I talked to MSG Jeff and practiced my best rapport building techniques when I tried to get a weekend during our off cycle, June to October. I wanted to have all the instructors sit down and go over everything before the next class. He wasn't very enthusiastic and told me he would have to 'look' at his training schedule and get back to me.

I knew what that meant so I went to the CSM and showed him my SOP, Handbook, and Counseling Forms. He was impressed and agreed that we should have a time to 'train the trainers.' I gave him a copy of my proposed training schedule for one of the weekends during the off cycle and told him I had spoken to MSG Jeff who wasn't very enthusiastic but hadn't said no.

2000: The Academy-The Ethical Dilemma Continues

The CSM said he would get with MSG Jeff and square things away. Not! The next month I asked him how it had gone with MSG Jeff and he said that the USAR had a full schedule each weekend until the next course started. Now I was getting mad again. I reminded the CSM that the USAR troops were getting paid for a unit training assembly each Friday night before drill and asked him if I could ask MSG Jeff for a couple of hours on Friday night during each drill weekend and he said OK but he didn't think that it would do any good. He was right.

MSG Jeff explained to me that yes, he gave them an extra day's pay each drill as a way of compensating them for having to drive to Austin for weekend drill but he didn't nor would he require them to do anything on those Friday nights. My brain was starting to boil. I looked up the regulation for what he was doing, and it was clearly illegal but who was going to do anything about it.

I had a mission to teach National Guard and Army Reserve NCOs how to be better NCOs, capable of running a platoon of soldiers. The training they were getting at this Academy was substandard and there was little I could do about it if I couldn't affect the way the instructors operated in the classroom. This started a small scale war between me and the USAR MSG who for his own petty reasons decided that he would do whatever it took to put me in my place and let things go on as they had for the past few years.

The instructors were in the middle and I soon lost respect for them because to a man they were more concerned about staying on the good side of MSG Jeff than they were about living by the Army Values, which in 1999 replaced our Army Ethics. It's amazing how all the good intentions of an Ethical Code can be negated by an unethical individual and a group of American fighting men and women who have been corrupted by their leaders.

Later during the off cycle, MSG Jeff gave a PT test to the instructors and several of them failed, I had my spies now. He pencil whipped the score cards and they all passed. I went up to the office that we allowed them to use in our facility and I saw one of the ones who failed grasping MSG Jeff's hand like he was shaking hands with the Pope and thanking him so profusely I wanted to vomit.

I asked my source later what that was all about and was told that MSG Jeff had just put the man's packet in for promotion after he failed his PT test. That was too much. This guy, when he got promoted, would become a BNCOC instructor so I went to the BNCOC Branch Chief and told him what was going on. He didn't like hearing about this but he had recently transferred from the USAR unit and didn't seem to want to engage MSG Jeff, so I took it to the CSM.

August 2000: Camp Beauregard, LA-NCOES Conference

In August, I was ordered to attend the annual NCOES Conference at the Louisiana National Guard Academy. They were our higher headquarters in the convoluted command and control arrangement of the Guard and the Army Reserve under the Sergeants Major Academy, and the Training and Doctrine Command (TRADOC). I was very unhappy with the way the last two classes had gone at the Texas Academy. I sat in an auditorium and listened to one briefing after another about how we needed to make little changes here and there to improve our greatly successful programs; how we were educating the NCOs of tomorrow by doing what we had always done, which was little.

I kept waiting for someone to address the glaring problems in Texas which were apparent to me but seemed to escape the observations of those from this headquarters who had inspected us recently and the rest of the higher ups in the NCOES system. I was humiliated back at SOD when all the NCOs I worked with told me that Texas had the worst

Academy they had ever seen and that they would rather miss an opportunity to go to a school and get a promotion than go to school in Texas.

I sat in my seat and boiled until someone brought up the sterling job that the Reserves were doing in providing experienced instructors for the Texas Academy. As soon as the speaker finished unloading that boat load of manure, I raised my hand and stood up to take umbrage with everything that had been said except for the fact that the instructors he was talking about were from the reserves. My main contention was that they were not experienced as squad leaders or platoon sergeants as USASMA required because the reserves didn't have squads and platoons of soldiers to lead.

As soon as the briefer could sit me down some of the Reserve people made a passionate defense of the instructors and the Reserve system and the NCOES system as it existed. Since I was a low ranking NCO who had no business saying anything except hooah! at the appropriate moment, people sitting near me began to lean away and after the break I returned to my seat and found I had plenty of room to stretch out because all my neighbors had moved. I was outraged at this point because I figured out that no one was or would do anything about the state of chaos at the Texas Academy, but I couldn't stop myself from approaching the TRADOC sergeant major at the next break.

I put the bottom-line up front and told him, "Sergeant Major, Texas is all fucked up!" The sergeant major was black and he looked at me carefully, as if he was looking at a crazy white man and now I realize that I may have given him that impression because no one had ever aired their dirty laundry like I just did to someone in his position.

He recovered and asked me what I meant by that and the floodgates opened. I tried to keep it short but there were so many problems competing for some face time with someone who might give a shit, I blabbed away until he cut me off because the next briefing was starting.

The rest of the conference was a waste of time and taxpayers' money as a far as I was concerned but it was useful for the maintenance of the system in place which I felt should be destroyed and rebuilt. We were all given handouts and floppy discs with the new NCOES Battalion SOP that was written in Louisiana and ignored in Texas except for the paperwork requirements that made the difference between success and failure as far as USASMA was concerned. As to the actual quality of the education, no one wished to discuss it.

The Academy-A Visit To The CSM

When we got back to Austin, during the next drill, my boss told me that I had screwed up big time and that I better learn to keep my mouth shut if I wanted to get anywhere. Then we were both told to report to the CSM. I walked into his office with my boss who closed the door behind us. The CSM's face was redder than usual and he stared at me for a moment before asking, "Sergeant Murry, did you tell the TRADOC Sergeant Major that Texas was all fucked up?"

I leaned over his desk and replied, "Sergeant Major, Texas is all fucked up!" My boss was looking real uncomfortable while the CSM sputtered about how we never mentioned internal problems to outsiders, etc. while I stood there quivering with anger and fear. The anger side won the fight within me and when the CSM stopped schooling me on the ways of a proper NCO, I went off and aired my list of correctable problems that had remained or gotten worse since I came here.

He listened for a while and when I took a breath, he held his hand up and told me he had heard enough and dismissed us. My boss took me outside and again tried to explain to me that I needed to learn how to get along with the system because there was nothing I could do about it. I just stood there silently, deciding not to waste my breath and wondering where I could transfer to now.

The Academy-Screw Up and Move Up

The Lord was watching out for me because instead of firing me, the CSM submitted my paperwork to the Sergeants Major Academy and I was certified as a small group leader for ANCOC. Seven days later I received certification for senior small group leader, ANCOC; my time on the podium had come to an end. Then, since my boss had moved on to greener fields, I was told I was the acting Branch Chief for ANCOC.

The USAR ANCOC instructors were not an easy bunch. Vern, an older paratrooper with time in the 82nd was skeptical about anything I had to say. George, a former helicopter pilot and then tank commander with the 124th Cav was playing his cards close to his chest. William was a cop in north Texas and was distracted by some shit storm in his department and Randolph was a local guy who was taking a wait and see attitude. Robert M was the senior small group leader and he wasn't interested in doing much more than the minimum necessary to get by.

I sat down with this group a few drills before the next course started. MSG Jeff gave me a half a day, so I tried to emphasize the most important aspects of teaching the course. For some reason we got into a short discussion on the Geneva Convention and Gerald, another USAR instructor just assigned to me, proceeded to inform us that the My Lai massacre was justified, and that Lt Calley had been screwed by the Army. This set me off and I didn't spare his feelings as I schooled the rest of the instructors who had varying degrees of knowledge about the event from none to a vague remembrance. After describing what had happened there, I told Gerald that he needed to stick with doctrine and policy and keep his stupid opinions to himself.

The last thing I did before we were through was give the instructors their beginning of the course counseling. Gerald refused to sign his because he was USAR and I was National Guard, so I had Robert M sign it and we were ready for the course in November.

CDTF Operations

Meanwhile back at CD, MSG Bob showed up one day from Bosnia, along with his wife and LTC L. Bob had been an Intel analyst in the beginning of the CD program and had done a stint in SOD where he gained the respect of the men there. LTC L was our original Intel officer when he was a captain and had been exiled to DPS for a number of years after his coup attempt against Bil.

Bil was dead now and since keeping an LTC at DPS doing the same work as an E-5 sergeant might be considered excessive, he was brought back to CD headquarters. We wondered where they would go and eventually, we heard that Bob and LTC L would be assigned to the J3.

Bob replaced the candidate and was now the NCOIC and LTC L was now the J3. The Operations section had gone from a captain and two sergeants to a LTC, a major, two captains, two lieutenants, four sergeants and a specialist. It was the fastest empire building operation I had ever seen, and Captain JT was its driving force.

The Academy-Acting ANCOC Branch Chief

As Branch Chief, I had to take care of all the administrative aspects of the course. We were being inspected by USASMA, so I reviewed all the student records several times to make sure we hadn't missed something. There were always problems with the students not having some document. I had to follow up on each one of these until all of them were taken care of. I would go from classroom to classroom twice a day and sit in for 15 minutes and observe. Robert M and I took turns counseling the instructors and making sure they had what they needed.

In December, I was told by Tony that I would have to appear before a promotion board at the end of January. "What?" I was astounded. Tony laughed and said, "The CSM is going to make you branch chief for ANCOC and promote you to Master Sergeant." He then told me some areas I needed to study and when I needed to show up for the board. The board was required by Texas regulations and there would be another sergeant who would be competing with me for the position. The CSM, MSG Tony, and MSG R would be on the board, and SFC Janie would be the recorder. I reported to the board two days after my birthday. I was 55 years old and I was being questioned by men twenty years younger than me. They each had a checklist and they were supposed to score me on my appearance, education, military background and experience, current events, awards, etc.

January 2001: The Academy

One morning I had to hunt down Robert M because he wasn't there for PT and found him still in bed at the Camp Mabry Motel. "I didn't know I had to be at PT," was what he had the audacity to tell me. These kinds of minor irritations kept me slightly angry and that motivated me to do the best I could until the end of the course.

When the course ended, I was happy to see that our success rate had improved; 91% of the students that took the final exam passed. We started with 58 students and graduated 45. Nine students had dropped out during the course or failed their PT test but only 4 students had failed the final exam twice which was a big improvement from the last class. The class average was 77 out of 100 so we still had a lot of work to do but we had improved.

After the course was finished, I started to get ready for the next course in November. I took care of the student boxes first. After getting access to the USASMA website, I was able to see what version of the ANCOC course we would be using and the required manuals. The manual library was in Building 17, an old warehouse with no AC so I would put on my PT uniform and bring some bottles of water and Gatorade and work in the lovely 120-degree warehouse filling student boxes.

I continued to badger the CSM about the troubles we were having with MSG Jeff and the USAR system. I also wrote several long and detailed memos to him about all the trouble I was having in making improvements to our operations. He knew I wasn't all mouth because of all the work I had done with the SOP and counseling so he took my side and began to look for a way to get National Guard instructors on the podium and to get rid of the USAR. I told him we should offer to let the good instructors transfer to the Guard and work at the Academy and he agreed. I felt that most of them would respond to good leadership and eventually return to the old honor code of: 'I will not lie, cheat, nor steal; nor will I tolerate anyone who does.'

CDTF Operations-Task Force Reorganization

Going to work at the J-3 was pretty routine, not much excitement. One afternoon when everybody was at lunch, I got a phone call from the New Mexico Counterdrug. One of their helicopters was following a suspected trafficker for their state police and they were getting close to El Paso. They wanted to continue the surveillance and our permission to enter Texas. There was no one around so I told them it was ok, but they would have to pay for anything they broke.

Around the same time, LTC Gladys, the CDTF commander, had her own set of problems to deal with. She passed her concerns down the chain of command and one day Captain JT stood in the center of the OPS section and said that there were too many cargo inspectors trying to move to Intel analyst positions. The reason they were trying to move was because they couldn't get promoted past E-6 while they were working cargo, but Intel analysts could go to E-8.

Captain JT and some of the other Intel qualified people like Bob felt that these people needed to have schools and qualifications before they could be an analyst. JT told me and Bob to work up some proposal that would establish some basic criteria for people who wanted to become Intel analysts. This little project grew and grew until it resulted in the reorganization of the entire Counterdrug Task Force.

The project kept me very busy. Bob kept telling me to chill out, that this wouldn't go anywhere, but I was a man with a mission. After some attempts to work together, Bob backed out of the main effort and I worked on much of it by myself. I would show Bob some of my ideas; while he thought some of them were ok, he generally felt that it wasn't worth the effort. This project kept me busy for two years.

We had been faced with three major problems. The first was the proposal that we fire everyone above E-6. That was a wonderful conundrum for all of us over that rank because we would be firing ourselves. Beside self-interest, which is hard to overcome, the other reason why we objected to this was that many of our best people who had been on the program for years would be sent home, greatly reducing our effective support to law enforcement. We had analysts who were the lead workers in major drug cases and the cops considered them essential in their operations. We also had some total losers out there who were ruining the reputation of CD at the same time.

This was the problem. I recommended that we give the higher-ranking people the additional responsibility of managing a 'cell' of CD personnel while performing their duties. This would include periodic visits, telephone contact, and face to face time with the law enforcement agency we were supporting. This was an attempt to evaluate individual performance, to see through the fog of war, and ensure that all our people were actually producing.

We also broke up the regions into threat areas, aligned with the high intensity drug trafficking areas in the state and planned to move higher ranking officers out of one area to another so that their additional responsibilities would be to manage the 'cell' leaders along with doing their own work.

Then there was the problem of the majors. According to Army regulations, a major approaching 15 years of commissioned service had to be discharged with no retirement benefits if he had not been promoted to LTC before his fifteenth year of commissioned service. This created some enormous pressure for these men. Since the CDTF restricted most positions to 0-4 or major, we were starting to have problems when several majors took promotions to LTC and then the CDTF took care of them by moving them to newly created positions in the TF or arranged for them to get jobs in some other part of the National Guard HQ.

This caused an uproar with the senior enlisted people who had been forced to turn down promotions when they were offered because their positions were capped at E-8 meaning that when they were offered a promotion to sergeant major, they couldn't take it because if they did they would be fired. We needed a pressure relief valve or there was going to be an explosion.

We attempted to solve the 'major' and the sergeant major problem by creating some LTC positions in the CDTF HQ and in the field. The personnel department at the Guard HQ agreed to give priority to our officers when a position opened in the full-time manning force, AGR. We also created seven E-9 positions in the CDTF. Since everyone on the Army side was required to retire when they got 20 years, we figured that would keep some upward mobility for the sergeants.

In order to support these changes, we were forced to develop hiring standards based on the scores a service member got when they enlisted. The test, the ASVAB, or Armed Services Vocational Aptitude Battery, would show whether a person had the mental acuity to perform certain types of work. An Army Intel analyst had to have a score of 105 in the ST category of test scores. The Air Force required a score of 55G on their convoluted scheme of mental evaluations.

The next thing was a method of developmental counseling and performance evaluation. This was a nightmare because the Army had a complete system for this, and the Air Force didn't. When we tried to press the Air Force to accept the logic of the Army way, they dug their heels in and said that it was unfair. I spent many hours searching through the Air Force websites trying to find their doctrines on leadership and their processes for personnel development.

They didn't have any because the Air Force is more like a corporation than a military organization. A very tiny number of Air Force people ever came near combat and most of them that did were officers. Most of them were pilots who if they flew with anyone else it was usually another officer. They were like knights of old, boring a hole in the sky once in a while, then returning to clean sheets and a martini before bed. They had people to take care of all the housekeeping work that supported the maintenance work that kept them in healthy steeds, that kept them flying and unconcerned about all the 'little people.'

The Academy-Promotion, Again

After the board, I went to Major Sue and told her that I was up for promotion to E-8. Nobody said anything to me about the results of the board but in May I was promoted to master sergeant. From January until May I never said a word to anyone else about my impending promotion, and when it came, I was pleasantly surprised. My promotion caused a problem at CD. My position in the J-3 was for an E-7. I gave her the information and she said not to worry about it. A week later she told me I could take the promotion because they had changed the TDA and now my position was for an E-8. I was torn between what was good for me and what was fair for the Task Force.

We were getting a lot of complaints from soldiers in the field who couldn't get promoted in their units because their position in CD was capped at a certain rank or pay grade. I asked her why and she said that the CD leadership decided that my position required an E-8 so I took the promotion and waited to see if anyone would complain. No one did except my wife who had to sew on yet another set of stripes on each of my uniforms, the third time in five years.

CDTF-Buckwheat Leaves

One day the word began to filter through the building, Buckwheat was going to retire. We had enjoyed a great deal of stability in our organization since he became the Task Force Commander in 1990. In 1995 he went out to DC to work as the deputy director of NGB Counterdrug. He returned after a year on the Potomac and became the CD Coordinator, Bil's old job. There he teamed up with LTC Tinsley and together they were the best command group I ever worked under.

Buckwheat faced out and took care of business with NGB and our own State Headquarters. It helped that the Adjutant General at that time was Major General Daniel James III. He was a fighter pilot with two tours in Southeast Asia and he and Buckwheat use to fly together as part of the 'fighter pilot mafia' that also operated as a force in the headquarters. We had it all our way when this group of professionals was running things. That would change after Buckwheat retired.

LTC Gladys Tinsley decided that we would have a big going away party for Buckwheat. I was directed to put together a short video highlighting Buckwheat's time in the service culminating with his leaving the Task Force. I took some clips from movies like "No Time for Sergeants" and "Dr. Strangelove" and it got a few laughs. I really appreciated his leadership and I wanted to wish him well.

Buckwheat was replaced by Colonel Russ Malesky. I had known him since he was a Captain and back-seater on the C-26. He was a decent guy who had been taking aerial photos and chasing crooks around while videotaping them for the cops. Unfortunately, none of that had prepared him for the game of hardball at the State Headquarters.

MG James left a year after Buckwheat retired. He was offered a ride to DC by our Governor who had been elected President of the United States. Now he was the head of the Air National Guard. Colonel Malesky showed up at the next directorates' staff meeting and was beaten up by all the CD haters who were no longer held in check by General James. They began to formulate policies that would be very detrimental to our efforts to support law enforcement if approved by the Adjutant General. Colonel Malesky would have to learn a lot in a short amount of time if he was to keep our operation going.

Home Front

My son Richard came to us one day and asked if he could go to public high school. Some of his church companions were going to the high school about a mile from our house and he wanted to go. My wife had been home schooling him from the time he was two years old and she had some reservations about sending her prize student to the public-school system. Since I had asked her to home-school Richard, I asked her to let him go to public school. She told me all the reasons why he shouldn't go until, when she didn't get the desired agreement from me, she said that he was retarded, and the other kids would laugh at him.

I hadn't spent as much time with him as I would have if I was working a normal job, but I was pretty sure that he was able to use the mind that God gave him to solve problems and finish projects. She was getting pretty emotional at this point, so I suggested that we have him tested by the school system to determine his capacity for education at the high school level.

His test scores came back showing that he was fully capable of be ing educated with the students of his age level as a high school freshman. My wife was not able to accept these test results and demanded a second opinion. I was starting to get tired of all this and told her, which didn't help. When she saw that I wasn't going to side with her she decided that she would apply for a teaching job at the high school so she could be nearby in case her little boy needed her. This was the last thing Richard wanted so we both tried to take care of this brokenhearted mom who had to face the fact that her son was growing up and this part of her life was over.

CDTF Operations-The Reorganization

I was totally involved with the reorganization efforts. All through this period we had to drag our work down to the conference room for in-progress reviews with the colonel and the XO. I was doing a lot of the work and Captain Scott, my boss, was trying to control me while at the same time keeping things moving toward some acceptable solutions for some insoluble problems.

The colonel and I had been flying together on photo recon missions since he was a captain so he would check with me informally once and a while about how things were going. When I showed him the military decision making process (MDMP) that we used to come to the conclusions we were recommending for the changes to the SOP that would hopefully keep our experienced people and would straighten out the problem children in our ranks, he said, "Gee, I like that. I wish we had something like that in the Air Force."

When I asked Colonel Malesky about how the Air Force did these things, he said that when a decision was to be made in the Air Force, the colonels or the generals got together over a cup of coffee and made one. I kind of liked that method myself but it required experienced men to make decisions that would be good ones and unfortunately you didn't always have them around so the Army tried to off-set the 'dumb-ass' potential by coming up with a method that might expose the weaknesses in someone's 'great idea' before it got someone killed.

We finally got everything the way we liked it and when the colonel finished 'fine tuning' it, he took it on the road and briefed the Guard HQ. The Adjutant General bought off on our reorganization plan. We had saved everyone's jobs for now, including our own. At the end of July, we completed the new SOP for the CDTF. This was labor intensive and I was right in the middle of it. I had written a manual for CDTF leaders and another for members and synched them with the new SOP. The Colonel told us we would implement the changes to the Task Force at the beginning of the new fiscal year on the 1st of October.

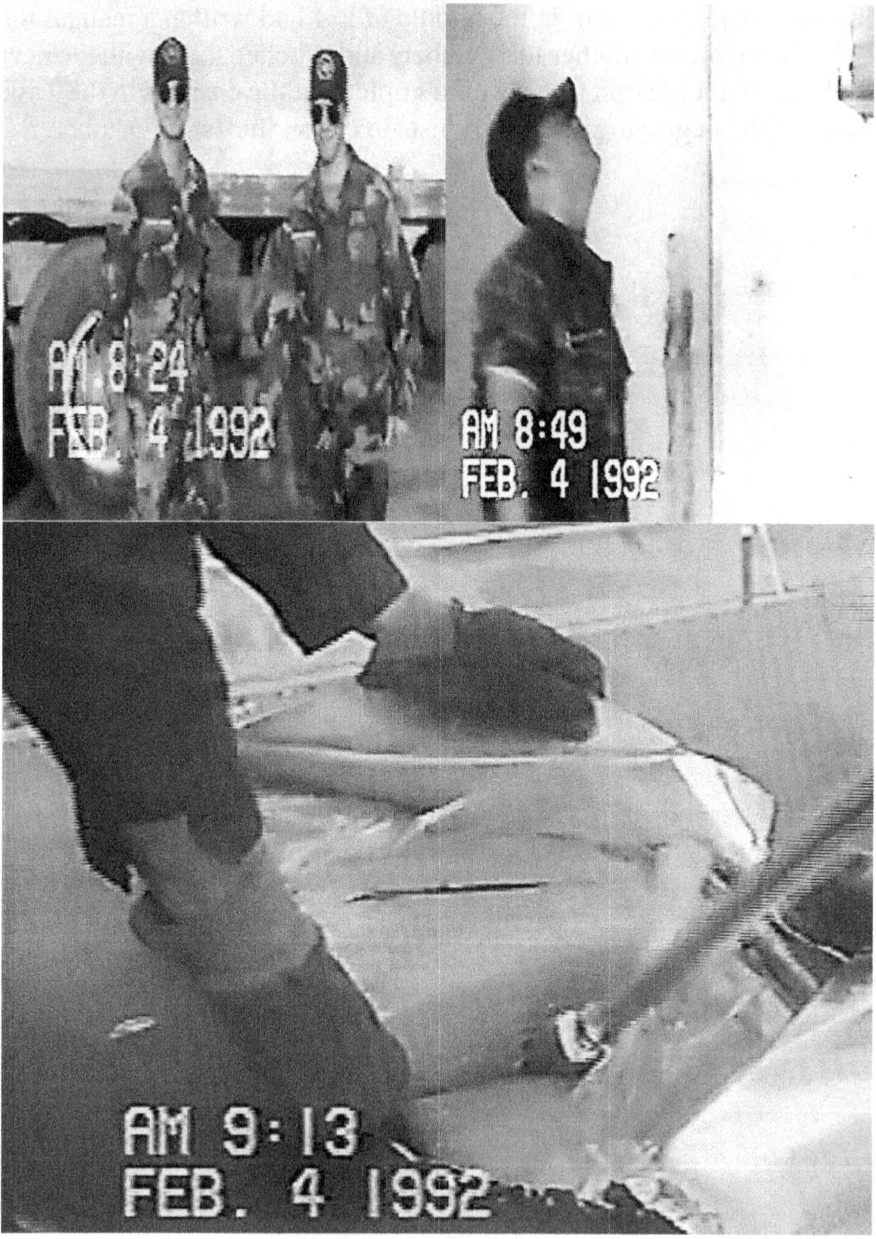

Cargo Inspectors at POE-Score!

Shall your brothers go to war, while you stay here?
Numbers 32:6b

Chapter 25-September 2001

911

Tuesday morning was the start of our work week. We worked four days a week, ten hours a day, minimum, and had three days off every weekend, unless something was going on. I got to work and was sitting at my desk when the phone rang. It was MSG Tony and he told me that a plane had hit the World Trade Center. He wasn't sure about what he had heard on the radio as he was driving to work but he wanted me to know something had happened in New York.

I figured it was the anniversary of the bombing in 1993 or something. Someone turned on the TV that we had on the wall in the office and there was the World Trade Center with smoke pouring out of one of the towers. As we sat there and watched, a plane appeared a long way off over the water. The news people were discussing the smoking tower situation and I watched this plane coming closer and closer until it flew right into the other tower, exploding into flame. That was a heart stopper. Everyone in the room was silent for a moment and I turned to Bob and said, "This is your Pearl Harbor."

I taped the news show as we watched replay after replay of the money shots. Then the buildings collapsed, and we watched the streams of survivors running down the street with the terrible clouds of dust pursuing them. The whole thing reminded me of the day I finished basic training and we heard that JFK had been shot in Dallas, TX and then we watched the flag that was flying over the main post at Fort Ord, CA lower to half-staff. It was a change of seasons for us all.

In a few hours SOD was pulling security around the perimeter of the Camp and they set up their remote cameras to cover areas that they couldn't watch due to their small numbers. The Texas National Guard mobilized itself to protect itself and the first thing they did was to surround the headquarters with concrete barriers, leaving the rest of us in the camp to protect ourselves.

I didn't think that we would be the next Al Qaeda target but as the airstrikes started in Afghanistan, I did consider that Texas, being the home of President Bush, could be a possible place for a terrorist attack. As I was driving to work one morning after that and as the downtown came into view I imagined a nuclear blast originating at the State Capitol and began to call on the Lord in whose hands I wanted to remain rather than go down the road to fear and worry.

Camp Mabry-A Rumor Of War

About a month after the war started, I was contacted by an old SOD comrade who asked me if I would be interested in going to Afghanistan with the Information Operations unit that was being formed in Texas. I was thrilled at the prospects of going to war. I said that I wanted in and that I would get back to him. The next day I talked to Don, a church elder who had been shepherding me for twenty years and asked him how he felt about my going to Afghanistan. He looked at me and said nothing for what seemed like forever before he said, "Greg, I would feel very lowly about that."

Although I was greatly disappointed, I said "Amen, Lord," and called my friend and told him I couldn't do it. I wanted to go because it was every soldier's duty to march to the sound of the guns. They weren't sending the National Guard to fight the only war in town but for some reason they did decide to send a National Guard Information Operations unit, whatever the hell that was. I had been invited to go but the Lord wasn't in it. I would have to wait for two years before I found out what he had in store for me.

Some of the SOD operators got called up with their units and several men that I knew were sent to guard a military installation in Arkansas. This turned out to be a nightmare tour because the Guard soldiers were not used to being on active duty for more than two weeks at a time. They were expected to guard this place like they were in Afghanistan and little by little, some of the NCOs and officers let things slide until they were having some serious discipline problems.

Service in Arkanistan was probably more difficult than in Afghanistan because try as you might it was hard to convince the troops that Al Qaeda might be coming through the wire that night. All the feedback I got from soldiers on these deployments made me more determined to teach the NCOs coming through the Academy what they needed to accomplish the mission and take care of their men.

The Academy-"Get Ready!"

In November, the next ANCOC class started. This time we were ready. George and Dave were my two senior small group leaders and they both belonged to the Guard. We spent long days getting all the student books ready. All the instructor materials was ready, and the Academy had installed video projectors and desk top computers in each classroom. I had one new instructor, Anthony, a USAR sergeant who had attended our last ANCOC class and had been promoted to SFC.

The other instructors were the same as the last cycle and I felt that they were responding to my exhortations to set the example and really help their students learn the material that might help them when they were deployed. I was able to spend a little more time with the instructors, giving them the Handbook and once again emphasizing the areas I thought were most important. The experiential learning cycle was a key element of the small group instruction method and I had the instructors sit with me in a classroom setting while each one led us through the ELC for several of the more difficult of the classes.

We also worked on an introduction program for the class. I used video clips of 'Saving Private Ryan' and 'A Few Good Men' to show the difference between physical and moral courage. I also showed a clip of President Bush telling the armed forces to "Get Ready" which became the class motto. I felt like we had to do a good job for the sake of these NCOs who were more than likely to deploy sooner or later because the President said that this would be a long war.

On Saturday morning of the first weekend we enrolled the students and gave them the briefing. When it was time for the students to leave the auditorium with their instructors I stood by the door and stared into each one's eyes. I think we impressed on them the seriousness of the course because we didn't have many of the usual problems that came up during the last class of the peacetime Guard.

CDTF Operations-A Major Change

The Department of Defense sent an undersecretary from the Counter Narcotics section down to the border to see what was going on. The new Defense Secretary, Donald Rumsfeld, didn't like a lot of what he found when he got to the Pentagon. He was on a war path and wanted the Armed Forces preparing for war, not doing police work.

"Dre" as we fondly referred to him, told us that he was doing away with the cargo inspection mission. We didn't think he could but a year later NGB told us that the cargo inspection mission would be closed by 2005. This left us with 150 people who would not have a job. Colonel Malesky told me that we needed to solve this problem on top of all the reorganization efforts. CPT JT's solution, off the top of his head, was to fire them. I made a strong argument against that, citing their consistent drug seizures while working in the hot sun at the ports of entry along the border, many of them for years.

The Colonel decided that we would make them all Intel analysts. On top of that, a few months later, "Oh, by the way," we found out that Colonel Malesky had promised all of these people to DEA as Intel people. The way things kept changing it was no wonder that Bob and some of the others threw up their hands and gave up on the reorganization.

Air Marshall Jim was the only one who consistently worked on the plan as part of his duties as the Task Force Training NCO. He was a former marine who had been a tank driver. He fought in Desert Storm and got out. Later he joined the Air National Guard and was assigned to the Weather Flight unit at Camp Mabry.

He did some research and came up with the idea that we would provide computer training and some computers for self-paced training at the ports for these men and women who had not used computers in their work or personal lives. He purchased a twenty-laptop classroom system that he would take to each port and conduct classes for the cargo inspectors to give them a jump start on the computer world.

CDTF-Texas DPS Narcotics

We briefed the Colonel on a plan to train the cargo inspectors as Intel Analysts. Colonel Malesky dropped another bomb, he was Air Force after all, and said that he wanted them trained to Texas DPS standards. He asked if we could and I got his permission to discuss this with the people at DPS Narcotics.

I contacted SGM Russ who I had known for a long time beginning with the early days of LRSD. He set up a meeting with Pat, the senior DPS Narcotics Analyst. We met at a Chinese Restaurant for lunch.

Colonel Malesky decided that he wanted to go and LTC L also went. Pat said that DPS didn't have an Intelligence Analyst Course but, she said, they always wanted to have one and now seemed to be the time to do it. Russ said he would be happy to work on it, but he was being deployed to Afghanistan.

By the time the lunch was over, I found myself being nominated for the task. I was interested in going over to DPS to see how the pros did it so I ignored some subtle warning signs, due to my "I'm an SOD operator and a surfer, I can do anything," attitude. I didn't know then that I was entering into a situation that was way beyond my abilities and in retrospect, I'm lucky that I passed through it all with my sanity intact. I can only thank the Lord Jesus and the prayers of my wife. Training our personnel to DPS standards was going to be hard because I soon discovered that DPS didn't have analyst training.

The next issue to arise was the General Counterdrug Intelligence Plan. This was a Drug Czar initiative which proposed that all Drug Intelligence be available to all agencies involved with Drug Law Enforcement. This was an issue which was at the heart of the argument between logic, self-interest, security, and necessity.

It was logical because in order to investigate and prosecute a multinational poly-drug trafficking organization, information if shared, could provide the probable cause needed for warrants, indictments, etc. Self-interest argues that if the local agency shares its information with the Feds, then they will make the case and take all the credit, asset seizures and publicity involved.

Security protects secrets. The past has shown that people can't keep secrets. There has been case after case where information had been leaked and investigations had been thwarted because the bad guys found out about informers in their midst. The informers were tortured and killed, and their handlers were devastated in some instances by the loss of people whom they had become close to in the course of working together in dangerous situations. Because of these past problems, the law enforcement culture held that an agent never reveals the identity of his informants to anyone except, maybe, to an agent whom he trusts, when he is reassigned or retires.

Necessity entered the picture when the Federal government tied funding to compliance. Each state was to develop their own method of data collection and the software to process the data. The data was then linked to nodes that connected various regions of the U.S. These nodes could be queried by an agent who was investigating someone and anything regarding the name or names of his suspects could come back as a 'hit.' The information wouldn't be there in many instances, but the name of an agent who had some information would come up with contact information.

Texas DPS chose a relatively inexperienced software company to develop the Texas reporting system that would comply with the General Counterdrug Intelligence Plan. By the time I showed up at DPS Narcotics, which is located at DPS Headquarters in Austin, the reporting system made strong men shudder when it entered conversations.

The system had three major components, the software, the hardware, and the people using the system. The software was an attempt to standardize the development of a Narcotics investigation. It covered every paperwork requirement for a DPS agent who was conducting an investigation. The hardware was unusual because it required a dedicated terminal that was connected to a special phone line. The terminal came from DPS and the agency that wanted to use the system was required pay for the phone line, which was expensive. The people that were required to use the system needed training. The basic training course was 40 hours and was taught at DPS Headquarters.

I took the basic course to see how we could integrate it into the Intelligence Analyst course. It was very difficult for me and I know that I am pretty slow when it comes to computer training, or any other training, for that matter. The instructor was a DPS Narcotics agent who came in from the field to teach the course. There were a number of Narcotics agents in the class. I sat next to one and became friendly with him. He told me that he was taking a refresher. I asked him what he thought of it. He didn't say anything. I assured him that I wouldn't mention his name, but I needed to know what the feeling of the agents in the field was. He took me outside and looking around, he told me that it was a nightmare in the field.

"Everyone has trouble because it is too complicated. The connection is extremely slow, and it takes forever to do any data entry. Most of the agents are way behind in their entries. They try to get the secretaries to do it. Many of the reports are kicked back because they were entered improperly. The lieutenants are always getting calls about these problems. Other than these problems, everyone loves it."

I began asking about the system whenever I talked to one of our guardsmen that was working at a federal, state or local law enforcement agency. The first place I checked was the Austin DEA office. They told me that they had sent their terminal back to DPS sometime before and said that nobody used it. The DEA wasn't going to put any of their information into it and they found that there wasn't anything of interest in the system. I found the same story everywhere. The federal and local cops weren't putting anything in the system. The DPS and TNCP Task Forces hated the system and were putting their information in but much of it was getting kicked back for incorrect entry.

2002: CDFT Operations-DEA

Colonel Malesky had promised the head of DEA Houston Intelligence that they would get all the cargo handlers that we were going to convert to Intel Analysts. I went with him to a meeting in Houston. The senior DEA people were there including the ASAC and senior analysts. The meeting theme was the DEA plan to attack the drug transportation organizations using intelligence analysis of all data regarding drug transportation.

CDTF-DPS Narcotics

After I had become familiar with the Texas reporting system, I was talking with Pat about integrating its training with the basic Intelligence Analyst Course. Since the Cargo Handlers who were transitioning to Intel were going to be in the pilot courses, I wondered out loud, to my regret, if maybe the DEA would use them to enter some minimum case information into the system. The next thing I heard about the subject was someone telling me that Pat had had a meeting with one of the Captain's where she indicated that the guardsmen that DPS was training, at no expense to the Guard, would probably be entering DEA case information into the Texas system.

This took on a life of its own in DPS Narcotics. They were frustrated because no one besides DPS was using it. Hadn't the President of the United States directed that the General Counterdrug Intelligence Plan would be used by everyone involved in Drug Law Enforcement? Why was it then that the DEA, the lead Federal Drug Law Enforcement agency, under the direct control of the President, how come they weren't following the orders of their commander in chief? This was something that the DPS guys professed not to understand. They were a paramilitary organization with a chain of command. They knew how to follow orders. They just didn't understand.

CDTF-DEA

In a matter of a few weeks the word got to Gary at DEA that DPS was expecting the Guard analysts to enter all the DEA case information into the DPS system. He called me one morning and told me that he would be in Austin that day and wanted to meet. I gave LTC L a heads up. When Gary showed up, he came right to the point. He heard that I had told DPS that DEA had agreed to use the Guard analysts trained by DPS to enter their case information into the DPS system; "That ain't gonna happen," he said.

I asked him who told him that I had done this, and he refused to tell me his source of information. I told him truthfully that I was aware that DPS was hoping to get other agencies to use their system and that I had said that I wondered if DEA would want to use the Guard analysts to enter information into it, but I never told DPS that DEA had agreed to have the Guard analysts enter data into their system.

Gary was a little hot under the collar so I moved my chair closer to him and looked him directly in the eye and said, "Sir I did not tell DPS that the Guard analysts were going to enter DEA data into their system." He stared me in the eye and finally blinked. "OK," he said, "I believe you."

I went to LTC L and told him what had happened. He was on short final for retirement and wasn't too interested in the day to day problems, but this was something that he could handle. He had worked for many years at DPS and knew everybody there. He called over there and arranged a meeting with the Narcotics leadership. He handled them in his own smooth way and at the end of the meeting, he convinced them that the Guard was not to blame for this fiasco and managed to shift the blame onto the DEA.

CDTF Operations-PowerPoint Briefings

Air Marshall Jim was supposed to prepare the annual refresher training for the entire task force and in-service training for the troops in the various activities. SOD conducted all their own training so, most of his time was spent in power point production for all the dog and pony shows for the command group.

The art of the briefing had risen to new heights since Generals Norman Schwarzkopf and Colin Powell and their televised briefings schooled the military during Desert Storm. Technology had rescued us from the posterboard briefing charts; I had to make a few of those in our early briefings to the Texas National Guard leadership before our first Unity missions. Next came a software program that helped you put together digital briefing slides that were made into 35mm color slides that you could use with a slide projector. That program was difficult to use but it was a step forward.

Finally, in the mid 90s, Power Point appeared on our computers and we all learned how to make briefing slides that looked good and made you look good. Around the same time, video projectors became affordable and when you connected your computer to one of them, you could put on quite a show with very little work involved. It took a while to figure out what to put on the slide. Many of the early slides had the entire text of what you were going to say. You would be speaking what was written on the screen.

This seemed redundant and the senior officers soon let everyone know that they wanted to see bullet comments on the screen that emphasized the main points that the briefer was trying to convey. Then came briefing training for NCOs in their training courses; NCOs soon found themselves giving Information briefings, Decision briefings, and many other types of presentations. This created a whole new skill set in the army. The term, Power Point Ranger, appeared and there were jokes about how soldiering had changed from attacking the enemy to dominating a dog and pony show for your own people.

CDTF Operations-Personnel Management

We still had all the problems we were going to correct with the reorganization of the Task Force. That was put on hold when the war started. One problem that gnawed away at our section was the knowledge that we had people in the field who needed a little supervision. I knew that most of our people in the field were doing good work but there were a few that were ruining the reputation of the good ones and the CD Task Force.

Texas was still divided up into five regions. Each region had an officer, the Region Commander, and an admin NCO to do the paperwork. The main function of the Region Commander had devolved to a daily accountability report that was required to be at the Task Force Headquarters by 10:00 AM. Every soldier or airman was required to call the Region Commanders office before 10:00 AM and state that he or she was present for duty. As long as everyone was at work, then everything was ok.

The region commanders were also supposed to get with the LEAs each year and go over the rules for employment of guardsmen but for many of them, this became a drill: make copies of last year's paperwork, use a little whiteout for dates, and fax exchanges for those all-important signatures.

Every year, the Operations section had to badger, and badger these officers to get their paperwork in by a certain date. Most of them never could because they weren't engaged with the LEAs so they didn't know what was going on in their region. Because they didn't visit these cops, they had no way of knowing if there were problems between the guardsmen and the cops.

The Academy-The Army Reserve

ANCOC finished and this time we had a 95% pass rate. I was still not satisfied that we were doing the best we could, but I felt like we had made progress. The instructor with the highest class average was given an Army Achievement Medal, and Anthony, the first time instructor had the lowest average, and was given a gentle chiding by me. We had the graduation ceremony in the camp chapel. The Academy commander, Colonel Rodriguez was the guest speaker. He must have been inspired by the stained-glass windows and the religious atmosphere because he told the graduates that as NCOs they were 'living rocks' for the building of the army. I had to keep from laughing; I'm sure he meant 'living stones,' because in current Army lingo a 'rock' was a less than intelligent soldier; or did he?

Shortly after graduation, I was told that MSG Jeff was getting promoted to Sergeant Major and would be moving on, some good news. I was full of hope that his replacement would work with me to improve the instructor development but that was not to be.

The replacement showed up the weekend after the course was finished. Saturday was busy with the instructors finishing their student files and turning in their instructor materials. He was a master sergeant who worked in a managerial position for the city of San Antonio. He was very friendly so after a greet and grin meeting I sat down with him and laid out the problems. He assured me that we would work together to solve them. I asked him when, he said "soon." I asked for more details as I was making the training schedule for the drills before the next class and he told me that he would get back with me 'soon.'

Sunday morning, after breakfast I got with him again and pushed a little harder for a training schedule. He asked me to meet him after lunch and we would work on it. When we got together after lunch, he told me that he had discussed the matter with SGM Jeff and there wouldn't be any time for instructor training before the next class.

This was too much, and I told him so. He said that he owed his loyalty to SGM Jeff because he had given him this position. I asked him about loyalty to our Army values which trumped any loyalty to a person who was disregarding our Army values. When I brought up the cheating on the PT tests and the sub-standard performance by some of the instructors, he sat there grinning at me like everything I was saying was great, but he wasn't going to do a thing about it. So much for a change with the USAR, I declared war at that point.

None of this was made any easier by the fact that he was black and I was white. Many people in the military were afraid to get into these kinds of confrontations because in the struggle for equal opportunity, the line between what was right and what was wrong was blurred by the recent history of our country.

I was as aware as anyone about the pathetic history of racism in this country and I considered the civil rights workers of the 50s and 60s to be the last American heroes, but I didn't think that they would want us to treat each other in a less than honorable way. I couldn't recompense everyone for the past behavior of some Americans, but I could make sure that I never treated anyone in my unit with less than the respect they deserved because they were operating within the same values that the Army demanded of all of us equally. I expected them to uphold those values and I expected them to hold me to the same standard.

I told him that if he didn't find some time for me and the SSGLs to get with the instructors before the next class I was going to ask the CSM to fire them all and we would train our own instructors. He was still grinning as I got up and left.

I went back to my office and started my memo to the CSM requesting that he do what I told the new MSG I was going to do. The CSM called me in for a face to face and I told him what had happened and why I wanted the USAR out of here. He surprised me by telling me that he agreed and would start making inquiries with the next higher headquarters to get the ridiculous agreement changed so that we could train, develop and evaluate our own instructors. Without unity of command or at least unity of effort it is foolish to try and do anything and this episode was a perfect illustration of why we have principals of war and what happens when we ignore them.

The Academy-ANCOC

The CSM did pressure the new MSG to give me a day with the instructors before the class started. SFCs George and Dave issued them their materials and went over the areas we had identified needing improvement, especially the experiential learning cycle. After I counseled George and Dave, they counseled their instructors, and we were ready to go.

The Academy-ANCOC

At the end of the course we added up the scores for each instructor. SFC Anthony's class had the highest average. I wrote a recommendation for an Army Achievement Medal for him and it was approved. Anthony was also black and I'm sure that the tension between his boss and me was a source of discomfort for him, but he was committed to his duty as an instructor and his class scores were a measurable way of showing that he had worked harder and smarter than his peers.

I also got one for SFC Robert who was retiring. We never became all buddy-buddy, but I do believe we reached a level of mutual respect and were able to operate together according to the Army Values.

One of the reserve instructors coached all the BNCOC classes with the exact areas that the test questions would come from. When they took their test, there was a 100% first time pass rate and most of the students had a 90% or better score. When the CSM started crowing about how good the BNCOC class did on their exam I mentioned that it was the most unusual showing on a final exam we had ever seen. He looked at me and said, "What do you mean, Sergeant Murry"?

I could see that he didn't want to hear anything negative about these wonderful results but I told him that not only was it the most unusual showing we had ever seen but it appeared to me that someone either gave the students the answers or told them what exact areas to study. He began to sputter about my saying something I knew nothing about. I pulled out the stats for the last three years and asked him how there could have been such a dramatic improvement this year

When the graduation day came, we had the Adjutant General as the guest speaker and when it was time to present the awards I stood on the stage and introduced SFC Anthony as the man who finished the year before in the outhouse but this year he was in the penthouse; it got a laugh out of General Marty.

The Academy-More Rumors Of War

That summer I heard a rumor from one of the full-timers that the Academy was going to Afghanistan. That was a joke. The Academy didn't have a wartime mission other than to go to Fort Hood and run the Army schools. I told the full-timer that he needed to stop smoking that stuff before the next piss-test. George and Dave and I started getting ready for the next class.

The CSM called me in and told me that he had done it and the Academy was now going to use Guard instructors. He directed me to tell the MSG to move out of our offices and find some other place to run their unit. There was an Army Reserve facility next to the camp so they wouldn't have to go too far. He told me I could offer positions to any of the instructors who wanted to come over to the Guard.

Over at the Academy, the Afghanistan rumor was still buzzing around like a mosquito in your bedroom at night. It's quiet for a long time and just when you're dropping off to sleep, here it comes again. There was starting to be some substance. Before it was like they were going to activate the Academy and send it after Al Qaeda in the Hindu Kush. That was too crazy; we were all old, overweight, and most of us couldn't shoot straight. Now, it was said, we were being considered for the job of running an Afghan Army training facility. I made inquiries about the source of these rumors and one of the full-timers nodded in the direction of the XO's office.

LTC Knight, the highest ranking full-timer, a West Point graduate, had recently joined our band of brothers and sisters. He had a knight in armor, about two feet high, standing on his desk. He wore jump wings and a CIB from the Gulf War.

Summer 2003: CDTF-Victory Through Reorganization

The CDTF published one of our sporadic newsletters and a good part was taken up by the colonel's Task Force reorganization announcement, which I wrote. When I got my copy, I waved it in front of MSG Bob who had been skeptical about our ability to do anything about the problems and gave him the old, "I told you so." He just smiled.

The Academy-Volunteers Of America

During the drill in August, my information source among the full-timers told me it was definite, some of the Academy was going to Afghanistan. "Who was going?' I asked. "It's a volunteer mission," he told me. This was a typical Guard maneuver. Fill a unit that was supposed to be a unit that trained together with people that didn't know each other, who were out of work and needed a job, who had nothing better to do, and send them to war; 'when will they ever learn…?'

He told me that LTC Knight was the guy heading this expedition, so I sought him out at the snack bar during lunch. He was standing in the food line waiting for a burger when I approached him and said, "Sir, I heard you were taking some people to Afghanistan." He looked at me in a typical officer way, then deigned to speak to me although he didn't have to, you understand, "I might be. What's it to you?" he asked.

“What are you going to be doing over there?” I asked. “We’re going to be advisors to the Afghan Army.” I looked at him and said, “The Counterdrug Task Force says I’m not allowed to volunteer for any deployment, but I will be extremely disappointed if you don’t take me, I was born for that mission.” He stepped back, “I wish that some of the other NCOs had your attitude.” Don’t forget, sir,” I told him, “I’ll be extremely disappointed if you don’t order me to go.” He nodded that he understood. A month later I got my orders, Afghanistan. I deployed in December 2003 with 136 Regional Training Institute (Forward) and returned to the Counterdrug Task Force in October 2004. Things had changed drastically at Counterdrug, but before I signed out I had one more trip to NGB in DC and I cover that episode in Book III. I remained in Counterdrug until I retired in 2005. That period of time is covered in my third book, “Content With My Wages, A Sergeant’s Story: Book III-Afghanistan.

Epilogue

When the war started on 911, everything changed for the soldiers among us. Most of the SOD operators ran to the sound of the guns. Tommy Reynolds got a head start. As a member of an FBI regional SWAT team, he deployed to Yemen and other places in Africa after the USS Cole was bombed in Aden. Several former LRSD men and SOD operators went to Afghanistan with an Information Operations unit. A number of operators served as security contractors in Iraq and Afghanistan. Captain Brian found a job as a combat advisor in Afghanistan. Then he served in Iraq and finally as a Lt. Colonel he commanded a unit in Afghanistan. In 2003, I deployed to Afghanistan with a good number of Counterdrug members including Roger Hernandez, a former SOD team leader. Chris Bolliger was a medical advisor there and we met in Kabul several times. Captain Smith went to Iraq and later retired as a Major General, (I told you that one bad OER didn′t matter.) Captain Krupp, who replaced Brian as SOD commander was with the combat advisor unit that replaced mine in August 2004

After the invasion of Iraq the National Guard went into high gear and many units deployed to both Iraq and Afghanistan, ending the longest break since 1953. Many more SOD operators and other Counterdrug Task Force members ran to the sound of the guns, so many, that the CO was forced to warn people that if they left CD voluntarily, they wouldn't have a job when they returned. That didn't have the desired effect but with so many deployments I think most people found a chair when the music stopped

All good things must end and I retired in 2006. When I left Vietnam, I had the address of one man in my platoon. I managed to find others by using a bag of dimes at phone booths. Since then, with the advent of the internet and social media I was able to find a few of the survivors. Now though, I have found many of the LRSD and SOD guys on Facebook along with many of the ones I was with in Afghanistan.

Author 1968 **Author 1998**

My Wife at the Range on Our First Date-The Day I fell in Love

My son and his family after his 5th jump

Annex A

MSG Bil Enney's
Words of Wisdom

**One Who Allows Their Heart and Emotions
To Guide Their Efforts Will Probably View Life as a Tragedy
One Who Is Guided by Thought Processes and Lessons Learned
Will Probably View Life as a Comedy**

1. Establish, Maintain, and Honor Rapport.

a. Establishment of rapport between counterparts can be difficult to achieve, even more difficult to maintain; but this rapport can vanish with no difficulty at all once the process is ignored.

b. Forget your counterparts and you can forget your mission.

c. Make no promises you can't keep, pay all bills on time, and burn no bridges that you could cross at a later date.

d. It is difficult to tune in, when you are constantly broadcasting.

2. Understand the Parameters at Hand

a. When sampling new and exotic tidbits, bite sized chunks may be advisable. You seldom choke on them and they can easily be concealed in a cocktail napkin if they are not to your liking.

b. Conventional methods don't always work in unique and unconventional situations.

c. Common Sense may be the reference guide you are forced to use if a regulation or manual doesn't exist to answer your question.

3. Know, Understand, and Never Exceed Your Own Capabilities

a. Know your subject matter, then talk!

b. Know your limitations, then exercise your ability.

c. You can be outranked, it's your fault if you are outsmarted.

d. It's hard to find fault with something that is legally, tactically, and morally correct.

e. Bonehead actions and statements will surely be in your counterpart's headquarters in less than 15 minutes.

f. Columbo always wins, J.R. lost half of Dallas and all of West Texas.

4. Know, Understand, and Never Over/Underestimate Your Audience
a. You may be neither politician nor philosopher; what if your counterpart is?
b. A protocol violated is an unpaid bill, your credit rating will surely suffer.

5. Agree Upon a Common Language
a. Acronym: A Creditless Relating Of Nonsensical Yankee Mishmash or perhaps A Creditable Revealing Of Noun Yielding Meaning.
b. Extensive use of acronyms in a "5" minute presentation can cause days of confusion and consternation. Simply adding 30 seconds to your presentation could save all that lost time.

6. Develop a Joint Plan that Complements and Enhances Existing Efforts--Unilateral decisions normally appease one party and displease the other.

7. Insure All Contingency Planning Is Complete
a. Failure or Surprise is no more than a contingency you haven't planned for.
b. Your plans may be secret, but be prepared to defend your actions when secrecy wanes!

8. Present One Plan To Those in Charge for Approval and Enforcement
a. What we need here is a plan we can live by, train by, accomplish our mission by, and "if necessary" prosecute by!
b. Implementers seldom have the final say.

9. Turn the Plan Loose and Stick With The Plan

10. Provide Quiet, Honest, and Professional Support

11. Cease Fire When Necessary
--It could be and perhaps should be your duty to yell "Cease Fire". You may incur someone's wrath, but you may save something or more importantly, someone.

12. Monitor Well the Effort

13. Debrief, Discuss, Analyze, and Document

Annex B

San Antonio Express News
Jul. 24, 1999
Dane Schiller

Laredo - The office of Gov. George W. Bush on Friday defended the use of National Guard patrols to search for drug traffickers along the Rio Grande and said the missions weren't an effort to "militarize the border." But a member of Congress called the missions a "prescription for disaster." Linda Edwards, spokeswoman for the governor, said Bush knew of and approved use of armed National Guard troops along the Texas-Mexico border. She said the National Guard troops haven't fired a shot in the 10 years the operations have been conducted.

However, a U.S. Marine on an anti-drug patrol fatally shot an 18-year-old herding goats near Redford in West Texas in May 1997. "This isn't militarization of the border. This is part of a drug-interdiction effort led by local law enforcement agencies," Edwards said in Austin. Critics, including U.S. Rep. Silvestre Reyes, D-El Paso, were surprised to learn of the guard patrols and said thy threatened border residents' safety. Reyes was once chief of the Border Patrol's El Paso Sector. After retiring, he was elected to Congress. The patrols, conducted by specially trained reconnaissance teams, gather intelligence, including spying on smuggling routes at the request of state and federal law enforcement agencies. "They could be disastrous with another Redford. From my perspective. I wouldn't be comfortable," Reyes said in Houston, where he was attending a congressional hearing on NASA.

In the Redford incident, Esequiel Hernandez Jr., who was carrying a .22-caliber rifle, was shot once in the ribcage by Marine Cpl. Clemente Banuelos. The Marine's four-member patrol was working with the Border Patrol. Banuelos later was discharged honorably from the service. That incident triggered a national debate over the use of military personnel along the border and led Defense Secretary William Cohen to cancel similar missions indefinitely. After the shooting, Bush said during a June 1997 trip to El Paso the Border Patrol "should be in charge of the border." "The military shouldn't be enforcing the law on the border," Bush said when asked about the shooting. Edwards insisted the governor's approval of the patrols is consistent with his El Paso statements.

"The governor has always said the military's mission is to fight the enemy; Mexico isn't the enemy" Edwards said. "Gov. Bush supports the effort involving the guard because the 30 guardsmen assigned to help law enforcement in the surveillance of drug traffickers are highly trained for the task and extremely helpful," she continued. The U.S. Drug Enforcement Administration, FBI, Customs Service and Texas Department of Public Safety are among the agencies using the National Guard reconnaissance teams. The missions, known as Listening Posts/Observation Posts, are approved by the Defense Department and the governors of states bordering Mexico. Bush is commander-in-chief of the Texas National Guard, and he and the Texas attorney general sign annual plans for the missions, Edwards said.

While the Border Patrol was the National Guard's chief customer before the Redford shooting, it no longer requests assistance, Col William Pettit, counter-drug coordinator for the guard said. "We aren't out there like a bunch of cowboys. We're there at the request of law enforcement," Pettit said in Austin. "Our safety record and training speaks for itself." As many as 30 anti-drug reconnaissance missions have been conducted this year, and two were near the border, he said. Other missions involved surveillance for marijuana crops. Petit declined to discuss the missions' specifics, including how often guardsmen are armed or where they're stationed along the 1,500-mile Texas-Mexico border.

Aaron Reed, spokesman for the Texas adjutant general's office, which oversees the National Guard, said security concerns prevent officials from discussing when and where guardsmen carry weapons. "We've got these guys out there, and there are some things we can't say because it would endanger them," he said. In recent years, guardsmen largely have shifted from hiding in the bushes on top of smuggling routes to observing from the distance with high powered scopes and other equipment, he said. As the sophistication of equipment available to guardsmen has increased, so have the challenges. Officials ever wary of well-financed international drug cartels estimate the majority of the cocaine and marijuana in the United States was smuggled across the Texas-Mexico border. The reconnaissance patrols are a small part of the National Guard's mission, with other efforts including helping inspect cargo traffic crossing international bridges. Reyes praised the guard's efforts at ports of entry as well as helping with analysis and operating equipment in counter-drug efforts. Still, Reyes said, Redford taught him mixing armed personnel with civilian populations is unwise. "I can see a prescription for disaster," he said. This may not be interesting to many, but thought some of you would like to hear local news on Gov. Bush, as opposed to the national media.

Nacho Guarache by Leo Ga

Made in United States
North Haven, CT
04 November 2026

21760291R00280